A Neuroscientist's Guide to Classical Conditioning

Springer
New York
Berlin
Heidelberg
Barcelona
Hong Kong
London
Milan
Paris
Singapore
Tokyo

John W. Moore
Editor

A Neuroscientist's Guide to Classical Conditioning

With 60 Figures

Springer

John W. Moore
Department of Psychology
University of Massachusetts, Amherst
Amherst, MA 01003
USA
jmmoore@cs.umass.edu

Cover illustration: Hermissenda are trained and tested in glass tubes on a turntable. The *Hermissenda*'s phototactic response to light is measured before training: the time the mollusk takes to reach a spot of light (darker shading at the center of the turntable) is recorded automatically when the animal reaches the photoelectric cell. Then, the *Hermissenda* are trained by being rotated while confined to the outer end of the tube and thus subjected to centrifugal force that is sensed by their statocysts. For one group, the rotation is precisely paired with a 30-s period of light; various control groups receive no training or are subjected to light and rotation in an explicitly unpaired manner. Finally, the *Hermissenda*'s velocity of movement toward the light is timed again to determine the effect of training.

Library of Congress Cataloging-in-Publication Data
A neuroscientist's guide to classical conditioning/editor, John W. Moore.
    p. cm.
  Include bibliographical references.
  ISBN 0-387-98805-X (softcover: alk. paper); ISBN 0-387-98787-8 (hardcover: alk. paper)
    1. Conditioned response.   I. Moore, John W. (John William), 1936–
QP416 .N48   2001
612.8—dc21   2001020201

Printed on acid-free paper.

© 2002 Springer-Verlag New York, Inc.
All rights reserved. This work may not be translated or copied in whole or in part without the written permission of the publisher (Springer-Verlag New York, Inc., 175 Fifth Avenue, New York, NY 10010, USA), except for brief excerpts in connection with reviews or scholarly analysis. Use in connection with any form of information storage and retrieval, electronic adaptation, computer software, or by similar or dissimilar methodology now known or hereafter developed is forbidden.
The use of general descriptive names, trade names, trademarks, etc., in this publication, even if the former are not especially identified, is not to be taken as a sign that such names, as understood by the Trade Marks and Merchandise Marks Act, may accordingly be used freely by anyone.

Production managed by Francine McNeill; manufacturing supervised by Erica Bresler.
Typeset by Impressions Book and Journal Services, Inc., Madison, WI.
Printed and bound by Maple-Vail Book Manufacturing Group, York, PA.
Printed in the United States of America.

9 8 7 6 5 4 3 2 1

ISBN 0-387-98787-8   SPIN 10715233   (hardcover)
ISBN 0-387-98805-X   SPIN 10715893   (softcover)

Springer-Verlag   New York Berlin Heidelberg
A member of BertelsmannSpringer Science+Business Media GmbH

*Dedicated to Isidore Gormezano
for his singular contributions to the field*

If someone quickly thrusts his hand against our eyes as if to strike us, even though we know him to be our friend, that he only does it in fun, and that he will take great care not to hurt us, we have trouble in preventing ourselves from closing them; and this shows that it is not by the intervention of our soul that they close, seeing that it is against our will, which is its only, or at least its principal activity; but it is because the machine of our body is so formed that the movement of this hand towards our eyes excites another movement in our brain, which conducts the animal spirits into the muscles which cause the eyelids to close.

—Volume I of *The Philosophical Works of Descartes* (1649), translated (in two volumes) by Elizabeth S. Haldane and G.R.T. Ross for Cambridge University Press, p. 388, corrected edition of 1967.

Preface

Why a Book About Classical Conditioning for Neuroscientists?

Most neuroscientists, whatever their home discipline, share some familiarity with classical conditioning. They recognize it as a form of learning and as a platform for investigating the nature of learning and memory at levels of analysis ranging from the cellular to the behavior of whole organisms. Furthermore, they have witnessed the steady growth in research on this topic. Hundreds of abstracts on classical conditioning have been listed in the *Society for Neuroscience Abstracts* in recent years, and the numbers are growing. As a paradigm of learning and memory, classical conditioning has become a tool for fundamental studies across a broad array of topics and specialties, ranging from computational modeling to motor control and from pharmacology and therapeutics to cognitive neuroscience.

At the same time that classical conditioning and other forms of behavioral learning have become increasingly important in the neurosciences, there has been an alarming contraction in the emphasis placed on these topics in training programs at all levels. Few undergraduate psychology programs offer laboratories in animal learning, and courses on the topic have been diluted to such a degree that few students have an opportunity to become facile with the terminology, concepts, and findings that make classical conditioning interesting and important for today's neuroscientist. Many neuroscientists are engaged in exciting work on classical conditioning using a variety of tools, but they generally lack an appreciation of where their work lies along a broad spectrum of current knowledge about behavioral learning and memory processes. What is needed is a comprehensive review of contemporary work in classical conditioning, and that is what this volume aims to provide. This volume should be the book of choice for neuroscientists who use classical conditioning in their research but who lack a sense of where their work fits into the larger picture. This book is not a primer on classical conditioning. Its objective is to provide a sophisticated overview of the field for working neuroscientists and their students. Readers who have no background familiarity with classical conditioning should consult the chapter by E. James Kehoe and Michaela Macrae (Chapter 6), the first part of which defines terms and illustrates basic phenomena.

Classical conditioning refers to a system of methods and findings about one of the brain's most important and ubiquitous functions, that of forming connections between correlated environmental events. A century ago, Pavlov noticed that the dogs used in his studies of digestive function would express anticipation of feedings in various ways, particularly by salivating. Pavlov's story is well known. Less well known to neuroscientists are the other forms of anticipatory reactions that have been examined in the laboratory. Pavlov's studies of classical conditioning of salivation have not survived into the modern era because of the expenses and difficulties in maintaining dogs as experimental subjects and because of the development of other preparations, such as the defensive eyeblink response. Nevertheless, almost all our terminology derives from Pavlov, and the basic phenomena, with very few exceptions, had been described by Pavlov by the time he died in 1936.

Regarding terminology, Pavlov and his English-language interpreters such as Hilgard and Marquis gave us the familiar terms that define the events that become conditioned (i.e., associated or connected) as a result of their correlation. They provided a lexicon to describe the behavioral and physiological consequences of this process. It is appropriate to review this terminology before proceeding further. The correlations that give rise to classically conditioned behaviors or reactions occur within the settings, contexts, or environments in which organisms live. They are part of the normal stream of events encountered from one moment to the next. Events that are correlated in time can become associated (i.e., "conditioned") in ways that have obvious utility to the organism, both as an individual and as a member of a species. Pavlov was quick to point out that conditioned responses (CRs), the learned expression of environmental correlations, were generally adaptive in nature with the function of preparing the organism for events it could not otherwise control. Pavlov brought this process of learning correlations (i.e., conditioning) and its behavioral consequences into the laboratory. The laboratory setting was made conducive for experimental work by limiting the questions to be asked to a relatively short list of issues, primarily those that address the factors that determine whether an environmental correlation between two events will result in learning, or as Pavlov expressed it, the acquisition of a CR. The first principle is that the events must be stimuli. That is, they must of themselves evoke some intrinsic or unconditioned reaction. Pavlov did not insist that the reactions of the organism to both correlated events be the same or even particularly obvious to an outside observer. It was only important that such observers witness the unconditioned reaction to the second event, because a measurable reaction was required to answer questions about the nature of the learning that occurs in classical conditioning. Thus, the first event, the conditioned stimulus or CS, may or may not evoke a reaction that the observer can readily see. A CS might evoke an orienting reflex if it is novel but no reaction at all if it is a familiar stimulus. By contrast, the second event, the unconditioned stimulus or US, is chosen to be biologically significant. That is, its intensity and functionality ensure a measurable reaction from the organism under study. Although the intensity or salience of the first stimulus, the CS, is not as crucial as that of the second stimulus, the US, Pavlov nevertheless knew that a strong CS would typically be more effective in promoting learning and expression of CRs than less intense or salient CSs.

In the language of classical conditioning, the basic unit of learning or experience is the trial. One joint occurrence of a CS and a US constitutes an acquisition trial. Because the extraction of a correlation between two events can occur in noisy environments, the development of a measurable CR typically requires many trials. A number of variables can influence the rate of CR emergence and its topographical features. Pavlov noted that the interval of time between the onset of a CS and the US is one important variable. If this interval is too long, conditioning may not occur. If the interval is less than 100 milliseconds (ms) or thereabouts, conditioning may not occur. There are intervals between the CS and US that are conducive to learning in some preparations but not in others. The interval between trials is another important variable influencing the rate of learning and the vigor of the CR. All else being equal, conditioning proceeds more efficiently when the interval between trials is longer (spaced) rather than shorter (massed).

Pavlov and his followers referred to the process of CR development as the *acquisition process*. The term has become synonymous with the term *conditioning* itself. In fact, the Pavlovian lexicon has become so familiar that the same terms are applied to the procedures (operations) used to produce a CR and the results of those procedures. Hence, the term *acquisition* can be applied to protocols in which a CS and US are arranged in trials in a manner designed to promote development of a CR or to the result of application of such protocols. In other words, I can say that acquisition consisted of a specified set of trials, and that as a consequence of this experience or training a CR has been acquired. The same sort of dual usage extends to other Pavlovian terminology. *Extinction* is the gradual diminution of a CR that occurs when the US no longer occurs with the CS, but the term also refers to the procedure of presenting a CS alone after a CR has been acquired. *Spontaneous recovery* refers both to the fact that CRs typically reappear despite extinction and to the procedure for testing for spontaneous recovery of CRs after a rest period. To extend this lesson on terminology further, an acquired CR can generalize from the original CS to other (usually related) stimuli. Pavlov called this process *irradiation,* but *stimulus generalization* is the term we would apply to any procedure in which an organism trained to make a CR to one stimulus is tested with an array of other stimuli.

Pavlov's lexicon was influenced by the neurophysiology of his day. The key concepts are *excitation* and *inhibition.* Excitation refers to hypothetical brain processes that promote CR acquisition and expression. Inhibition refers to those counteracting processes that reduce or eliminate a CR. As far as the CR is concerned, excitation is a property of the CS. A CS that produced a robust CR is said to express excitatory conditioning. A CS that does not produce a robust CR might have insufficient excitation because the CS and US have not been paired together often enough or because of some overlying and counteracting inhibitory process such as extinction.

Pavlov's followers, for example, Hilgard and Marquis, were responsible for the idea that the role of the US is to reinforce learning. All that was intended by this usage of the term *reinforcement* is that the development of a CR requires pairing a US with a CS. Further pairings strengthen (reinforce) the connection between the CS and the US, and withholding the US causes extinction. Although the meaning of reinforcement in classical conditioning is straightforward, the term can cause confusion among neuroscientists because the term can mean reward or punishment of behaviors emitted by organisms in circumstances in which there is no clearly delineated triggering event, such as a CS. These days, *reinforcement learning* refers to instances of learning that involve goal-directed actions. This is the sort of learning that experimental psychologists refer to as *instrumental* or *operant conditioning*. It differs from classical conditioning in that reinforcement depends on actions of the organism. In classical conditioning, by contrast, the reinforcing events are not under the control of the organism. This learning involves adaptation to circumstances that, although predicted by a signaling stimulus such as a CS, are beyond the organism's power to change.

Experimental psychologists generally agree that different things are learned in classical conditioning and in instrumental or operant conditioning. In classical conditioning, what is learned is a connection (statistical association) between a CS and a US. Neurobiologists are interested in the physical manifestations of this association, as expressed by evidence of reorganization in neural pathways that connect brain regions concerned with the CS and those concerned with the US and expression of the CR. Instrumental conditioning is similar, but the statistical associations of interest are those between actions and outcomes. Favorable outcomes strengthen these associations, and neurobiologists are concerned with how this happens in brains. Experimental psychologists agree that both forms of reinforcement operate in most real-world instances of behavioral learning. Furthermore, the empirical relationships and the descriptive and theoretical language of classical and instrumental conditioning overlap a great deal. It is beyond the scope of this Preface to pursue this issue, except to note that a sophisticated knowledge of the facts and theories of classical conditioning would provide a solid, perhaps essential, grounding for the study of more complex and elaborate forms of behavioral learning based on principles of reinforcement.

Contents and Contributors

The lead chapter is by Stephen R. Coleman, a former doctoral student of Isidore Gormezano and currently Professor of Psychology at Cleveland State University. Coleman has published extensively on the history of classical and instrumental (operant) conditioning. His chapter is entitled "Circumstances and Themes in the History of Classical Conditioning." His chapter offers a diagnosis of the way in which American psychologists selectively appropriated Pavlov's work by subordinating it to the indigenous research on habit formation (through instrumental conditioning procedures) and demoting for several decades the role of neurophysiological questions in classical conditioning research. The result has been a belated renaissance of interest in the subject of this book.

The second chapter, by Bernard G. Schreurs and Daniel L. Alkon, is entitled "Cellular Mechanisms of Classical Conditioning." In addition to providing a detailed snapshot of fast-moving developments in the cellular and molecular biology of classical conditioning, the authors discuss important fundamental issues, such as the use of control groups to separate associative processes that can effect a CR from nonassociative processes unrelated to learning. In addition, the chapter introduces readers to several model systems employed in contemporary research. The most widely used model system is eyeblink conditioning, and the chapter considers how synaptic plasticity within the cerebellum contributes to this learning.

The third chapter, by Ray W. Winters, Philip M. McCabe, and Neil Schneiderman, is entitled "Functional Utility and Neurobiology of Conditioned Autonomic Responses." Autonomic CRs are those mediated by the autonomic nervous system and key brain structures such as the amygdala, which includes the conditioning of fear and its physiological substrates. In addition, this chapter describes research on attentional processes that underlie classical conditioning and discrimination learning.

The fourth chapter, by Germund Hesslow and Christopher H. Yeo, is entitled "The Functional Anatomy of Skeletal Conditioning." This chapter presents a comprehensive overview of the anatomical substrates of learning and performance of CRs, such as the eyeblink, at the neural systems level of analysis. The chapter emphasizes recent findings on the role of the cerebellum and hippocampus in the acquisition and expression of the classically conditioned eyeblink response.

The fifth chapter, by Paul R. Solomon, is entitled "Classical Conditioning: Applications and Extensions to Clinical Neuroscience." This chapter demonstrates that the classical eyeblink conditioning preparations in humans and animals have contributed insights about mechanisms that are responsible for impairments that result from aging, disease, and exposure to toxins.

The sixth chapter, by E. James Kehoe and Michaela Macrae, is entitled "Fundamental Behavioral Methods and Findings in Classical Conditioning." This chapter provides a comprehensive review of the paradigms and experimental designs, and findings (principally in eyeblink conditioning) that challenge emerging computational models of classical conditioning.

The seventh chapter, by Susan E. Brandon, Edgar H. Vogel, and Allan R. Wagner, is entitled "Computational Theories of Classical Conditioning." This chapter caps the volume with an extensive review of formal models of learning and conditioning in the associationist tradition of the Hull–Spence behavior theory and the Rescorla–Wagner model. Contemporary computational models strive to account for all fundamental behavioral data in paradigms reviewed in the penultimate chapter. Successful models operate in real time and can describe response topography, the effects of stimulus compounding (i.e., credit assignment with multiple CSs), and the underlying motivational support for conditioning.

Dedication

It is fitting to dedicate a book about classical conditioning for neuroscientists to Professor Isidore Gormezano, who retired as the Kenneth W. Spence Professor of Psychology at the University of Iowa in 1997 after a remarkable career of teaching, research, and service to his field and profession. His influence among the current generation of researchers is reflected in the list of contributors. Coleman (Iowa, 1972), Kehoe (Iowa, 1976), Moore (Indiana, 1962), Schneiderman (Indiana, 1965), and Schreurs (Iowa, 1985) received their doctoral training under Gormezano's supervision. Gormezano received a B.A. degree from New York University (University Heights College) in 1952; he spent 2 years in the Army Medical Corps before entering the University of Wisconsin for his graduate training. Gormezano received his Ph.D. in experimental psychology in 1958 from the University of Wisconsin. His Ph.D. mentor was David Grant, then one of the leaders in the field of classical conditioning and experimental psychology generally. Gormezano went to Indiana University in 1958 and quickly rose to rank of professor. He left Indiana in 1966 to assume the position vacated by Kenneth W. Spence when Spence moved to the University of Texas.

In 1994, Gormezano received the Howard Crosby Warren Medal from the Society of Experimental Psychologists "for his fundamental and long-term contributions to our experimental methods and systematic knowledge of learning, especially classical conditioning, and for his contributions to neuroscience and psychopharmacology," thereby joining a pantheon of Warren Medal recipients who have contributed to the field of learning: Karl Lashley (1937), Elmer Culler (1938), Ernest Hilgard (1940), B.F. Skinner (1942), Clark Hull (1945), Kenneth W. Spence (1953), Neal Miller (1954), Harry Harlow (1956), Donald Hebb (1958), Carl Hovland (1960), James Olds (1962), William Estes (1963), Benton Underwood (1964), Richard Solomon (1968), Delos Wickens (1973), Leo Postman (1974), John Garcia (1978), Abram Amsel (1980), Eric Kandel (1984), Gordon Bower (1986), Richard Thompson (1989), Robert Rescorla and Allan Wagner (1991), and M.E. Bitterman (1997).

Gormezano's publications include 11 articles describing studies of human eyeblink conditioning. These writings are primarily concerned with the methodological problem that the eyeblink is partially under voluntary control and the possibility that instrumental learning contingencies may operate in eyeblink conditioning. More than 70 research articles concern classical conditioning of nictitating membrane extension or jaw movement in rabbits. Many of these studies are seminal parametric investigations. Others exploit these preparations in studies involving anatomical, pharmacological, or physiological manipulations. Gormezano has also been coeditor of two books on classical conditioning and learning: *Classical Conditioning,* Third Edition, with W.F. Prokasy and R.F. Thompson (Erlbaum 1987), and *Learning and Memory: The Biological Substrates,* with E.A. Wasserman (Erlbaum 1992). He has published 10 review articles, 6 in Spanish, 24 chapters, some in textbooks, and 13 articles on methodology and instrumentation.

Gormezano prided himself on being a blue-collar experimentalist with the skills of a machinist, carpenter, and electronics savant to build and develop instrumentation solutions for work in the field. Perhaps the most noteworthy of his methodological innovations is the "Gormezano box" used to restrain rabbits in experiments in classical conditioning. This device is described in Gormezano's authoritative chapter in Sidowski's 1966 book on methods and instrumentation.

The Gormezano restrainer allows rabbits to sit snugly but comfortably in a natural posture. Padded pinna clamps and an adjustable neck yoke contributed to stability. This device is suitable for all conditioning protocols, including neuronal recording with single-unit isolation. These restraining boxes have been used in almost all rabbit labs.

Gormezano was among the first to use digital technology to record nictitating membrane responses waveforms, and he and his associates developed the most elaborate early data-acquisition and -analysis systems based on stand-alone microprocessors. The Apple/FIRST system (based on Apple II computers and the powerful FIRST operating system) is still in use in some labs, even as powerful PCs and workstations have come on line during the past 20 years. Gormezano appreciated the power and efficiency of running several subjects at a time. Gormezano's other methodological innovations include (1) a goldfish restrainer and analog transducer for measuring tail flicks in response to electrical stimulation in as many as 12 fish simultaneously; (2) a pneumatically driven noise-free tactile stimulater; (3) a computer-controlled I.V. drug delivery system; and (4) a system for delivering gaseous anesthetics to as many as 8 rabbits simultaneously while recording response parameters.

Gormezano is especially proud of his contributions to what he refers to in his C.V. as *scientific developments*. These include the following: (1) rabbit jaw movement response and its laws of classical conditioning; (2) rabbit eyelid, nictitating membrane, eyeball retraction, and their laws of classical conditioning; (3) frog nictitating membrane response and its reflexology; and (4) contingent yoked control design in classical conditioning.

Gormezano has given more than 120 invited lectures and addresses at meetings, and as a colloquia and symposia participant. In addition to invitations from colleges and universities in the United States, invitations to lecture have come from many countries. In 1975, he gave 15 lectures at Australian universities over a span of weeks. He gave a series of 9 lectures in Spain in 1981 and a series of 4 lectures in Russia in 1991. Other foreign venues include England, Czechoslovakia, France, Norway, Mexico, Canada, Hungary, and Germany. He has lectured extensively in Mexico, Spain, and Russia, further enhancing his international reputation and influence.

Acknowledgments

In addition to the contributors, I wish to thank two individuals without whose help my responsibilities as editor would have been overwhelming. Stephen R. Coleman helped with proofreading and fine-tuning several chapters, and he helped with the creation of the indexes. Jordan S. Marks, a graduate student in neuroscience and behavior, also helped with editing, but more importantly, he served as Webmaster and digital savant responsible for formatting textual and graphic materials into digitized documents.

Amherst, Massachusetts   *John W. Moore*

Contents

Preface		ix
Contributors		xix
1.	Circumstances and Themes in the History of Classical Conditioning *Stephen R. Coleman*	1
2.	Cellular Mechanisms of Classical Conditioning *Bernard G. Schreurs and Daniel L. Alkon*	14
3.	Functional Utility and Neurobiology of Conditioned Autonomic Responses *Ray W. Winters, Philip M. McCabe, and Neil Schneiderman*	46
4.	The Functional Anatomy of Skeletal Conditioning *Germund Hesslow and Christopher H. Yeo*	86
5.	Classical Conditioning: Applications and Extensions to Clinical Neuroscience *Paul R. Solomon*	147
6.	Fundamental Behavioral Methods and Findings in Classical Conditioning *E. James Kehoe and Michaela Macrae*	171
7.	Computational Theories of Classical Conditioning *Susan E. Brandon, Edgar H. Vogel, and Allan R. Wagner*	232
Name Index		311
Subject Index		313

Contributors

Daniel L. Alkon, Laboratory of Adaptive Systems, National Institute of Neurological Disorders and Stroke, National Institutes of Health, Bethesda, MD 20892, USA

Susan E. Brandon, Department of Psychology, Yale University, New Haven, CT 06520, USA

Stephen R. Coleman, Department of Psychology, Cleveland State University, Cleveland, OH 44115, USA

Germund Hesslow, Department of Physiological Sciences, University of Lund, Lund SE-221 84, Sweden

E. James Kehoe, School of Psychology, The University of New South Wales, Sydney, NSW 2052, Australia

Michaela Macrae, School of Psychology, The University of New South Wales, Sydney, NSW 2052, Australia

Philip M. McCabe, Department of Psychology, University of Miami, Coral Gables, FL 33124, USA

John W. Moore, Department of Psychology, University of Massachusetts, Amherst, MA 01003, USA

Neil Schneiderman, Department of Psychology, University of Miami, Coral Gables, FL 33124, USA

Bernard G. Schreurs, Department of Physiology and Blanchette Rockefeller Neurosciences Institute, West Virginia University School of Medicine, Morgantown, WV 26506, USA

Paul R. Solomon, Department of Psychology, Williams College, Williamstown, MA 01267, USA

Edgar H. Vogel, Department of Psychology, Yale University, New Haven, CT 06520, USA

Allan R. Wagner, Department of Psychology, Yale University, New Haven, CT 06520, USA

Ray W. Winters, Department of Psychology, University of Miami, Coral Gables, FL 33124, USA

Christopher H. Yeo, Department of Anatomy and Developmental Biology, University College London, London WC1E 6BT, UK

1
Circumstances and Themes in the History of Classical Conditioning

STEPHEN R. COLEMAN

What we now call classical (or Pavlovian) conditioning was examined systematically in the early twentieth century by a digestive system physiologist named Ivan P. Pavlov. His visibility to psychologists at that time and his enduring fame as a pioneer in twentieth-century psychology have obscured the differences that eventually estranged him from American pioneers in conditioning research and theory. Those differences turned the subsequent history of conditioning in the United States away from Pavlov's cherished objectives (e.g., Pavlov, 1932). Like Wilhelm Wundt, who is routinely identified as the founder of experimental psychology (but see Blumenthal, 1975), Pavlov came to be another misunderstood but frequently cited ancestor of contemporary psychology. Because of a very selective appropriation of Pavlov's work by American psychologists, some later projects in American conditioning can best be understood as efforts to reconnect with Pavlov (Dewsbury, 1997; Gormezano et al., 1983; Kline, 1961; Rescorla, 1988).

Pavlov

Pavlov's discoveries were made at the Military Medical Academy in Petrograd (St. Petersburg) in an institute for physiological research of which he was Director. His training in physiology and his research on nervous system control ("nervism": Babkin, 1949, p. 225) of the digestive system eventually led him to an interest in questions about a simple kind of learning that he called signalization (Pavlov, 1927/1960, p. 17) and which we now identify as classical conditioning. Historically, questions about learning belonged to the philosophical province of association psychology, but Pavlov was convinced that a physiological inquiry into those questions would be more fruitful than speculation or introspection. However, in the period before the development of electronic amplifiers after World War I, technical obstacles to recording from the brain limited the ways in which Pavlov could address those questions (Hilgard, 1987, pp. 446–447). His strategy for handling that limitation resulted in a less clear-headed reception from American psychologists than he would have liked (Pavlov, 1932).

Pavlov's strategy for investigating signalization in dogs resembled the methodology advocated in the early seventeenth century by Francis Bacon. That is, Pavlov accumulated quantitative (tabled, individual subject) data from carefully conducted laboratory research (Pavlov, 1927/1960, pp. 20–21, 25–27) to refine an overall theory of the brain mechanisms of signalization. A small number of hypotheses about some aspect of the theory would be narrowed down by comparing their predicted outcomes against the result of an appropriately designed experiment. Hypotheses were thereby either retained or rejected, and the overall theory was repeatedly modified in piecemeal fashion in an ongoing program of systematic experimentation. As a result of Pavlov's Baconian investigative strategy, his "inferential physiology" (see Kimble, 1961, pp. 32–35) was constantly changing, and it resisted easy summarization (Loucks, 1937; Morgulis, 1914). Although Pavlov did not provide a summary appraisal of the merits of his canine salivary-conditioning preparation (Pavlov, 1927/1960, lect. 2), we offer the following description as a modern equivalent that captures his investigative aims:

> The mammalian order exhibits the greatest amount of evolutionary development of the cerebral hemispheres, and the dog is a representative mammal. The salivary system supplies representative behavioral data that offer advantages in ease and precision of measurement and provide less encouragement to "anthropomorphic" interpretations than do other food-relevant actions of the dog (Pavlov, 1927/1960, p. 18). Because the components of the digestive system obey the principle of nervism, the canine salivary-conditioning preparation will surely reveal important phenomena of signalization learning. Nevertheless, the principal value of these phenomena consists in helping us develop an empirically responsive account of the cortical mechanisms that we must assume if we are to account for such phenomena and rule out alternative explanations.

Let us enlarge the perspective so as to see what factors influenced the reception of Pavlov's work by psychologists in the United States. In the late nineteenth century, experimental psychology was often identified as the New Psychology because it was gradually replacing an established philosophical psychology as an academic discipline in American colleges and universities (Evans, 1984). That older, entrenched psychology was a speculative system of mental faculties derived from Thomas Reid's philosophy of the human mind (Reid, 1785/1971). The resulting complex and lengthy process of disciplinary revision has been placed under such historical themes as psychology's quest for disciplinary independence from academic philosophy and psychology's search for identity as a natural science, perhaps as a biological science, as many psychologists in the early twentieth century urged (Angell, 1907).

Psychology's disciplinary aspirations were fundamental to its agenda in the history of American psychology in the period of 1890 to 1930. This period witnessed such developments as the founding of psychology laboratories in the late 1800s and the subsequent conflict among antagonistic proposals (i.e., such schools of psychology as functionalism, behaviorism, etc.) for a disciplinary identity for psychology in the first third of the twentieth century. Questions about psychology's status were the most pressing concerns when the writings of Pavlov and others (Kimble, 1967, p. 44) came to the attention of American psychologists in the period 1910–1930.

Psychology's quest for scientific status from 1890 to 1930 inspired commitments by psychologists to such creeds as objectivism, positivism, mechanism, and behaviorism. Early behaviorists were firmly rooted in the nineteenth-century Zeitgeist of mechanism (see Williams, 1976, pp. 163–169), a worldview that promoted the extension of a scientific attitude to human beings. This extension was opposed by the Old Psychology, as well as by academic philosophy (Kuklick, 1977), by established religions and their distinguished educational institutions in the United States, and by folk belief and perennial superstition (Leahey, 1997; for contemporary flavor, see White, 1896).

Early behaviorists were sure that human beings can be adequately understood only from a framework of materialism, naturalism, and biological evolution. They distrusted and tried to avoid concepts and ideas that belonged to a traditional, religiously supported doctrine of human nature (Watson, 1930, pp. 1–5). Finally, they regarded causal explanation as the only legitimate mode of explanation. When early behaviorists spoke of causality, however, they meant "mechanical necessity." That is, antecedent physical conditions (e.g., eliciting stimuli) bring about (compel or provoke) the effect (e.g., response) invariably, inevitably, "necessarily." Examples of this particular interpretation of causality can be found throughout the writings of Watson (1913, 1925), Hull (1937; also 1943, especially ch. 1; see Smith, 1986, pp. 158–162), and Skinner (1931). As a result, when Pavlov came to the attention of American psychologists, what he put forth as a representative nineteenth-century mechanist (Pavlov, 1927/1960, lect. 1 and 2) was entirely in step with the mechanistic leanings of his audience.

The work of Marshall Hall and other nineteenth-century physiologists (see Fearing, 1930) had brought the concept of reflex fully within this framework of analogy and metaphor that we call mechanism (Fearing, 1930; Skinner, 1931). As a result, the concept of reflex had great appeal to early behaviorists, independent of its established role in nervous system physiology, the science on which the New Psychology had been most dependent in the effort to bring Mind into a scientific framework. The reception of Pavlov's work by American psychologists occurred within this Zeitgeist.

In broad terms, Pavlov's findings substantiated much that was already well known, for example, that a great variety of organisms can learn to anticipate future events. But Pavlov's essential contribution was that he had objectively demonstrated many features of anticipation and had empirically identified the circumstances on which anticipation depends. He had thereby shown that anticipation of the future is an entirely natural, biological property of living organisms. That a laboratory scientist like Pavlov could provoke the admiration of such public intellectuals as H.G. Wells (Wells, November 13, 1927) may seem a bit surprising, but one should keep in mind that Pavlov's careful demonstrations served as examples of how questions about behavior, mentality, or almost any topic could be addressed empirically and given reliable and definite answers. Indeed, the appeal of such related doctrines as objectivism, positivism, and scientism owed much to the presumption that objective information is always preferable to religious authority, traditional verities, superstition, and mere opinion (Ross, 1991). Pavlov's reception by the New Psychology in the early twentieth century reflected this climate of esteem for science, and classical conditioning has subsequently enjoyed a reputation as a specialty whose solid empirical foundation has fostered steady progress for about a century.

Pavlov's reception in the United States was affected not only by such all-encompassing attitudes as we have just described but also by specific, accidental, historical circumstances. In this regard, it is noteworthy that John B. Watson espoused behaviorism (Watson, 1913) long before he became a Pavlovian (Watson, 1925; Watson & Rayner, 1920). Behaviorism was also promoted by E.R. Guthrie (Smith & Guthrie, 1921) well before he developed a theory of learning that assimilated Pavlov's findings (Guthrie, 1930). This delay between commitment to behaviorism and utilization of Pavlovian ideas resulted mostly from a language barrier. Because Russian was not a world scientific language at the time, Americans had to rely on brief English-language reviews (Morgulis, 1914; Yerkes & Morgulis, 1909) and Pavlov's short speeches (Pavlov, 1906) for information about his work.

After Pavlov

A thorough description of Pavlov's procedures, findings, and theory was not available in English until 1927, when G.V. Anrep published an authorized (by Pavlov) translation of lectures that Pavlov had delivered in 1924 on the subject of conditioning (*Conditioned Reflexes:* Pavlov, 1927/1960). By that time, behaviorism was already in the full bloom of its early period and, with only a sketchy picture of classical conditioning, was using and extending Pavlovian concepts. Indeed, Watson's (1925) theorizing had stretched the concept of the Pavlovian "conditioned reflex" (CR) to include overt activity sequences as well as inferred bodily states, such as emotions. This conceptual loosening ran counter to qualities that had made the conditioned reflex an appealing construct for early behaviorists, especially its close tie to laboratory procedures and results.

The conceptual loosening of "conditioned reflex," which we have just described, was part of a more pervasive development. By the mid-1920s, psychologists had replaced "reflex "with terms such as "response" or "reaction," as well as "conditioned stimulus-response (S-R) connections," "associations," or "relations" (Allport, 1924, p. 41; Smith & Guthrie, 1921; Watson & Rayner, 1920).

This seemingly minor terminological change reflected the concern for disciplinary identity that was a preoccupation of American psychology in the early twentieth century. Substitution of "response" for "reflex" contributed to a loosening of the historical dependence of American psychology on physiology (Skinner, 1938, ch. 12). This terminological change also supported an elaboration of the significance of conditioning in normal and abnormal development of personality (Burnham, 1924; Watson, 1925; Watson & Rayner, 1920) in humans engaged in complex daily activities (Tolman, 1932, ch. 1). Pavlov's work on such apparently insignificant behavior as the salivation of dogs would have more obvious relevance to human psychology if the Pavlovian CR could be freed from the restrictiveness of the physiological concept of "reflex" and if "conditioned response" could be extended to a larger range of learned activities, of which Watson's (1925, 1930) speculative extensions were the most ambitious.

We have emphasized the importance of physiology in the development of the laboratory-based New Psychology in the United States. Such a relation could conceivably have shaped an American conditioning enterprise that closely resembled Pavlov's in making the discovery of the brain mechanisms of classical conditioning a priority. Indeed, several physiological inquiries into classical conditioning were undertaken from the late 1920s into the early 1940s in the United States. For example, to discover what conditions are necessary or sufficient for conditioning, and what kind of association (a stimulus-stimulus, or S-S, association or stimulus-response, or S-R, association) is developed during conditioning, investigators blocked the expression of the unconditioned response (UR) by drugs (Crisler, 1930) or evoked it by electrical stimulation as an unconditioned stimulus (US) (Loucks, 1935). This line of research engaged the interest of talented investigators in the 1930s and early 1940s (Culler, Girden, Harlow, Hilgard, Shurrager, and others; see Hilgard & Marquis, 1940, pp. 317–326, and Kimble, 1961, pp. 6–8), but these promising inquiries never moved from a marginal to a primary status in the classical conditioning enterprise. Many circumstances contributed to this missed opportunity; the following are examples of such influences.

First, physiologically oriented research in conditioning clashed with the growing conviction that behavioral psychology could (and, moreover, should) be carried out as an independent discipline and not become sidetracked by questions about the physiological substrate of learning (Tolman, 1932, ch. 1; see especially Skinner, 1938, ch. 12, for a spirited promotion of this position).

Next, the objectivism and positivism of some scientific psychologists inclined them to see Pavlov's physiological inferences as entirely speculative (Guthrie, 1930; Loucks, 1933, 1937). Even after a renewal of Western interest in Pavlov during the Cold War period (Kline, 1961), such writers as Kimble (1961) and Spence (1961) never missed a chance to remind readers of the inferential status of Pavlov's physiology. Third, Pavlov's physiology was contrary to the Sherringtonian system of neuron and synapse (Denny-Brown, 1932), which had become established in neurophysiology by the time *Conditioned Reflexes* appeared in 1927. Fourth, although the neurophysiological investigation of some movement systems had made enormous progress in the analysis of animal posture and locomotion (Creed et al., 1932; Magnus, 1924; Sherrington, 1906), neurophysiology in the 1930s had little to offer for an understanding of even the simplest forms of learning (Hilgard & Marquis, 1940, ch. 13; Skinner, 1938, ch. 12; Spence, 1956).

Thus far, we have focused on what American psychology failed to do with Pavlov. That is, Pavlov's use of conditioning as a vehicle for the study of brain mechanisms was not given the welcome he had expected (Pavlov, 1932) and did not establish itself as the primary aim of the study of conditioning in the United States. We now take note of what was done with Pavlov's work in the United States by describing in general terms two programs of conditioning research.

We have already mentioned the first of these two American conditioning programs, namely, the speculative extension of conditioning as a scientific refinement of the philosophical concept of association (see Gormezano et al., 1983, pp. 198–201) or as the fundamental process in all varieties of learning, as Watson claimed (Watson, 1925, 1930). Watson's work belongs to a first stage of behaviorism, as we noted earlier, whereas the publication of *Conditioned Reflexes* in 1927 provided material for influential publications (Guthrie, 1930; Hull, 1929) that began a second phase of behaviorism, typically called neobehaviorism.

In this second phase, the CR was saddled with a theory-serving role that Pavlov could never have anticipated, because it resulted from a sharp distinction of classical or Pavlovian conditioning from instrumental or Thorndikean learning that was made during the 1930s and solidified in succeeding years (Hilgard & Marquis, 1940, pp. 68–74). The distinction was not symmetrical, because theorists tended to regard instrumental learning through reinforcement and punishment as the more important in the life of organisms (e.g., Skinner, 1938, pp. 19, 22, 438), and because instrumental conditioning dominated as a vehicle for learning theory debates among Hull, Tolman, and Guthrie from the 1930s into the 1950s.

A widely endorsed version of the two-types formula assigned a subordinate but important role to Pavlov-type conditioning as a hypothetical mechanism for the anticipation of events (Hull, 1937) or as an inferred source of conditioned motivation (Amsel, 1958; Miller, 1951; Mowrer, 1960; Spence, 1956). This theoretical use of conditioning called for hypothetical CRs to replace presumptive inner, so-called mental or psychic, determinants of behavior, thus extending the scope and power of behaviorist theory (Hall, 1989, pp. 18–22). In place of mental states, theorists made use of theoretical, so-called covert CRs to represent the organism's acquired anticipation of future events. According to this use of conditioning, environmental cues of future events influence behavior by eliciting inferred CRs that mediate the influence of future events by motivating, inhibiting, prompting, or otherwise affecting the performance of instrumental actions. There were many versions of such a mediating response model in conditioning and in other types of learning (Osgood, 1953, pp. 342–412; see Leahey, 1997, pp. 405–407, 413–415), often with a role for hypothesized Pavlovian CRs (Hull, 1952; Mowrer, 1960; Spence, 1956).

It is obvious that the postulation of hypothetical CRs that mediate the effects of environmental arrangements on overt instrumental behavior was consonant with Watson's earlier loosening of the denotation of the concept of CR. However, use of CRs as hypothetical mediators also diminishes the empirical content of CR and thus constitutes a departure from the doctrine of objectivism, which had been a touchstone of early behaviorism and a hallmark of Pavlov's employment of the fundamental concepts of conditioning. So, again we have encountered features of the American conditioning enterprise that would have disturbed Pavlov.

The success of mediation-based theory invited questions concerning the reality status of postulated mediating CRs. Some influential theorists declared that such questions are extraneous to the enterprise of theory development and testing (Spence, 1956). Others were less sure of this separation (Lachman, 1960). Still others regarded the mediating CR as a hypothetical construct, a potentially measurable behavioral mechanism, and so they treated the question of the reality status of mediating CRs as an empirical question (Williams, 1965). Indeed, the objectivist dogma in the history of conditioning encouraged investigators to overcome technical difficulties to direct recording of presumptive CRs in animals moving freely in instrumental-conditioning preparations (Miller & DeBold, 1965). The very terminology of learning theory, a terminology of stimuli and responses, encouraged an assumption of measurability that a mentalistic language of thoughts, experiences, and feelings would not have encouraged.

At this point, it is sufficient to say that direct efforts to substantiate or validate such mediational models of conditioning ultimately failed, so that the influential review of two-factor mediation theory by Rescorla and Solomon (1967) was mostly a critique of the direct measurement strategy, a diagnosis of the problem, and a call for theorizing of a more cognitive style. The appeal of such theorizing was tremendously enhanced by the discovery of so-called informational effects in classical conditioning (Kamin, 1969; Rescorla, 1967; cf. an alternative model proposed by Rescorla & Wagner, 1972), and cognitive-style theory became a sort of Zeitgeist in the last quarter-century (Mackintosh, 1974, 1983; see Hilgard's diagnosis in his 1987, pp. 221–225).

In describing developments in the first of two American uses of conditioning, we have continued our theme of American departures from Pavlov's program. Earlier, we noted that Pavlov's physiological theory failed to take hold in America. In describing American behavior theory, we saw that the empirical character of conditioning concepts was attenuated as CRs were assigned a theoretical role. We now describe a more thoroughly empirical use of conditioning in the United States.

Pavlov's salivary-conditioning arrangement was not much more successful than his physiological theory in winning American converts. The salivary-conditioning procedure was employed by a few productive American researchers, such as W.H. Gantt, but its wider acceptance was stalled by technical obstacles (see Gormezano, 1966, pp. 389–397). In addition, American investigators worked in the wake of Watson's promotion of conditioning in human psychology and, therefore, were particularly interested in the possibility and the pervasiveness of human conditioning (Hamel, 1919; Lashley, 1916). From these concerns arose quite a variety of human conditioning preparations (see Hilgard & Marquis, 1940, pp. 30–35; Hull, 1934, pp. 392–416; Kimble, 1961, p. 51).

As the classical conditioning enterprise grew, the various human conditioning arrangments competed for dominance as the preferred conditioning preparation, a competition in which one would expect technical and methodological virtues to be deciding factors. The human eyeblink arrrangement, first demonstrated by Hulsey Cason (1922), was numerically dominant by the 1950s and 1960s, thanks to methodological refinements in CR definition (Grant, 1943a, b, 1945), improvements in recording eyeblink CRs (Spence & Taylor, 1951), and use of this arrangement in the productive conditioning programs of Kenneth Spence at Iowa and David Grant at Wisconsin (Coleman, 1985, pp. 106–109).

Other vehicles for studying conditioning did not adhere so closely to the Pavlovian practice of directly measuring the CR as did the salivary and eyeblink preparations. Although CR-based mediation theory ultimately floundered in the 1960s, as we have just seen, the idea of inferred CRs in instrumental-learning preparations survived in the conditioned emotional response (CER) preparation that Estes and Skinner (1941) introduced and that Watson and Rayner (1920) had anticipated. In the CER preparation, the learning that results from CS-US presentations (without concurrent measurement of CRs) is assessed later by examining the effect of superimposing the CS on an established baseline of instrumental-response (IR) performance. Reduction in response rate below the baseline is an indicator of the degree of conditioning of fear that the CS had undergone. During the 1960s and 1970s, there emerged several Pavlovian preparations that Gormezano and Kehoe (1975) would identify as CS-IR preparations: in a CS-IR preparation, one infers CR strength (during a separate test phase after conditioning) from change in instrumental-response (IR) strength from a CS-free baseline. By contrast, such CS-CR preparations as salivary and eyeblink provide direct, on-trial measurement of CR during its development. Although such CS-CR arrangements as human eyeblink conditioning were dominant in the 1940s and 1950s, as we noted, CS-IR arrangements became more prevalent in the 1970s, 1980s, and 1990s.

This methodological development was sharply at variance with the logic of Pavlov's insistence that learned salivation should be considered a reflex and, therefore, should be annexed to physiology on the grounds that he could identify the stimulus circumstances that attended measured changes in salivation from occasion to occasion (Pavlov, 1927/1960, pp. 20–26; 1928, pp. 264–265). Pavlov's stance was quite consistent with an important property of the conditioning arrangement he employed, namely, that it afforded the possibility of a close specification of the stimulus antecedents of variations in the response (measured dependent variable). CS-IR preparations do not permit this specification but provide the convenience of a (later) indirect assessment and inference of the degree of previously established conditioning. Consequently, even in laboratory procedures, American researchers departed from the example that Pavlov set.

A sizeable amount of classical conditioning research simply obeyed the functionalist injunction to ascertain the empirical relations (Hilgard, 1956, ch. 10) that conditioning exhibits. Such work could be done without commitments to ambitious theorizing. Indeed, functionalism was a viable, laboratory-based alternative to behaviorism. Gormezano was well versed in a line of functionalist thinking that stemmed from Raymond Dodge (at Wesleyan), through E.R. Hilgard (at Yale, then at Stanford), to David Grant (at Wisconsin), with whom Gormezano trained from 1955 to 1958. The metapsychological commitments of this lineage (see Hilgard, 1956, ch. 10) dissuaded Gormezano from adopting an identifiable allegiance to behaviorism—at least not to either of the two varieties (Hull–Spence and Skinnerian) that were dominant when he was in graduate school—even though his conceptual tough-mindedness resembled the behaviorist attitude of "objectivism."

The same training background made Gormezano uncomfortable with the tendency of later cognitive-style conditioning theorists (Dickinson, 1980; Mackintosh, 1974; Rescorla, 1967) to conflate theory and description. Finally, the cautious empiricism of his training background inclined him away from higher-level formulae, such as evolutionary schemata that were promoted in the "biological predisposition" movement that closely followed interest in selective associations (Garcia & Koelling, 1966) and autoshaping (Brown & Jenkins, 1968).

Gormezano's training made it possible for him, early in his career, to distinguish himself as a methodologist (Gormezano, 1965, 1966). He followed up by laboriously refining the rabbit-eyeblink and nictitating-membrane response (NMR) preparations, and published demonstrations that they had all the methodologically desirable qualities that a classical conditioning preparation should have (Gormezano et al., 1962; Schneiderman et al., 1962).

His subsequent functionalist projects, which aimed to map, in definitive fashion, the influence of CS-US interval (Smith et al., 1969), CS effects (Gormezano, 1972), and US factors (Tait et al., 1983) were predicated upon the success of his efforts to show that the NMR preparation met the highest methodological standards among conditioning preparations. This preparation figures prominently in the chapters of this book.

Finally, with a methodologically superior preparation, one could address questions of the biological substrate of conditioning with greater optimism than is afforded by other preparations (Cegavske et al., 1976). Gormezano's work along these lines intensified in the late 1970s and became his primary interest in the second half of his career (roughly 1979–1999). Gormezano benefited from a windfall of technical advances in techniques of electrical and chemical stimulation and CNS inactivation that were developed in the last 20 years. With such tools, Gormezano was able to provide experimental demonstrations, whereas Pavlov, facing unsolvable technical obstacles to recording from the brain, had to fall back on a strategy of inference, with consequences that we have described at length earlier. One might go so far as to say that Gormezano eventually accomplished a technically improved reconnection with Pavlov's use of conditioning as a vehicle for exploring brain mechanisms of learning. Good scientific work is cumulative, and Gormezano thus made it possible for others to make further progress along that path.

References

Allport, F.H. (1924). *Social psychology.* Boston: Houghton Mifflin.
Amsel, A. (1958). The role of frustrative nonreward in noncontinuous reward situations. *Psychological Bulletin, 55,* 102–119.
Angell, J.R. (1907). The province of functional psychology. *Psychological Review, 14,* 61–91.
Babkin, B.P. (1949). *Pavlov: a biography.* Chicago: University of Chicago Press.
Blumenthal, A.L. (1975). A reappraisal of Wilhelm Wundt. *American Psychologist, 30,* 1081–1086.
Brown, P.L., and Jenkins, H.M. (1968). Autoshaping of the pigeon's key-peck. *Journal of the Experimental Analysis of Behavior, 11,* 1–8.
Burnham, W.H. (1924). *The normal mind.* New York: Appleton-Century-Crofts.

Cason, H. (1922). The conditioned eyelid reaction. *Journal of Experimental Psychology, 5,* 153–196.

Cegavske, C.F., Thompson, R.F., Patterson, M.M., and Gormezano, I. (1976). Mechanisms of efferent neuronal control of the reflex nictitating membrane response in the rabbit. *Journal of Comparative and Physiological Psychology, 90,* 411–423.

Coleman, S.R. (1985). The problem of volition and the conditioned reflex. Part I: Conceptual background, 1900–1940. *Behaviorism, 13,* 99–123.

Creed, R.S., Denny-Brown, D., Eccles, J.C., Liddell, E.G.T., and Sherrington, C.S. (1932). *Reflex activity of the spinal cord.* London: Oxford University Press.

Crisler, G. (1930). Salivation is unnecessary for the establishment of the salivary conditioned reflex induced by morphine. *American Journal of Physiology, 94,* 553–556.

Denny-Brown, D. (1932). Theoretical deductions from the physiology of the cerebral cortex. *Journal of Neurology and Psychopathology, 13,* 52–67.

Dewsbury, D.A. (1997). In celebration of the centennial of Ivan P. Pavlov's (1897/1902) *The Work of the Digestive Glands. American Psychologist, 52,* 933–935.

Dickinson, A. (1980) *Contemporary animal learning theory.* Cambridge: Cambridge University Press.

Estes, W.K., and Skinner, B.F. (1941). Some quantitative properties of anxiety. *Journal of Experimental Psychology, 29,* 390–400.

Evans, R.B. (1984). The origins of American academic psychology. In J. Brozek (Ed.), *Explorations in the history of psychology in the United States* (pp. 17–60). Lewisburg, PA: Bucknell University Press.

Fearing, F. (1930). *Reflex action.* Philadelphia: Williams and Wilkins.

Garcia, J., and Koelling, R.A. (1966). Relation of cue to consequence in avoidance learning. *Psychonomic Science, 4,* 123–124.

Gormezano, I. (1965). Yoked comparisons of classical and instrumental conditioning of the eyelid response; and an addendum on "voluntary responders." In W.F. Prokasy (Ed.), *Classical conditioning* (pp. 48–70). New York: Appleton-Century-Crofts.

Gormezano, I. (1966). Classical conditioning. In J.B. Sidowski (Ed.), *Experimental methods and instrumentation in psychology* (pp. 385–420). New York: McGraw-Hill.

Gormezano, I. (1972). Investigations and defense and reward conditioning in the rabbit. In A.H. Black, and W.F. Prokasy (Eds.), *Classical conditioning II: current research and theory* (pp. 151–181). New York: Appleton-Century-Crofts.

Gormezano, I., and Kehoe, E.J. (1975). Classical conditioning: some methodological-conceptual issues. In W.K. Estes (Ed.), *Handbook of learning and cognitive processes,* Vol. 2. *Conditioning and behavior theory* (pp. 143–179). Hillsdale, NJ: Erlbaum.

Gormezano, I., Schneiderman, N., Deaux, E., and Fuentes, I. (1962). Nictitating membrane: classical conditioning and extinction in the albino rabbit. *Science, 138,* 33–34.

Gormezano, I., Kehoe, E.J., and Marshall, B.S. (1983). Twenty years of classical conditioning research with the rabbit. *Progress in Psychobiology and Physiological Psychology, 10,* 197–275.

Grant, D.A. (1943a). The pseudo-conditioned eyelid response. *Journal of Experimental Psychology, 32,* 139–149.

Grant, D.A. (1943b). Sensitization and association in eyelid conditioning. *Journal of Experimental Psychology, 32,* 201–212.

Grant, D.A. (1945). A sensitized eyelid reaction related to the conditioned eyelid response. *Journal of Experimental Psychology, 35,* 393–402.

Guthrie, E.R. (1930). Conditioning as a principle of learning. *Psychological Review, 37,* 412–428.

Hall, J.F. (1989). *Learning and memory* (2nd ed.). Boston: Allyn and Bacon.
Hamel, I.A. (1919). A study and analysis of the conditioned reflex. *Psychological Monographs, 27* (No. 118).
Hilgard, E.R. (1956). *Theories of learning* (2nd ed.). New York: Appleton-Century-Crofts.
Hilgard, E.R. (1987). *Psychology in America: a historical survey.* San Diego, CA: Harcourt Brace Jovanovich.
Hilgard, E.R., and Marquis, D. G. (1940). *Conditioning and learning.* New York: Appleton-Century-Crofts.
Hull, C.L. (1929). A functional interpretation of the conditioned reflex. *Psychological Review, 36,* 498–511.
Hull, C.L. (1934). Learning: II. The factor of the conditioned reflex. In C. Murchison (Ed.), *A handbook of general experimental psychology* (pp. 382–455). Worcester, MA: Clark University Press.
Hull, C.L. (1937). Mind, mechanism, and adaptive behavior. *Psychological Review, 44,* 1–32.
Hull, C.L. (1943). *Principles of behavior.* New York: Appleton-Century-Crofts.
Hull, C.L. (1952). *A behavior system.* New Haven, CT: Yale University Press.
Kamin, L.J. (1969). Predictability, surprise, attention, and conditioning. In B.A. Campbell and R.M. Church (Eds.), *Punishment and aversive behavior* (pp. 279–296). New York: Appleton-Century-Crofts.
Kimble, G.A. (1961). *Hilgard and Marquis' conditioning and learning.* New York: Appleton-Century-Crofts.
Kimble, G.A. (Ed.) (1967). *Foundations of conditioning and learning.* New York: Appleton-Century-Crofts.
Kline, N.S. (Ed.) (1961). Pavlovian conference on higher nervous activity. *Annals of the New York Academy of Sciences, 92* (Art. 3), pp. 813–1198.
Kuklick, B. (1977). *The rise of American philosophy, Cambridge, Massachusetts, 1860–1930.* New Haven: Yale University Press.
Lachman, R. (1960). The model in theory construction. *Psychological Review, 67,* 113–129.
Lashley, K.S. (1916). Reflex secretion of the human parotid gland. *Journal of Experimental Psychology, 1,* 461–493.
Leahey, T.H. (1997). *A history of psychology* (4th ed.). Upper Saddle River, NJ: Prentice-Hall.
Loucks, R.B. (1933). An appraisal of Pavlov's systematization of behavior from the experimental standpoint. *Journal of Comparative Psychology, 15,* 1–47.
Loucks, R.B. (1935). The experimental delimitation of structures essential for learning: the attempt to condition striped muscle responses with faradization of the sigmoid gyrus. *Journal of Psychology, 1,* 5–44.
Loucks, R.B. (1937). Reflexology and the psychobiological approach. *Psychological Review, 44,* 320–338.
Mackintosh, N.J. (1974). *The psychology of animal learning.* New York: Academic Press.
Mackintosh, N.J. (1983). *Conditioning and associative learning.* Oxford: Oxford University Press.
Magnus, R. (1924). *Die Körperstellung.* Berlin: Springer.
Miller, N.E. (1951). Learnable drives and rewards. In S.S. Stevens (Ed.), *Handbook of experimental psychology* (pp. 435–472). New York: Wiley.
Miller, N.E., and Debold, R.C. (1965). Classically conditioned tongue-licking and operant bar pressing recorded simultaneously in the rat. *Journal of Comparative and Physiological Psychology, 59,* 109–111.

Morgulis, S. (1914). Pavlov's theory of the function of the central nervous system and a digest of some of the more recent contributions to this subject from Pavlov's laboratory. *Journal of Animal Behavior, 4*, 362–379.

Mowrer, O.H. (1960). *Learning theory and behavior.* New York: Wiley.

Osgood, C.E. (1953). *Method and theory in experimental psychology.* New York: Oxford University Press.

Pavlov, I.P. (1906). The scientific investigation of the psychical faculties or processes in the higher animals. *Science, 24*, 613–619.

Pavlov, I.P. (1928). *Lectures on conditioned reflexes.* (W.H. Gantt, Ed. and Trans.). New York: International.

Pavlov, I.P. (1932). The reply of a physiologist to psychologists. *Psychological Review, 39*, 91–127.

Pavlov, I.P. (1960). *Conditioned reflexes.* (G.V. Anrep, Ed. and Trans.). New York: Dover. (Original work published 1927.)

Reid, T. (1971). *Essays on the intellectual powers of man.* New York: Garland. (Original work published 1785.)

Rescorla, R.A. (1967). Classical conditioning and its proper control procedures. *Psychological Review, 74*, 71–80.

Rescorla, R.A. (1988). Pavlovian conditioning: it's not what you think it is. *American Psychologist, 43*, 151–160.

Rescorla, R.A., and Solomon, R. L. (1967). Two-process learning theory: relationships between Pavlovian conditioning and instrumental learning. *Psychological Review, 74*, 151–182.

Rescorla, R.A., and Wagner, A.R. (1972). A theory of Pavlovian conditioning: variations in the effectiveness of reinforcement and nonreinforcement. In A.H. Black and W.F. Prokasy (Eds.), *Classical conditioning II: current research and theory* (pp. 64–99). New York: Appleton-Century-Crofts.

Ross, D. (1991). *The origins of American social science.* Cambridge: Cambridge University Press.

Schneiderman, N., Fuentes, I., and Gormezano, I. (1962). Acquisition and extinction of the classically conditioned eyelid response in the albino rabbit. *Science, 136*, 650–652.

Sherrington, C.S. (1906). *The integrative action of the nervous system.* New Haven, CT: Yale University Press.

Skinner, B.F. (1931). The concept of the reflex in the description of behavior. *Journal of General Psychology, 5*, 427–458.

Skinner, B.F. (1938). *The behavior of organisms.* New York: Appleton-Century-Crofts.

Smith, L.D. (1986). *Behaviorism and logical positivism: a reassessment of the alliance.* Stanford, CA: Stanford University Press.

Smith, M.C., Coleman, S.R., and Gormezano, I. (1969). Classical conditioning of the rabbit's nictitating membrane response at backward, simultaneous, and forward CS-US intervals. *Journal of Comparative and Physiological Psychology, 69*, 226–231.

Smith, S., and Guthrie, E.R. (1921). *General psychology in terms of behavior.* New York: Appleton.

Spence, K.W. (1956). *Behavior theory and conditioning.* New Haven, CT: Yale University Press.

Spence, K.W. (1961). Discussion: Part VI. *Annals of the New York Academy of Sciences, 92* (Art. 3), 1187–1189.

Spence, K.W., and Taylor, J. (1951). Anxiety and strength of the UCS as determiners of the amount of eyelid conditioning. *Journal of Experimental Psychology, 42*, 183–188.

Tait, R.W., Kehoe, E.J., and Gormezano, I. (1983). Effects of US duration on classical conditioning of the rabbit's nictitating membrane response. *Journal of Experimental Psychology: Animal Behavior Processes, 9,* 91–101.

Tolman, E.C. (1932). *Purposive behavior in animals and men.* New York: Appleton-Century.

Watson, J.B. (1913). Psychology as the behaviorist views it. *Psychological Review, 20,* 158–177.

Watson, J.B. (1925). *Behaviorism.* New York: Norton.

Watson, J.B. (1930). *Behaviorism* (rev. ed.). Chicago: University of Chicago Press.

Watson, J.B., and Rayner, R. (1920). Conditioned emotional reactions. *Journal of Experimental Psychology, 3,* 1–14.

Wells, H.G. (1927, November 13). Mr. Wells appraises Mr. Shaw. *New York Times Sunday Magazine,* pp. 1, 16.

White, A.D. (1896). *A history of the warfare of science with theology in Christendom* (2 vols.). New York: Appleton.

Williams, D.R. (1965). Classical conditioning and incentive motivation. In W.F. Prokasy (Ed.), *Classical conditioning* (pp. 340–357). New York: Appleton-Century-Crofts.

Williams, R. (1976). *Keywords: a vocabulary of culture and society.* New York: Oxford University Press.

Yerkes, R.M., and Morgulis, S. (1909). The method of Pawlow in animal psychology. *Psychological Bulletin, 6,* 257–273.

2
Cellular Mechanisms of Classical Conditioning

BERNARD G. SCHREURS AND DANIEL L. ALKON

The search for the cellular mechanisms of learning and memory has fostered the development of a large number of model systems, preparations that demonstrate learning and are tractable to biological analyses. We examine briefly the history and status of classical conditioning as a means of studying the biological basis of associative learning. Several model systems, including *Aplysia, Hermissenda, Drosophila,* and the rabbit nictitating membrane response, are examined for their ability to demonstrate associative learning when classical conditioning procedures are employed. The cellular mechanisms of learning and memory are assessed in light of the types of behavior change that these model systems exhibit.

The modern study of learning and memory has been characterized, in part, by a search for the biological basis of behavior change. This search has fostered the development of a large number of model systems, preparations that demonstrate learning and which are tractable to biological analysis. Model systems that have been studied include *Aplysia* gill and siphon withdrawal (Hawkins et al., 1998a), bee navigation and proboscis extension (Menzel & Muller, 1996), cat short-latency eyeblink (Gruart et al., 1995; Kim et al., 1983; Woody, 1970), crayfish escape responses (Krasne & Teshiba, 1995; Zucker, 1972), dog salivation (Kulitka, 1992) and leg flexion (Aleksandrov et al., 1992), *Drosophila* odor avoidance (Dubnau & Tully, 1998) and proboscis extension (Holliday & Hirsch, 1986), fly proboscis extension (Zawistowski & Hirsch, 1984), *Hermissenda* phototaxis and foot contraction (Alkon, 1983, 1989), *Limax* taste aversion (Gelperin, 1994), leech escape responses (Sahley, 1994), insect leg movement (Harris, 1991), monkey choice behavior (Mishkin, 1982), octopus discrimination learning (Michels et al., 1987), pigeon heart rate changes (Cohen, 1980), rabbit heart rate changes (Schneiderman et al., 1987) and nictitating membrane/eyelid response (Gormezano et al., 1983), rat eyelid response (Skelton, 1988) and spatial maze learning (Biegler & Morris, 1993), squid discrimination learning (Allen et al., 1985), turtle eyeblink responses (Keifer et al., 1995), and human eyelid conditioning (Molchan et al., 1994; Solomon et al., 1989; Woodruff-Pak, 1997).

The utility of model systems for increasing our understanding of learning and memory depends, in large part, upon the type of learning that is studied (Schreurs, 1989). A useful distinction that might be made between different types of learning is the distinction between nonassociative learning (e.g., habituation, sensitization) and associative learning (e.g., classical conditioning, instrumental conditioning). Examples of nonassociative learning such as habituation and sensitization provide interesting information about how an organism responds to repeated presentations of a single stimulus (Brown, 1998). Perhaps more compelling, however, is information about how an organism responds to a number of events that have become associated. Studying examples of associative learning like classical conditioning can best provide this information.

Despite the growing body of research employing both vertebrate and invertebrate preparations as model systems, it remains unclear whether all the vertebrate and invertebrate preparations being subjected to classical conditioning procedures demonstrate associative learning (Schreurs, 1989). The purpose of the present chapter is to (1) examine briefly the history and status of classical conditioning as a means of demonstrating associative learning; (2) assess the ability of a cross section of model systems to demonstrate associative learning when subjected to classical conditioning procedures; and (3) explore the biological mechanisms that have been proposed to be involved in the associative learning demonstrated by these model systems.

Associative Learning

Associative learning may be defined as a relatively permanent change in behavior that results from the temporal conjunction of two or more events. Despite a number of important caveats such as "silent" learning in which learning occurs even when observable behavior is blocked (Mackintosh, 1983; Krupa et al., 1996), the definition of associative learning as stated continues to be of heuristic value. It seems clear, however, that not every two events that occur together are associated. Aristotle proposed that previous contiguity, as well as similarity and contrast, determined which events would be recalled together. Subsequently, the British Empiricists postulated the Laws of Association as a means of specifying formally which events would become associated. The Law of Contiguity stated that an associative connection between two events would be formed only if they occurred in spatial and temporal proximity to one another. Secondary Laws of Association dealt with the frequency with which events occurred in contiguity, the duration of events, their intensities, the number of other associations in which the two events had been involved, the similarity of the association to other associations, and the abilities, emotional state, and bodily state of the person experiencing the events (Gormezano & Kehoe, 1981).

Ebbinghaus (1885), who documented his own ability to learn and then relearn pairs of nonsense syllables, conducted the first empirical assessment of the laws of association in 1885. However, the study of associations received its first major impetus in Russia where, in the early 1900s, classical conditioning was proposed as a prototypical example of associative learning (Pavlov, 1927). By the 1920s, researchers in the United States viewed the classical conditioning procedure and the resulting formation of a conditioned response to an originally indifferent stimulus as an "almost ideal example" of associative learning. Just a few years later, the notion of classical conditioning as merely an example of associative learning (as is the current view) was abandoned and classical conditioning became synonymous with associative learning.

Classical Conditioning

Defining Characteristics

The proposal that classical conditioning was synonymous with associative learning and the ensuing use of classical conditioning as a means of studying associative learning necessitated the identification of the defining characteristics of classical conditioning and the specification of its appropriate control procedures (Gormezano & Kehoe, 1975). A recent statement of the defining characteristics of classical conditioning includes (1) the presentation of an unconditioned stimulus (US) that reliably elicits an unconditioned response (UR); (2) the use of a conditioned stimulus (CS) that has been shown by test to produce a response that initially does not resemble the UR; (3) the repeated presentation of the CS and US with a specified order and temporal interval; and (4) the emergence of a new response to the CS, the conditioned response (CR), which is similar to the UR (see Figure 2.1).

A broader conceptualization of classical conditioning has emerged in the past 20 years that has been characterized as the study of the *relations* among stimuli in the environment (Rescorla, 1988). In a broad description of classical conditioning, the CS is said to signal the US during pairings, and the question is whether exposure to the relation between the CS and US modifies the behavior of the organism in a detectable way. For example, if an organism shows an augmented response to the CS as a result of being exposed to the relationship between the CS and US, then an association has been suggested by some to be formed between the two events.

Control Procedures

The need to specify appropriate control procedures acknowledges the fact that although repeated pairings of a CS and US may, under appropriate conditions, lead to the emergence of a CR, the occurrence of a response to the CS may result from nonassociative as well as associative processes (Gormezano & Kehoe, 1975). The first of the nonassociative factors that may influence responding is the baseline level of activity that occurs in most response systems. The second factor is the elicitation by the CS of URs in untrained animals in the target response system or in other response systems. For example, a bright light can elicit a reflexive blink in many species including humans (Grant, 1943, 1944). A bright light may also

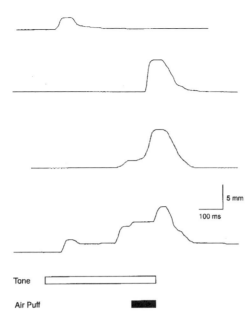

FIGURE 2.1. Sample of human eyelid responses to the tone and air puff during the course of a classical conditioning experiment. From *top* to *bottom:* an unconditioned response to the tone (alpha response), an unconditioned response to the air puff (UR), a conditioned response during pairing of the tone and air puff (CR), a combination of alpha response and conditioned response during pairing, and duration and timing of the tone conditioned stimulus and air puff unconditioned stimulus.

produce changes in heart rate, in respiration, and in the electrodermal response. If the eyeblink is the target response, the occurrence of a reflexive or unconditioned blink to the light has been termed an "alpha" response (see the top trace in Figure 2.1). Changes in the reflexive blink as a result of CS–US pairings have been termed "alpha conditioning" (Grant & Adams, 1944). If the change in heart rate is the target response, then the reflexive blink to light is a response to the CS in an effector system other than that containing the target response. The third potential nonassociative contributor to responding is the sensitizing effect that presentation of a US may have on the frequency of baseline responding or CS-elicited responses. To assess the contribution of nonassociative processes to responding, researchers have adopted control procedures incorporating US-alone, CS-alone, and explicitly unpaired presentations of the CS and US.

A broader approach to control procedures, epitomized by the "truly random" control, is based on manipulating the degree to which there is a relationship between the CS and US (Rescorla, 1988). For example, in the explicitly unpaired control procedure the CS and US never occur together and, thus, there is a negative relationship between the CS and US. In fact, the CS is thought of as perfect predictor of the absence of the US. Conversely, the truly random control procedure consists of independently programmed occurrences of the CS and US that are presented in an attempt to ensure that there is no consistent relationship between the CS and US. In the truly random control, there is an equal probability rate of US occurrence in both the presence and absence of the CS (Rescorla, 1988). Experiments designed to examine the effects of a truly random control procedure have found that some conditioning to the CS does occur as a result of fortuitous CS–US pairings that occur when there is an equal probability of US occurrence in the presence and in the absence of the CS (Ayers et al., 1975; Benedict & Ayers, 1972).

Emergence of a New Response

Initial experiments using the classical conditioning procedure identified a number of responses, including salivation, leg flexion, finger withdrawal, knee jerk, and eyelid closure, that could be conditioned to a variety of stimuli. The major outcome of these early classical conditioning experiments, as well as many current experiments, was the emergence of a new response to the CS that in some way resembled the response to the US. The emergence of a new response occurred even when the CS elicited a response of its own at the start of conditioning (see Figure 2.1). For example, Pavlov (1927) reported an experiment by Erofeeva in which skin shock was paired with food. Although the shock initially elicited defensive responses, after repeated pairings of shock and food, the dog elicited a conditioned salivary response to skin shock without showing any of the previously observed defensive responses. More recently, groups of rabbits were presented with CS–US pairings in which water in the mouth (CS) was paired with an air puff to the eye (US) or air puff to the eye (CS) was paired with water in the mouth (US). In the former case rabbits elicited conditioned nictitating membrane extension to water and in the latter case rabbits elicited conditioned jaw movement to air puff (Gormezano & Tait, 1976). Despite unconditioned jaw movement to the water CS and unconditioned nictitating membrane extension to the corneal air puff CS, rabbits were able to acquire a new response to the CS in the effector system elicited by the US. Indeed, acquisition of the new response by animals given paired training reached a level of conditioning that was substantially and significantly higher than the responses of animals given explicitly unpaired presentations of the CS and US. In another recent experiment, rabbits were given two blink-eliciting electrical pulses (Schreurs & Alkon, 1990); the first served as the CS and the second served as the US. After repeated pairings of the two response-eliciting electrical pulses, rabbits began to elicit a new blink that emerged close to the US and moved forward toward the CS as a function of the training. Importantly, there were no significant changes in the original blink to the CS (i.e., no alpha conditioning). This result confirmed that classical conditioning led to the emergence of a CR without any changes in the UR to the CS.

The emergence of a new response to the CS in the effector system activated by the US has historically been used as the index of associative learning. Moreover, it continues to be advocated as the hallmark of associative learning when classical conditioning procedures are employed. The emergence of a CR that resembles the UR has led to the postulation that the CS becomes a substitute for the US. Support for the stimulus substitution theory was derived from a number of observations including Pavlov's report that a previously conditioned CS could support conditioning when paired with a new CS (i.e., second-order conditioning). However, it soon became clear that a number of the responses that emerged as a result of CS–US pairings were different from the UR. For example, conditioning of pupillary change and heart rate change produced CRs that were in the opposite direction to those produced by the US (i.e., compensatory responses). At a more prosaic level, many CRs differed from URs in terms of latency, amplitude, and recruitment.

As a result of the lack of identity between a CR and UR, some researchers abandoned classical conditioning procedures as the means of studying associative learning. Other researchers have adopted a less stringent specification of CRs and have begun to study behaviors that are "related to" the CS and/or US (Rescorla, 1988). These responses have been variously described as instrumental approach behaviors, sign-directed behaviors, and goal-directed behaviors and are used as evidence that an association has been formed between two events (Boakes, 1977; Hearst & Jenkins, 1974). In brief, investigators have proposed that the nature of the CR is determined, in part, by the type of CS or US employed in any given experiment. For example, in autoshaping, a pigeon may peck at a key light with "eating" or "drinking" motions depending on whether food or water has been paired with the lighted key (Jenkins & Moore, 1973). A light may elicit the orienting response of rearing in a rat, whereas a tone may elicit the orienting response of head turning (Holland, 1980). The frequency of these behaviors has been shown to change as a function of training.

Although the emergence of goal-directed behaviors may be a measure of associative learning, the utility of employing preparations that study goal-directed behavior as model systems may be problematic because of the very nature of the responses that are studied. Specifically, the ability to identify the biological basis of different target responses is hindered by the fact that the responses are recognized in terms of their outcome. That is, an approach response or the depression of a bar or key can occur in any number of different ways that may change from moment to moment, from presentation to presentation, and from animal to animal. The variation in responses and complexity of any single response makes it difficult to identify a unique neural output pathway for the target response. Without an identified motor output pathway that is consistent from presentation to presentation and from animal to animal, it would be difficult to study the complete neural circuitry and identify the neural substrates involved in these instances of associative learning. However, the isolation of discrete responses or the use of restrained subjects may assist in the identification of potential motor output pathways and thus make preparations that display the emergence of goal-directed behavior a powerful addition to the search for the biological basis of learning.

Model Systems

Despite the problems inherent in different definitions of associative learning and differences in what is considered to be a CR, a number of model systems have been exposed to classical conditioning procedures to study the neural substrates of associative learning. For the purposes of exposition, we review here several model systems that employ classical conditioning procedures to study behavior change and examine these model systems in light of a traditional definition of associative learning.

Aplysia

In experiments with *Aplysia* gill and siphon withdrawal, Kandel and his associates have found that a touch of the siphon or mantle elicits mild siphon and gill withdrawal, whereas tail shock elicits strong siphon and gill withdrawal (Kandel, 1976). Examination of the *Aplysia* model system has provided valuable insights into the nonassociative learning phenomena of habituation and sensitization. With repeated stimulus presentations, the siphon and gill withdrawal reflex can be habituated or decreased. The response decrement or habituation is not the result of fatigue, and it can be dishabituated or restored with a stimulus of different strength. The same withdrawal reflex can also be sensitized or enhanced. In other words, the response will become larger if it is elicited after a series of stronger stimulations (e.g., repeated shocks) to the tail.

The circuitry involved in siphon and gill withdrawal consists of neurons in both the peripheral and central nervous systems (Glanzman, 1995). Nevertheless, interest has focused on the abdominal ganglion where there are about 25 identified sensory neurons of the siphon that synapse with interneurons and motor neurons which control withdrawal.

One electrophysiological correlate of habituation is a progressive reduction in the amplitude of excitatory postsynaptic potentials in the interneurons and motor neurons that control the withdrawal reflex. The biophysical basis for this reduction has been suggested to be a reduction in calcium current and is being further examined. During repeated presentations of tail shock (the sensitization procedure), interneurons are excited and release increased amounts of transmitter that, in turn, presynaptically facilitate the sensory neuron connections. This synaptic facilitation has been shown to produce reduction in potassium current, enhanced slow depolarization, broadening of sensory neuron action potentials, and increases in synaptic potentials. Cyclic AMP-mediated phosphorylation has been implicated in production of the persistent synaptic enhancement associated with sensitization (Hawkins & Kandel, 1984).

Serotonin has been proposed as the critical neurotransmitter that initially reduces a nonvoltage-dependent potassium current (the 'S' current) and ultimately activates gene transcription through a factor known as 'CREB' (cyclic AMP response element binding protein; see Abel & Kandel, 1998, for a recent review). Although the neurochemical, biochemical, and molecular biological techniques used to analyze this cascade for nonassociative behavioral modification have been most impressive (Abel & Kandel, 1998), many significant issues remain to be resolved. Serotonin, for example, has been identified within synaptic terminals on the critical sensory neurons but it has been unequivocally ruled out as the transmitter released by the L29 interneuron onto these same sensory neurons. However, L29 is the interneuron that on excitation in cultured neuron or ganglia preparations produces sensitization. In the cultured neuron configuration there are no serotonergic neuron or synaptic endings, but sensitization nevertheless occurs. Moreover, it is the cultured neuron configuration that has been most commonly analyzed for long-term changes of protein synthesis. It should be noted that repeated application of serotonin to cultured sensory neurons causes increased CREB-mediated synthesis of a vast number of proteins, raising some question as to the specificity of the serotonergic model for nonassociative synaptic change.

Although the electrophysiological phenomena thought to underlie habituation and sensitization are postulated to take place at the level of the synaptic connections between the sensory neurons, motor neurons, and interneurons, they have only been observed to take place at the level of the cell body. However, structural alterations in terminal branching have now been correlated with habituation and sensitization (for review, see Baily & Kandel, 1993).

Following the examination of habituation and sensitization, Kandel and colleagues have sought to develop procedures to demonstrate that *Aplysia* is capable of being classically conditioned (Carew et al., 1981; Hawkins et al., 1986). In an initial series of experiments, unrestrained, freely moving *Aplysia* were presented with a CS consisting of a tactile stimulus delivered by a single nylon bristle from a paintbrush held firmly by a hemostat and a strong electrical US delivered by the manual application of spanning electrodes to the tail. The CS was applied by inserting the bristle into the funnel of the siphon and briskly moving it upward a single time. The duration of contact with the inner surface of the siphon was estimated to be approximately 0.5 s. Both the bristle CS and the electric shock US always elicited siphon withdrawal, and in the former case a stopwatch measured the response.

The acquisition and extinction of the conditioned siphon withdrawal response was studied using six groups, a paired CS–US group and five control groups: strictly alternating CS and US, programmed random occurrence of CS with respect to US, US alone, CS alone, and naive (untreated). During acquisition, the US was presented every 5 min and, for the paired group, the CS preceded the US by 30 s. Siphon withdrawal was tested immediately after the first CS presentation and then after every 5 CS presentations. Extinction was conducted by delivering the CS alone for 10 presentations immediately after training. Acquisition in the paired group was characterized by a positively accelerated increase in the duration of siphon withdrawal from an initial mean of 8.6 s to a terminal level of 32.5 s, whereas the control groups did not change significantly during the course of training from their initial duration of 5 s. During extinction the paired group showed a significant decrease in the duration of siphon withdrawal from the first CS presentation (32 s) to the 10th CS presentation (15 s), whereas the control groups showed no systematic change from their initial values of siphon withdrawal (5 s).

In other experiments, researchers have attempted to differentially condition the siphon withdrawal response in *Aplysia* (Carew et al., 1983). In a typical experiment, subjects were presented with a US consisting of tail shock and CSs consisting of nylon bristle stimulation of the siphon and electrical stimulation of the mantle. In two groups of animals, the US followed CS+ (siphon or mantle stimulation, counterbalanced) by 0.5 s and CS– (mantle or siphon stimulation, counterbalanced) occurred at the midpoint of the 5-min interval between CS presentations. The results revealed that at 15 min and 24 h after training there was a greater increase in the duration of siphon withdrawal to CS+ than to CS–. The increase in siphon withdrawal to CS– was attributed to sensitization and the increase in siphon withdrawal to CS+ was attributed to associative learning. Examination of the effects of the interval between CS+ and the US (interstimulus interval) revealed that the siphon withdrawal response of greatest duration occurred at an interstimulus interval of 0.5 s with a response of shorter duration at an interval of 1 s and little or no increase in duration at longer (2-, 5-, 10-s), shorter (0-s), or backward (–0.5, –1, –1.5 s) interstimulus intervals.

Recently, Kandel and colleagues have used a number of simplified preparations including an isolated mantle organ preparation and cell culture to replicate and study the behavioral and biophysical effects of both nonassociative and associative procedures (Bao et al., 1998; Hawkins et al., 1998a, b).

The *Aplysia* model system involves a response that occurs to the CS from the outset of training. The response to the CS, siphon withdrawal, is the same as the response to the US and thus may be characterized as an alpha response. As a result of a pairings operation, increases in the duration of the alpha response have been observed relative to a number of nonassociative control groups. Interestingly, there are also several vertebrate preparations, including human eyelid conditioning, in which the CS (see Figure 2.1) elicits an alpha response. In these cases, the characteristics of the alpha response, for example, onset latency and amplitude, are noted and responses that fall within the latency range of an alpha response are eliminated from consideration as conditioned responses. When alpha responses are eliminated from consideration, conditioned responses can be observed to emerge and display characteristic features such as an onset latency that first occurs at or near the point of US onset and then moves forward in time as a function of continued training (Gormezano, 1966). To date, there has been no report of the emergence of a new siphon withdrawal response that occurs outside the latency range of the unconditioned siphon withdrawal (alpha) elicited by the bristle CS.

Because the bristle CS elicits a response quite similar to the siphon withdrawal response elicited by the shock US, the *Aplysia* training procedure may be conceptualized as a procedure in which a weak US is followed by a strong US. However, in other cases where two USs have been paired, such as the rabbit experiments described earlier, although an unconditioned response to the first US has been observed, there has also been a new response that emerges to the first US (the 'CS') that resembles the UR to the second US.

The fact that only changes in the *Aplysia* alpha response have been reported to date may reflect limitations in the currently employed measurement techniques rather than any inherent limitation in the organism's ability to show the emergence of a conditioned response. For example, high-gain mechanical transduction of the siphon withdrawal response (e.g., a force transducer has been used in an in vitro *Aplysia* gill withdrawal preparation) may help resolve small contractions or relaxations of the siphon not detectable by visual inspection. Moreover, mechanical transduction may allow precise specification of response latency, amplitude, and duration providing a number of dependent variables that could be examined for the emergence of a conditioned response. To overcome some of these limitations, Hawkins et al. (1998b) have recently used a surgically reduced siphon, gill, and mantle preparation that allows precise control of stimulus presentation through computer-controlled stimulators and automated response measurement through a movement transducer. This more precise control and quantification of behavior have produced essentially the same results as those obtained in the original experiments; there was an increase in the amplitude of the response to the CS following paired training (Hawkins et al., 1998b). However, the possibility exists that the "identified" CS-US pathway used in *Aplysia* learning experiments thus far simply cannot support classical conditioning. There is evidence that the gill withdrawal reflex can occur in the absence of the abdominal ganglion and may be controlled to a significant extent by the peripheral nervous system (Leonard et al., 1989). Finally, even the gill withdrawal reflex itself is not a simple stereotypic movement but a number of complex and variable action patterns suggesting a more elaborate underlying circuitry than previously assumed (Leonard et al., 1989).

Drosophila

Classical conditioning procedures have been used to study learning in the fruit fly, *Drosophila melanogaster*. In an appetitive procedure, presentation of a sucrose solution (US) to individual, hungry *Drosophila* elicits the unconditioned response of proboscis extension. Hirsch and colleagues have shown that pairings of a salt solution (CS) and a sucrose solution (US) result in anticipatory proboscis extension to the salt solution (Holliday & Hirsch, 1986; Lofdahl et al., 1992). Appropriate controls have included unpaired presentations of the CS solution and US solution, CS-alone presentations, and US-alone presentations (Holliday & Hirsch, 1986). One major concern in determining the associative nature of conditioned proboscis extension has been the presence of a form of sensitization termed the "central excitatory state" (Vargo & Hirsch, 1982). The central excitatory state consists of increased responding to a CS when preceded by sucrose stimulation independently of CS–US pairings. The central excitatory state may be reduced by the presentation of another stimulus during the intertrial interval (Holliday & Hirsch, 1986; Vargo & Hirsch, 1982). Selective breeding over 25 generations has produced one population of flies that show good conditioning and another population in which very few, if any flies learn to associate saline with sucrose (Lofdahl et al., 1992).

A second form of stimulus pairings used to study learning in individual *Drosophila* involves the pairing of sucrose and an aversive stimulus such as shock or quinine solution (Brigui et al., 1990; De Jianne et al., 1985; Medioni & Vaysse, 1975). Repeated pairings of sucrose and quinine produce suppression of proboscis extension to the sucrose. Control groups have been included to assess the level of sensitization, pseudoconditioning (unpaired), and habituation (Vaysse & Medioni, 1976). Although flies in these groups did show suppression of proboscis extension, the level of suppression was significantly lower than the level of suppression shown by flies that received pairings of sucrose and quinine. The conditioned suppression model has been used most recently to study questions of aging (Fresquet & Medioni, 1993) and the effects of hypergravity (Minois & Le Bourg, 1997).

A third form of stimulus pairings used to study learning in *Drosophila* involves training and testing of groups of flies. Specifically, a group of flies (approximately 100) receives presentation of an odor (e.g., 3-octanol) followed by exposure to a shock grid (Tully & Quinn, 1985). On a subsequent trial, a second odor (e.g., 4-methylcyclohexanol) is presented alone. The group of flies is then placed at a choice point in a T-maze that contains both odors. The number of animals that avoids the odor previously presented with shock is divided by the total number of flies in the T-maze and the resulting ratio is used as an index of learning. Thus, learning is inferred from group performance and, although individual flies are said to have learned, the animals are only tested as a group (for detailed discussion, see Ricker et al., 1986; Hirsch & Holliday, 1988). There are several difficulties in making inferences about the ability of individual flies to learn; not the least of these difficulties is that in a group training procedure different flies may receive different experiences. For example, although all flies may be exposed to the odor, not all flies will necessarily receive the shock because they may not all be in contact with the shock grid (Wallace & Sperlich, 1988). A second difficulty concerns individual differences in reactivity to odor, and in fact there is evidence for considerable individual and strain differences in reactivity to shock (Preat, 1998; Wallace & Sperlich, 1988). For example, Preat (1998) has shown that odor avoidance in the *Drosophila* mutants *amnesiac, dunce,* and *rutabaga* (see following) is greatly decreased after exposure to electric shock, and he concluded that the putative memory deficits in these mutants may result from the effect of the electric shock US.

Despite the problems inherent in inferring individual learning from experiments limited to testing groups, the odor avoidance procedure has been used to study the effects of mutations on learning (Dubnau & Tully, 1998; Tully, 1997). For example, mutants such as *dunce, rutabaga, radish, turnip, cabbage, amnesiac, latheo, linotte, nalyot,* and *golovan* have been found that show lower acquisition or retention of odor avoidance than wild-type flies (Tully, 1996). Indeed, these mutants have been identified to have disruptions of genes encoding among other things an adenylcyclase, a subunit of a G-protein, a phosphodiesterase, and a subunit of a cAMP-dependent protein kinase (Goodwin et al., 1997). As a result, the argument has been made that learning in *Drosophila* involves CREB-mediated genetic changes localized to the mushroom bodies of the *Drosophila* nervous system akin to the CREB-mediated changes hypothesized to occur in *Aplysia* (Milner et al., 1998). However, the aforementioned difficulties in establishing the associative nature of behavioral changes confound possible interpretations of the CREB-knockout flies (see following). The gene knockout may, in fact, affect a fly's responsiveness to shock rather than the fly's associative learning ability.

Hermissenda

Classical conditioning in the marine snail *Hermissenda* has been demonstrated in a procedure in which 3 s of light CS are presented together with 2 s of rotation US (Alkon, 1983). Usually, rotation of the animal elicits clinging with the body musculature contracted and the foot gripping the bottom of the enclosure. Presentations of the light CS alone to animals placed in the dark during the light portion of their diurnal cycle produce the phototropic response of movement toward the light source. As a result of light–rotation pairings, classically conditioned animals have a significantly longer latency to move toward the light than control animals that received rotation alone, random but separate presentation of light and rotation, light alone, or strictly alternating presentations of light and rotation (Figure 2.2). The learned increase in latency to move toward light is a function of the number of CS–US pairings and CS–US interval. The conditioned clinging can be extinguished and is specific to the light CS; it displays savings and is sensitive to CS/US contingencies. In addition to learning-specific changes in movement to light, trained animals also show a new response to light, foot contraction, which emerges as a function of training.

Observations of *Hermissenda* in the dark revealed a 15%–20% shortening in the length of the foot (the single organ of locomotion) in response to rotation. It was also noted that there was a small lengthening of the foot in response to light before training. After pairings of light and rotation, the foot contraction elicited as an unconditioned response to rotation emerged as a new response to light in all the

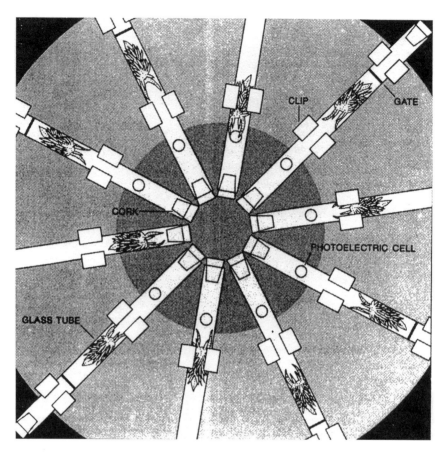

FIGURE 2.2. *Hermissenda* are trained and tested in glass tubes on a turntable. The *Hermissenda*'s phototactic response to light is measured before training: the time the mollusk takes to reach a spot of light (*darker shading* at the center of the turntable) is recorded automatically when the animal reaches the photoelectric cell. Then, the *Hermissenda* are trained by being rotated while confined to the outer end of the tube and thus subjected to centrifugal force that is sensed by their statocysts. For one group, the rotation is precisely paired with a 30-s period of light; various control groups receive no training or are subjected to light and rotation in an explicitly unpaired manner. Finally, the *Hermissenda*'s velocity of movement toward the light is timed again to determine the effect of training.

trained animals. The new response of foot contraction to the light did not occur in untrained animals nor did it occur in animals that received random presentations of light and rotation (Lederhendler et al., 1986). Simultaneous measurement of both locomotion and foot contraction revealed that there is a rapid decrease in locomotion and a gradual increase in foot contraction as a function of training (Matzel et al., 1990a). Finally, foot contraction was also shown to be a function of the number of CS–US pairings and of the interval between the CS and US (Matzel et al., 1990b).

The sensory inputs involved in *Hermissenda* associative learning include the visual system and a primitive vestibular system (Alkon, 1983). The visual system consists of two relatively primitive eyes, each containing a lens, pigment cells, and five photoreceptors (three type B cells and two type A cells). The vestibular system consists of two statocysts that contain 12 hair cells and a cluster of crystals called stataconia that brush up against the hair cells when there are changes in the direction of gravitational force. In *Hermissenda,* there is convergence between the visual and vestibular pathways at the level of the sensory cells themselves. There is also convergence of the sensory inputs on interneurons that, in turn, connect to the motor outputs. The motor output neurons involved in classical conditioning of foot contraction include motoneurons controlling the muscles of the foot and turning of the animal toward a light source.

A detailed analysis of the networks within and between the visual and vestibular pathways has been possible. Knowledge of this neural organization has allowed a step-by-step tracking of training-elicited signals as they flow through the identified pathways (Alkon, 1987).

Pairings of light and rotation lead to an increase in the excitability of the type B photoreceptor and a decrease in the excitability of the type A receptor. The changes in excitability are the net effect of interactions of several inputs to the sensory circuitry of *Hermissenda*. First, light excites the B photoreceptors and rotation excites the hair cells, which, in turn, inhibit the B photoreceptors. When rotation stops, hair cell activity is reduced even below unstimulated levels so that the B photoreceptor is released from hair cell inhibition. Second, when light and rotation are paired, excitation of the B photoreceptor allows the inhibitory signals of the hair cell on the B photoreceptor to be "shunted," which lessens the effect of rotation-induced inhibition on that B photoreceptor. Third, when light and rotation are stopped, there is an increase in the excitatory feedback from the second-order visual cells of the optic ganglion onto the B photoreceptor. Fourth, a transmitter released by the hair cells (γ-aminobutyric acid, GABA) onto the B photoreceptor causes prolonged depolarization and potassium current reduction. These effects of GABA are enhanced by B-photoreceptor depolarization, which is maximized when light is paired with rotation during training. The cumulative effect of these four sources of increased excitation is an increase in the overall excitability of the B photoreceptor so that its response to light is enhanced and prolonged. One of the results of increased B-photoreceptor excitability is the inhibition of a chain of neurons responsible for motoneuron impulses that drive the muscles, which cause turning of the foot. Another result is an increased excitation of interneurons driven by hair cells that then excite motoneurons controlling foot contraction.

The pairing-specific excitability of the type B photoreceptor has been shown to result from an increase in intracellular calcium which interacts with an elevation of diacylglycerol (a membrane-bound lipid) and probably arachidonic acid to induce movement of the calcium-sensitive enzyme protein kinase C (PKC) from the cell cytoplasm to the cell membrane. In the cell cytoplasm, PKC increases potassium ion flow, which in turn decreases membrane excitability. In the cell membrane,

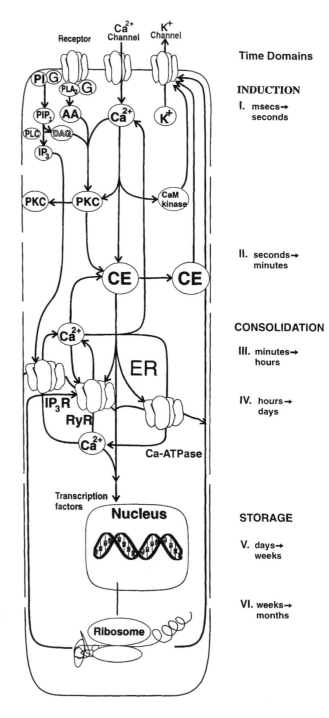

PKC reduces potassium ion flow, which in turn increases membrane excitability (Figure 2.3). Agents that block the movement of PKC from the cytoplasm to the membrane also block the decrease in potassium ion flow. Translocation of PKC to the inner surface of the neuronal membrane allows it to interact directly with protein targets for phosphorylation (Alkon, 1989).

In *Hermissenda* (and later in the rabbit hippocampus; see following) a particular calcium- and guanosine triphosphate-(GTP-) binding protein known as calexcitin was shown to undergo phosphorylation only during classical conditioning procedures. Recently, calexcitin has been cloned and sequenced completely. This protein has been shown to be a high-affinity substrate for the alpha-isozyme of PKC. The alpha-isozyme of PKC, although not limited to neuronal tissue, is by far the most abundant of the PKC isozymes in the brain. PKC-mediated phosphorylation of calexcitin also causes the latter to translocate to neuronal membranes of three principal types: the outer wall membrane, the membrane of the endoplasmic reticulum (ER), and the nuclear membrane. These translocation targets are consistent with the recently revealed pleiotropic (i.e., multiple) functions of calexcitin, functions that are consistent with and apparently required for synaptic modification during learning (Alkon et al., 1998). At the outer wall membrane, calexcitin directly inactivates voltage-dependent potassium currents just as occurs during classical conditioning to cause increased excitability (and thus increased synaptic weight). At the nuclear membrane, calexcitin has been found to increase turnover of mRNA, presumably for a number of learning-related proteins.

Other studies had demonstrated previously that there is a close correlation of *Hermissenda* classical conditioning with increased turnover of mRNA for a variety of protein species. At the ER membrane, calexcitin has now been shown to bind the ryanodine receptor (RYR) with high affinity (Nelson et al., 1999). Direct measurements of efflux at isolated ER membrane vesicles in fact demonstrated that calexcitin is a specific endogenous ligand necessary for calcium-mediated calcium release via the RYR. Calexcitin therefore serves as a signaling molecule

◄─────────────────────────

FIGURE 2.3. Possible roles of calcium and calcium-dependent kinases in the pathway from depolarization to long-term increases in excitability. Calmodulin-dependent protein kinase type II (*CaM kinase*) and protein kinase C (*PKC*) may act synergistically to integrate different signal modalities (e.g., depolarization and neurotransmitter influx, fast and slow time courses) in response to an influx of calcium. PKC phosphorylates calexcitin (*CE*), which in turn blocks potassium channels, induces increased RNA synthesis, and inhibits retrograde axonal transport. CE is translocated to particulate fractions after phosphorylation by PKC, which places it in closer proximity to potassium channels. PKC is also activated and translocated to the membrane after learning. Autophosphorylation of PKC and CaMKII render the kinases calcium independent, prolonging the response after calcium is removed. Both PKC and CaM kinase may also phosphorylate ion channels (not shown). *ER*, endoplasmic reticulum, *RyR*, ryanodine receptor. (Figure courtesy of T.J. Nelson.)

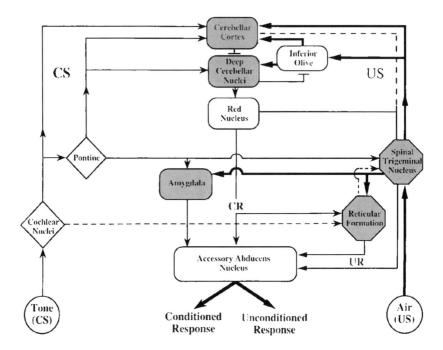

FIGURE 2.4. Simplified circuit diagram of pathways involved in classical conditioning of the rabbit nictating membrane response (NMR). Conditioned stimulus (CS) inputs are shown with *thin lines* and unconditioned stimulus (US) inputs with *thick lines;* unconditioned response (UR) and conditioned response (CR) pathways to the final common motor output (accessory abducens nucleus) are shown with *lines of intermediate thickness.* *Dashed lines* represent proposed connections that have not been verified. Possible sites of learning-related plasticity (e.g., cerebellum, amygdala, spinal trigeminal nucleus, reticular formation) are *shaded.* These sites receive both CS and US inputs and may be capable of affecting US processing and/or UR elicitation.

that amplifies calcium elevation in response to learning-associated synaptic transmitters that initiate second messengers such as diacylglycerol, arachidonic acid, and calcium itself. The crucial role of the RYR was also recently confirmed with gene fingerprinting technology that revealed prolonged learning-specific enhancement of RYR mRNA expression for many hours after rat spatial maze learning. Enhanced RYR expression was further confirmed by quantitative reverse transcriptase-polymerase chain reaction (RT-PCR), Northern blots, and other molecular biological techniques such as in situ hybridization.

Rabbit Eyeblink/Nictitating Membrane Response

Classical conditioning of the rabbit nictitating membrane response (NMR) and eyeblink was first reported by Gormezano and his colleagues when they paired a tone CS with a corneal air puff US (Gormezano et al., 1962; Gormezano, 1966). Usually, corneal air puff elicits extension of the nictitating membrane and closure of the outer eyelids (eyeblink), whereas the tone does not elicit that response. Animals given paired CS–US presentations show the emergence of a nictitating membrane response to tone and a progressive increase in the frequency of conditioned nictitating membrane extension across days of training to a level of 95%. In marked contrast, animals given CS-alone, US-alone, or explicitly unpaired CS-alone and US-alone presentations never exceeded a frequency of 6% membrane extension on any single day and averaged a level not appreciably higher than the base rate of 2%–3%. Extensive behavioral experiments have shown that acquisition of the conditioned NMR displays all the hallmarks of classical conditioning, including sensitivity to the number of CS–US pairings, CS-US interval, CS and US intensity, CS specificity, and savings.

The sensory inputs of the rabbit NMR include the brain stem auditory pathways and corneal inputs that travel via the spinal trigeminal nucleus to the inferior olive and cerebellum and to the motor output of the accessory abducens nucleus that controls eyeball retraction and the resultant sweep of the nictitating membrane (Figure 2.4). In addition to these primary pathways, CS and US information also travels to many other parts of the brain including the hippocampus, cerebellum, and cortex. Extensive lesion and recording research, initiated, in large part, by Thompson and his colleagues, has revealed the important role played by the hippocampus and cerebellum (Thompson, 1986). For example, neural activity that mimics and precedes conditioned nictitating membrane responding has been recorded in the hippocampus. Electrodes placed in the region of the CA1 pyramidal cells of the hippocampus show that the CA1 cells respond to a tone CS at a higher frequency and shorter latency as a consequence of classical conditioning of the NMR than as a consequence of unpaired stimulus presentations.

Intracellular recording of CA1 cells in a slice of hippocampus obtained from classically conditioned rabbits revealed a reduction of the flow in potassium ions through the cell membrane in much the same way as in the *Hermissenda* type B photoreceptor. In fact, voltage-clamp recordings in CA1 cells showed that potassium ion currents active in the presence of calcium were modified as a function of classical conditioning (Sanchez-Andres & Alkon, 1991). This current is similar to one of those implicated in the associative learning demonstrated by *Hermissenda*. Measurements of PKC after classical conditioning indicate that there is an increase in membrane-associated PKC near the CA1 cell bodies even 1 day after all training has been completed (Olds et al., 1989). Three days after training (i.e., well into the period of memory retention), maximal PKC labeling moved from the cell bodies to the region of the CA1 dendrites. Interestingly, the movement of PKC in CA1 cells can be artificially induced by the drug phorbol ester, which also causes the same potassium ion flow reduction that takes place during conditioning (Alkon, 1989). Translocation of PKC to the CA1 membrane by phorbol ester also causes enhanced summation of excitatory postsynaptic potentials (EPSPs) elicited by activation of presynaptic fibers known as Schaeffer collaterals. This

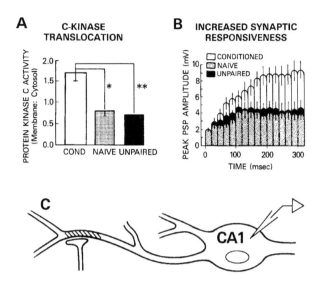

FIGURE 2.5. Enhanced synaptic responsiveness parallels PKC translocation after classical conditioning of the rabbit NMR. The proportion of PKC associated with the plasma membrane (**A**) is greatest in regions surrounding and including the CA1 pyramidal cells microdissected from classically conditioned rabbits compared to controls (*naive* and *unpaired*). Summation of postsynaptic potentials (*PSP*) is greatest in responses of CA1 pyramidal cells from conditioned rabbits (**B**) compared to cells from control animals. A model of a *CA1* pyramidal cell shows the position of a recording electrode in the soma (**C**). The *hatched area* represents postsynaptic regions in proximal dendrites where specific binding of radioactive phorbol ester was localized by autoradiographic analysis.

same enhanced EPSP summation was demonstrated to occur only in rabbits previously trained with a classical conditioning procedure (Figure 2.5).

These remarkable biophysical and molecular parallels between *Hermissenda* and rabbit classical conditioning also included a learning-specific increase in phosphorylation of calexcitin in both these very different species. Thus, conservation of a molecular sequence for memory storage was implicated by parallel cellular events in molluscan and mammalian memory paradigms. This conservation was given further support by findings of memory-specific translocation of PKC in the hippocampus and other brain structures for other learning paradigms such as cue and platform rat spatial maze learning and rat olfactory discrimination learning. Furthermore, different laboratories using different paradigms and different methodologies (e.g., monoclonal antibodies for PKC) have independently confirmed the role of PKC in associative memory (for review, see Van der Zee et al., 1997).

Neural recording in the cerebellum during classical conditioning of the rabbit NMR has shown both increased and decreased extracellular activity in the dentate/interpositus deep nuclei and cortex correlated with the CS, US, CR, and UR (for review, see Hesslow & Yeo, this volume; Thompson & Kim, 1996). Lesions of the deep nuclei of the cerebellum have been shown to abolish CRs in trained rabbits and prevent the acquisition of CRs in naive rabbits. Lesions of the cerebellar cortex, particularly lobule HVI, have severely impaired CRs and bilateral lesions of HVI have abolished CRs.

A number of years ago, we adopted the cerebellar slice preparation to identify the cellular correlates of these conditioning-specific changes in Purkinje cell activity (Schreurs et al., 1991, 1997a, 1998). In a typical experiment, rabbits are given sessions of paired or explicitly unpaired presentations of a tone and electrical stimulation around the right eye. Slices are prepared of lobule HVI from the right side of the cerebellum 24 h later, and Purkinje cell electrophysiological properties are measured including membrane excitability (measured as threshold for dendritic spikes; Schreurs et al., 1991, 1997a, 1998), synaptic excitability (measured as current required to elicit an EPSP; Schreurs et al., 1997a), and potassium channel function (measured as the effects of potassium channels antagonists, Schreurs et al., 1998).

Our experiments show that membrane excitability was higher in Purkinje cell dendrites in lobule HVI of rabbits given paired stimulus presentations than of rabbits given unpaired stimulus presentations (Figure 2.6) (Schreurs et al., 1991, 1997a, 1998). This excitability was indexed, in part, by a lower minimum current required to elicit dendritic calcium spikes in cells from paired animals than in cells from unpaired animals. A second index of a change in excitability was a decrease in the size of a potassium channel-mediated transient membrane hyperpolarization in cells from paired rabbits (Schreurs et al., 1998). A third index of excitability was a decrease in the threshold current required to elicit EPSPs and Purkinje cell spikes, which was found to be lower in cells from paired animals than in cells from unpaired control animals (Schreurs et al., 1997a). A fourth index of excitability was an increase in membrane-bound protein kinase C, which was specific to lobule HVI of paired animals and was not found in lobule HVI of unpaired or sit control animals, nor was it found to be different in the dentate/interpositus nuclei of any group (Freeman et al., 1998).

The increase in Purkinje cell excitability was found to be highly correlated with the acquisition of CRs and was preserved for as long as 1 month following conditioning (Schreurs et al., 1998). In addition, the ability to induce a pairing-specific form of long-term depression (LTD; see following) in Purkinje cells of lobule HVI was occluded by classical conditioning (Schreurs et al., 1997a). These data suggest that the conditioning-induced increase in Purkinje cell excitability prevents LTD and that cerebellar LTD may not be the mechanism underlying conditioning of the rabbit NMR. Pharmacological experiments implicating a specific potassium channel (Schreurs et al., 1997a) suggest that PKC calexcitin-mediated changes in potassium channels of the type found in *Hermissenda* B photoreceptors and rabbit hippocampal CA1 cells may be the mechanism responsible for the observed increase in Purkinje cell dendritic excitability.

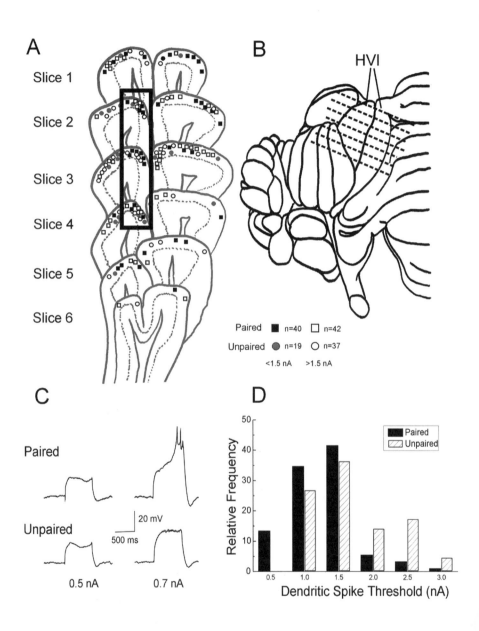

A Specific Molecular Cascade Common to Several Model Systems

Although most of the molecular steps involved in associative learning and memory were first identified and defined in *Hermissenda,* many of the same steps have now been confirmed in mammalian model systems (Alkon et al., 1998). Thus, long-term changes of voltage-dependent potassium currents have been demonstrated with current- or voltage-clamp measurements in the rabbit hippocampus following eyeblink conditioning (Sanchez-Andres & Alkon, 1991) and inferred from dendritic recordings from cerebellar Purkinje cells in the same paradigm. Persistent translocation of PKC has been observed in rabbit hippocampus and cerebellum following NMR conditioning, in the rat hippocampus following spatial maze learning, and in the rat olfactory cortex following odor discrimination learning (as well as in single identified neurons following conditioning in *Hermissenda*). Furthermore, other laboratories have also found evidence of PKC translocation or increased levels of PKC isozymes following associative learning paradigms (for review, see Van der Zee et al., 1997). Evidence of increased synthesis of the type II ryanodine receptor was found in specific regions of the rat hippocampus following rat spatial maze learning (see Alkon et al., 1998). This increase in ryanodine receptor synthesis is thought to mediate learning-associated increases in calcium release (Alkon et al., 1998).

FIGURE 2.6. Recording sites for individual Purkinje cells located on sequential parasagittal slices from Larsell's lobule HVI (**A**), an anterior view of the right cerebellum (**B**) indicating the area from which the slices were cut (*dashed lines*). The slices are depicted so that the rabbit's left folium of lobule HVI is shown on the *left* and the rabbit's right folium is shown on the *right.* The cell locations for Paired (*squares*) and Unpaired (*circles*) animals are based on visual inspection (12 × magnification) of the electrode position in the slice at the time of recording. Locations are divided between those below (*filled*) or at and above (*open*) the mean dendritic spike threshold for cells from Paired animals (1.5 nA). The individual slices were obtained by tracing cresyl violet-stained parasagittal sections cut through the right cerebellar hemisphere. The *dotted line* within each slice indicates the border between the molecular and granular layers. Individual intradendritic recordings (**C**) and relative frequency distributions of local dendritic calcium spike thresholds (**D**) for Purkinje cells from slices of lobule HVI taken from the right cerebellum 24 h after 3 days of Paired training or Unpaired stimulus presentations. Individual recordings show depolarized responses to 700-ms current injections of 0.5 and 0.7 nA for a cell from a Paired animal with dendritic spikes occurring at 0.7 nA and for a cell from an Unpaired animal that remains below threshold at 0.7 nA. The relative frequency distribution shows a clear shift in thresholds to the left for Paired animals with only cells from Paired animals in the lowest threshold bin (0.5–0.9 nA).

All these studies collectively implicate a cascade that responds to profound elevation of intracellular calcium in association with other second-messenger signals such as diacylerglycerol or arachidonic acid. Thus, on the cellular level, the temporal association of synaptic signals transduces and transmits the temporal association of sensory stimuli received by the animal during behavioral training. Associated synaptic signals are then transduced and transmitted in the postsynaptic dendrites by molecular messengers that are themselves temporally associated. There is a striking parallel between this molecular cascade of associative memory and the cascade responsible for muscular contraction (see Alkon et al., 1998). Furthermore, recent observations suggest that the calexcitin-activated molecular cascade can induce structural rearrangements of neuronal processes. These findings have prompted the theory that permanent representations of consistently encountered stimulus associations involve changes in the structural geometries of dendrites that respond to the calcium–calexcitin–ryanodine receptor cascade in a manner analogous to muscle contraction responses to similar cascades (Alkon et al., 1998).

Ultimately, the most important generality of the conserved molecular mechanisms of associative memory so far demonstrated is to the human brain's biochemistry of memory. One way of beginning to test such generality is through the study of neuronal pathophysiology in Alzheimer's disease. This disorder is the one human dementia that most specifically involves loss of associative memory early in the course of the disease's natural history. Recent studies have in fact revealed a remarkably specific disruption of molecular events in cells from Alzheimer's (AD) patients (versus age-matched controls and non-AD dementias). Thus, diagnostic changes of voltage-dependent potassium currents (Etcheberrigaray et al., 1994a), calexcitin, PKC isozymes, and inositol triphosphate-mediated release of intracellular calcium have all been repeatedly observed in AD versus control patients. Perhaps equally as important, exposure of control cells to 10 nM soluble beta-amyloid (comparable to endogenous levels) induces many of the same molecular changes that are so characteristic of the AD cellular phenotype (Etcheberrigaray et al., 1996). Such observations (consistent with more recent findings of other laboratories) lend support to the view that early AD involves human memory loss because there is dysfunction of the same molecular events that have been implicated as important for animal associative learning and memory (Etcheberrigaray et al., 1994b).

Cellular Correlates Versus Causes of Learning and Memory

The cellular mechanisms discussed thus far are correlates of actual learning and memory in behaving animals. In a few cases, elimination of a particular enzyme or a particular membrane channel in a living animal has been achieved with molecular biological interventions to demonstrate necessity for, and not just correlation with, the learning that takes place. For example, temporary blockade of mRNA expression of the Kv 1.1 late potassium channel (with a technique called "antisense") eliminated memory retention of mouse instrumental fear conditioning and rat spatial maze learning. This antisense "knockdown" of a late potassium channel produced memory deficits without affecting a number of measures of sensory and motor activity for both species tested (Meiri et al., 1997).

"Knockouts" via transgenic changes of the mouse or rat genome have also been used to assess the role of substrates such as calcium/calmodulin II kinase, the mGluR1 glutamate receptor, the N-methyl-D-aspartate (NMDA) receptor, and the gamma isozyme of PKC. Although the first three knockouts do impair learning and memory, their impact on sensory, motor, attentional, and arousal functions have not yet been assessed thoroughly. In the case of the alpha-isozyme of calcium/calmodulin II kinase, when functional assessments were performed the generalized motor activity was so elevated as to make the animals epileptic. Until the possibility of nonlearning impairments are examined systematically for each "knockout," the specific and direct necessity for learning and memory of the factor that was "knocked out" will remain uncertain. The knockout technique also introduces the uncertainty of developmental defects that alter the actual structure of important brain regions such as the hippocampus. Developmental abnormalities could also obscure the requirement of a particular cellular event for memory storage in adult, fully differentiated brain.

Another genetic manipulation, overexpression of the NMDA receptor (Tang et al., 1999), produced behavioral results that have been interpreted as "enhanced learning." Although modest increases in some forms of learned behaviors were demonstrated, controls for arousal or sensitization were not included in the study. Thus, it was not possible to rule out NMDA-induced enhanced arousal or sensitization that could increase behavioral changes in general and not just increase learning. Furthermore, the NMDA overexpression did markedly enhance extinction of the "learned" behavior, a finding consistent with generalized enhanced behavioral changes. If true, generalized enhanced behavioral responding would not be expected to improve net learning because loss of memory and interference by successive training experiences would minimize any net gain of permanent stored information.

Cellular Models of Learning and Memory

Correlations and specific lesions (be they structural, genetic, or pharmacological) within memory paradigms require the behavioral assessment of living animals. However, there has been a growing trend in the use of synaptic models of cellular processes such as long-term potentiation (LTP) and LTD that are hypothesized to underlie learning and memory. Neither LTP nor LTD has ever been found to occur in response to in vivo stimulation of *intact sensory pathways* in either living animals or reduced preparations of animal brain. However, the prolonged and marked changes of synaptic response amplitude induced by high-frequency stimulation of presynaptic fibers with either LTP or LTD protocols have suggested that LTP or LTD may be related to and therefore serve as models for learning-induced synaptic changes.

LTP

LTP is an increase in EPSP amplitude and decrease in threshold for a postsynaptic action potential induced by high-frequency stimulation of presynaptic fibers such as the Schaeffer collaterals that synapse on CA1 pyramidal cells (Bliss & Lomo, 1973). Typically, several trains of 100-Hz stimulation are sufficient to cause LTP that lasts for 1 h or more. Many other stimulation paradigms, including temporally related pre- and postsynaptic stimulation, have also been used to induce LTP. Furthermore, LTP has been induced in different regions of the hippocampus as well as in different regions of the brain including the cortex (for review, see Buonomano & Merzenich, 1998). Available evidence suggests that the mechanisms for LTP induction vary for different regions. For example, although both pre- and postsynaptic mechanisms contribute to hippocampal CA1 LTP, postsynaptic mechanisms appear to predominate in hippocampal dentate gyrus LTP (Bliss & Collingridge, 1993). In no case does LTP appear to involve long-lasting changes of voltage-dependent potassium channels as have been implicated in learning and memory of alive, behaving animals. In vivo learning does appear to share at least one major mechanism with LTP, namely, activation of PKC. Activation of calcium/calmodulin II kinase, originally implicated in *Hermissenda* associative learning, also seems to contribute to LTP (Bliss & Collingridge, 1993).

LTD

Long-term depression is found in both in vivo and in vitro preparations and is characterized by a reduction in EPSPs as a function of direct electrical stimulation of presynaptic pathways (Linden & Connor, 1995). In the cerebellar cortical slice, LTD describes a reduction in the size of Purkinje cell EPSPs that results from conjoint stimulation of the climbing fiber and parallel fiber inputs to the Purkinje cell. The conditions for Purkinje cell stimulation are critical to the induction of LTD because it only occurs if climbing fiber stimulation occurs before parallel fiber stimulation in the presence of a GABA inhibitor such as picrotoxin (Schreurs & Alkon, 1993). An extensive investigation of LTD has shown it to be calcium mediated and dependent on PKC (Freeman et al., 1998), two factors that also appear to be important for associative learning (see earlier). The fact that Purkinje cells of the cerebellum are capable of synaptic plasticity and that the cerebellar cortex is important for classical conditioning has led to the suggestion that LTD underlies classical conditioning. However, as noted previously, if slices of cerebellar cortex are examined after classical conditioning, Purkinje cells actually increase in excitability rather than decrease and a normally inducible form of LTD appears to be occluded (Schreurs et al., 1997a). Consequently, LTD in the cerebellar cortex may not be a mechanism involved in the cerebellar-dependent form of classical conditioning.

The relevance of LTP and LTD mechanisms for associative learning and memory has come under increasing scrutiny and, in a number of cases, called into question. As mentioned, the hippocampal Kv1.1 (potassium channel) "knockdown" experiment eliminated memory of the spatial maze. However, this same Kv1.1 knockdown of postsynaptic potassium channels had no effect on LTP in hippocampal CA1 pyramidal cells or LTP in the dentate gyrus. Conversely, knockdown of the Kv1.4 presynaptic potassium channel in the rat hippocampus had no effect at all on spatial maze learning and memory but entirely eliminated both short- and long-term LTP in hippocampal CA1 pyramidal cells. Taken together, these experiments demonstrate that LTP in CA1 cells is neither necessary (Kv1.4) nor sufficient (Kv1.1) for rat spatial maze learning and memory. More recently, knockout of the postsynaptic α-amino-3hydroxy-5-methyl-4-isoxazolepropionate(AMPA) receptor confirmed the lack of necessity of LTP in CA1 or dentate gyrus for spatial maze learning. Specifically, the AMPA knockout, as expected from many previous studies, eliminated LTP but in this case the knockout had no effect on rat spatial maze learning and memory (Zamanillo et al., 1999).

Summary and Conclusions

With the large number of model systems currently being examined, the potential for insights into the neural mechanisms underlying learning and memory is substantial. When research using model systems is focused on associative learning, there are two model systems that have already begun to yield insights into the cellular and molecular mechanisms underlying learning. In fact, research with the marine snail *Hermissenda* and the albino rabbit suggests that both the behavioral and neural mechanisms involved in associative learning may be conserved across these seemingly disparate species. In both species, the animal learns to associate an innocuous signal with a significant event within a precisely circumscribed temporal interval, and the result of this associative learning is a persistent change in the membrane properties of specific target cells in the neural systems involved in each association.

Our evidence suggests that classical conditioning involves a molecular cascade that responds to profound elevation of intracellular calcium in association with other second-messenger signals such as diacylerglycerol or arachidonic acid. The temporal association of synaptic signals transduces and transmits the temporal association of sensory stimuli received by the animal during behavioral training. Associated synaptic signals are then transduced and transmitted in the postsynaptic dendrites by molecular messengers that are themselves temporally associated. The result is a reduction in the flow of potassium ions through the cell membrane, leading to an increased excitability of the cell. Finally, it should be noted that the changes that have been observed in these model systems are only examples of changes that may occur in many areas of the brain. Evidence from human position emission tomography (PET) experiments suggests that functional connections are formed between many brain areas during tasks as relatively simple as eyeblink conditioning (Schreurs et al., 1997b). Each of these areas may be responsible for storing different kinds of information about what is apparently a simple learning task.

References

Abel, T., and Kandel, E. (1998). Positive and negative regulatory mechanisms that mediate long-term memory storage. *Brain Research Reviews, 26,* 360–378.

Aleksandrov, A.V., Vasil'eva, O.N., Ioffe, M.E., and Frolov, A.A. (1992). Certain methods of biomechanical description of various postural adjustment patterns during motoric learning in dogs. *Neuroscience and Behavioral Physiology, 22,* 503–512.

Alkon, D.L. (1983). Learning in marine snails. *Scientific American, 249,* 70–84.

Alkon, D.L. (1987). *Memory traces in the brain.* New York: Cambridge University Press.

Alkon, D.L. (1989). Memory storage and neural systems. *Scientific American, 261,* 42–50.

Alkon, D.L., Nelson, T.J., Zhao, W., and Cavallaro, S, (1998). Time domains of neuronal Ca^{2+} signaling and associative memory: steps through a calexcitin, ryanodine receptor, K^+ channel cascade. *Trends in Neuroscience, 21,* 529–537.

Allen, A., Michels, J., and Young, J.Z. (1985). Memory and visual discrimination by squids. *Marine Behavior and Physiology, 11,* 271–282.

Ayers, J.J.B., Benedict, J.O., and Wichter, E.S. (1975). Systematic manipulation of individual events in a truly random control in rats. *Journal of Comparative and Physiological Psychology, 88,* 97–103.

Baily, C.H., and Kandel, E.R. (1993). Structural changes accompanying membrane storage. *Annual Review of Physiology, 55,* 397–426.

Bao, J.X., Kandel, E.R., and Hawkins, R.D. (1998). Involvement of presynaptic and postsynaptic mechanisms in a cellular analog of classical conditioning at Aplysia sensory-motor neuron synapses in isolated cell culture. *Journal of Neuroscience, 18,* 458–466.

Benedict, J.O., and Ayers, J.J.B. (1972). Factors affecting conditioning in the truly random control procedure in the rat. *Journal of Comparative and Physiological Psychology, 78,* 323–330.

Biegler, R., and Morris, R.G. (1993). Landmark stability is a prerequisite for spatial but not discrimination learning. *Nature, 361,* 631–663.

Bliss, T.V.P., and Collingridge, G.L. (1993). A synaptic model of memory: long-term potentiation in the hippocampus. *Nature, 261,* 31–39.

Bliss, T.V.P., and Lomo, T. (1973). Long-lasting potentiation of synaptic transmission in the dentate area of the anesthetized rabbit following stimulation of the perforant path. *Journal of Physiology (London), 232,* 331–356.

Boakes, R.A. (1977). Performance on learning to associate a stimulus with positive reinforcement. In H. Davis and H.M.B. Hurwitz (Eds.), *Operant Pavlovian interactions* (pp. 67–97). Hillsdale, NJ: Erlbaum.

Brigui, N., Le Bourg, E., and Medioni, J. (1990). Conditioned suppression of the proboscis-extension response in young, middle-aged, and old *Drosophila melanogaster* flies: acquisition and extinction. *Journal of Comparative Psychology, 104,* 289–296.

Brown, G.D. (1998). Operational terminology for stimulus exposure (SE) conditioning. *Behavioural Brain Research, 95,* 143–150.

Buonomano, D.V., and Merzenich, M.M. (1998). Cortical plasticity: from synapses to maps. *Annual Review of Neuroscience, 21,* 149–186.

Carew, T.J., Hawkins, R.D., and Kandel, E.R. (1983). Differential classical conditioning of a defensive withdrawal reflex in *Aplysia. Science, 219,* 397–400.

Carew, T.J., Walters, E.T., and Kandel, E.R. (1981). Classical conditioning in a simple withdrawal reflex in *Aplysia californica. Journal of Neuroscience, 1,* 1426–1437.

Cohen, D.H. (1980). The functional neuroanatomy of a conditioned response. In R.F. Thompson, L.H. Hicks, and V.B. Shvyrkov (Eds.), *Neural mechanisms of goal-directed behavior and learning* (pp. 283–302). New York: Academic Press.

De Jianne, D., McGuire, T.R., and Pruzan-Hotchkiss, A. (1985). Conditioned suppression of proboscis extension in *Drosophila melanogaster. Journal of Comparative Psychology, 99,* 74–80.

Dubnau, J., and Tully, T. (1998). Gene discovery in *Drosophila:* new insights for learning and memory. *Annual Review of Neuroscience, 21,* 407–444.

Ebbinghaus, H. (1885). *Uber das gedachtnis.* Leipzig: Duncker and Humbolt. [Translated by H.A. Ruger and Clara Bussenins (1913) in *Memory.* New York: Columbia.]

Etcheberrigaray, R., Ito, E., Kim, C.S., and Alkon, D.L. (1994a). Soluble β-amyloid induction of Alzheimer's phenotype for human fibroblast K^+ channels. *Science, 264,* 276–279.

Etcheberrigaray, R., Gibson, G.E., and Alkon, D.L. (1994b). Molecular mechanisms of memory and the pathophysiology of Alzheimer's disease. *Annals of the New York Academy of Sciences, 747,* 245–255.

Etcheberrigaray, R., Payne, J.L., and Alkon, D.L. (1996). Soluble beta-amyloid induces Alzheimer's disease features in human fibroblasts and in neuronal tissues. *Life Sciences, 59,* 491–498.

Freeman, J.H., Jr., Scharenberg, A.M., Olds, J.L., and Schreurs, B.G. (1998). Classical conditioning increases membrane-bound protein kinase C in rabbit cerebellum. *NeuroReport, 9,* 2669–2673.

Fresquet, N., and Medioni, J. (1993). Effects of aging on visual discrimination learning in Drosphila melanogaster. *Quarterly Journal of Experimental Psychology, 46,* 399–412.

Gelperin, A. (1994). Nitric oxide mediates network oscillations of olfactory interneurons in a terrestrial mollusc. *Nature, 369,* 61–63.

Glanzman, D.L. (1995). The cellular basis of classical conditioning in *Aplysia californica*—it's not as simple as you think. *Trends in Neuroscience, 18,* 30–36.

Goodwin, S.F., Del Vecchio, M., Velinzon, K., Hogel, C. Russell, S.R.H., Tully, T., and Kaiser, T. (1997). Defective learning in mutants of the *Drosophila* gene for regulatory subunit of cAMP-dependent protein kinase. *Journal of Neuroscience, 17,* 8817–8827.

Gormezano, I. (1966). Classical conditioning. In J.B. Sidowski (Ed.), *Experimental methods and instrumentation in psychology* (pp. 385–420). New York: McGraw-Hill.

Gormezano, I., and Kehoe, E.J. (1975) Classical conditioning: some methodological-conceptual issues. In W.K. Estes, (Ed.), *Handbook of learning and cognitive processes, Vol. 2. Conditioning and behavior theory* (pp. 143–179). Hillsdale, NJ: Erlbaum.

Gormezano, I., and Kehoe, E.J. (1981). Classical conditioning and the law of contiguity. In P. Harzem and M.D. Zeiler (Eds.), *Advances in analysis of behavior. Vol. 2. Predictability, correlation, and contiguity* (pp. 1–45). Sussex, England: Wiley & Sons.

Gormezano, I., and Tait, R.W. (1976). The Pavlovian analysis of instrumental conditioning. *Pavlovian Journal of Biological Sciences, 11,* 37–55.

Gormezano, I., Kehoe, E.J., and Marshall, B. (1983). Twenty years of classical conditioning research with the rabbit. In J.M. Sprague and A.N. Epstein (Eds.), *Progress in psychobiology and physiological psychology, Vol. 11.* (pp. 197–275). New York: Academic Press.

Gormezano, I., Schneiderman, N., Deaux, E.G., and Fuentes, I. (1962). Nictitating membrane: classical conditioning and extinction in the albino rabbit. *Science, 138,* 33–34.

Grant, D.A. (1943). Sensitization and association in eyelid conditioning. *Journal of Experimental Psychology, 32,* 201–212.

Grant, D.A. (1944). A sensitized eyelid reaction related to the conditioned eyelid response. *Journal of Experimental Psychology, 35,* 393–404.

Grant, D.A., and Adams, J.K. (1944). 'Alpha' conditioning in the eyelid. *Journal of Experimental Psychology, 34,* 136–142.

Gruart, A., Blazquez, P., and Delgado-Garcia, J.M. (1995). Kinematics of spontaneous, reflex and conditioned eyelid movements in the alert cat. *Journal of Neurophysiology, 74,* 226–248.

Harris, C.L. (1991). An improved Horridge procedure for studying leg-position learning in cockroaches. *Physiology & Behavior, 49,* 543–548.

Hawkins, R.D., and Kandel, E.R. (1984). Is there a cell-biological alphabet for simple forms of learning? *Psychological Review, 91,* 375–391.

Hawkins, R.D., Carew, T.J., and Kandel, E.R. (1986). Effects of interstimulus interval and contingency on classical conditioning of *Aplysia* siphon-withdrawal reflex. *Journal of Neuroscience, 6,* 1695–1701.

Hawkins, R.D., Greene, W., and Kandel, E.R. (1998a). Classical conditioning, differential conditioning, and second-order conditioning of the *Aplysia* gill-withdrawal reflex in a simplified mantle organ preparation. *Behavioral Neuroscience, 112,* 636–645.

Hawkins, R.D., Cohen, T.E., Greene, W., and Kandel, E.R. (1998b). Relationships between dishabituation, sensitization, and inhibition of the gill- and siphon-withdrawal reflex in *Aplysia californica:* effects of response measure, test time, and training stimulus. *Behavioral Neuroscience, 112,* 24–38.

Hearst, E., and Jenkins, H.M. (1974). *Sign tracking: the stimulus reinforcer relation and directed action.* Monograph. Austin, TX: Psychonomic Society.

Hirsch, J., and Holliday, M. (1988). A fundamental distinction in the analysis and interpretation of behavior. *Journal of Comparative Psychology, 102,* 372–377.

Holland, P.C. (1980). Influence of visual conditioned stimulus characteristics on the form of Pavlovian appetitive conditioned responding in rats. *Journal of Experimental Psychology: Animal Behavior Processes, 6,* 81–97.

Holliday, M. and Hirsch, J. (1986). Excitatory conditioning of individual *Drosophila melanogaster. Journal of Experimental Psychology: Animal Behavior Processes, 12,* 131–142.

Jenkins, H.M., and Moore, B.R. (1973). The form of the autoshaped response with food or water reinforcers. *Journal of the Experimental Analysis of Behavior, 20,* 163–181.

Kandel, E.R. (1976) *Cellular basis of behavior: an introduction to behavioral neurobiology.* San Francisco, Free Press.

Keifer, J., Armstrong, K.E., and Houk, J.C. (1995). In vitro classical conditioning of abducens nerve discharge in turtles. *Journal of Neuroscience, 15,* 5036–5048.

Kim, E.H.-J., Woody, C.D., and Berthier, N.E. (1983). Rapid acquisition of conditioned eye-blink responses in cats following pairing of an auditory CS with a glabellar-tap US and hypothalamic stimulation. *Journal of Neurophysiology, 49,* 767–779.

Krasne, F.B., and Teshiba, T. (1995). Habituation of an invertebrate escape reflex due to modulation by higher centers rather than local events. *Proceedings of the National Academy of Sciences of the United States of America, 92,* 3362–3366.

Krupa, D.J., Weng, J., and Thompson, R.F. (1996). Inactivation of brainstem nuclei blocks expression but not acquisition of the rabbit's classically conditioned eyeblink response. *Behavioral Neuroscience, 110,* 219–227.

Kulitka, E.F. (1992). Influence of superficial polarization of the cerebral cortex of the dog on extinctive inhibition. *Neuroscience & Behavioral Physiology, 22,* 56–58.

Lederhendler, I., Gart, S., and Alkon, D.L. (1986). Classical conditioning of *Hermissenda*: origin of a new response. *Journal of Neuroscience, 6,* 1325–1331.

Leonard, J.L., Edstrom, J., and Lukowiak, K. (1989). Reexamination of the gill withdrawal reflex of Aplysia californica Cooper (Gastropoda: Opisthobranchia). *Behavioral Neuroscience, 103,* 585–604.

Linden, D.J., and Connor, J.A. (1995). Long-term depression. *Annual Review of Neuroscience, 18,* 319–357.

Lofdahl, K.L., Holliday, M.J., and Hirsch, J. (1992). Selection for conditionability in Drosophila melanogster. *Journal of Comparative Psychology, 106,* 172–183.

Mackintosh, N.J. (1983). *Conditioning and associative learning.* Oxford: Oxford University Press.

Matzel, L.D., Schreurs, B.G., and Alkon, D.L. (1990a). Pavlovian conditioning of distinct components of *Hermissenda*'s response to rotation. *Behavioral and Neural Biology, 54,* 131–145.

Matzel, L.D., Schreurs, B.G., Lederhendler, I., and Alkon, D.L. (1990b). Acquisition of conditioned associations in *Hermissenda*: additive effects of contiguity and the forward interstimulus interval. *Behavioral Neuroscience, 104,* 597–606.

Medioni, J., and Vaysse, G. (1975). Suppression conditionnelle d'un reflexe chez la Drosphile Drosophila melanogaster): acquistion et extinction. *Comptes Rendus des Seances de la Societe de Biologie, 169,* 1386–1391.

Meiri, N., Ghelardini, C., Tesco, G., Galeotti, N., Dahl, D., Tomsic, D., Cavallaro, S., Quattrone, A., Capaccioli, S., Bartolini, A., and Alkon, D.L. (1997). Reversible antisense inhibition of shaker-like Kv1.1 potassium channel expression impairs associative memory in mouse and rat. *Proceedings of the National Academy of Sciences of the United States of America, 94,* 4430–4434.

Menzel, R., and Muller, R. (1996). Learning and memory in honeybees: from behavior to neural substrates. *Annual Review of Neuroscience, 19,* 379–404.

Michels, J., Robertson, J.D., and Young, J.Z. (1987). Can conditioned aversive tactile stimuli affect extinction of visual responses in octopus? *Marine Behavior & Physiology, 13,* 1–11.

Milner, B., Squire, L.R., and Kandel, E.R. (1998). Cognitive neuroscience and the study of memory. *Neuron, 20,* 445–468.

Minois, N., and Le Bourg, E. (1997). Hypergravity and aging in *Drosophila melanogaster.* 9. Conditioned suppression and habituation of the proboscis extension response. *Aging Clinical and Experimental Research, 9,* 281–291.

Mishkin, M. (1982). A memory system in the monkey. *Philosophical Transactions of the Royal Society of London, 298,* 85–95.

Molchan, S.E., Sunderland, T., McIntosh, A.R., Herscovitch, P., and Schreurs, B.G. (1994). A functional anatomical study of associative learning in humans. *Proceedings of the National Academy of Sciences of the United States of America, 91,* 8122–8126.

Nelson, T.J., Zhao, W.Q., Yuan, S., Favit, A., Pozzo-Miller, L., and Alkon, D.L. (1999). Calexcitin interaction with neuronal ryanodine receptors. *Biochemical Journal, 341,* 423–433.

Olds, J.L., Anderson, M., McPhie, D., Staten, L., and Alkon, D.L. (1989). Imaging memory-specific changes in the distribution of protein kinase C within the hippocampus. *Science, 245,* 866–869.

Pavlov, I.P. (1927). *Conditioned reflexes*. (Translated by G.V. Anrep.) London: Oxford University Press.

Preat, T. (1998). Decreased odor avoidance after electrical shock in *Drosophila* mutants biases learning and memory tests. *Journal of Neuroscience, 18,* 8534–8538.

Rescorla, R.A. (1988). Pavlovian conditioning. It's not what you think. *American Psychologist, 43,* 151–160.

Ricker, J.P., Hirsch, J., Holliday, M., and Vargo, M.A. (1986). An examination of claims for classical conditioning as a phenotype in the genetic analysis of Diptera. In J.L. Fuller and E.C. Simmel (Eds.), *Perspectives in behavior genetics* (pp. 155–200). Hillsdale, NJ: Erlbaum.

Sahley, C.L. (1994). Serotonin depletion impairs but does not eliminate classical conditioning in the leech *Hirudo medicinalis*. *Behavioral Neuroscience, 108,* 1043–1052.

Sanchez-Andres, J.V., and Alkon, D.L. (1991). Voltage-clamp analysis of the effects of classical conditioning on the hippocampus. *Journal of Neurophysiology, 65,* 796–807.

Schneiderman, N., McCabe, P.M., Haselton, J.R., Ellenberger, H.H., Jarrell, T.W., and Gentile, C.G. (1987). Neurobiological bases of conditioned bradycardia in rabbits. In I. Gormezano, W.F. Prokasy, and R.F. Thompson (Eds.), *Classical conditioning (3rd ed.)* (pp. 37–63). Hillsdale, NJ: Erlbaum.

Schreurs, B.G. (1989). Classical conditioning of model systems: a behavioral review. *Psychobiology 17,* 145–155.

Schreurs, B.G., and Alkon, D.L. (1990). US-US conditioning of the rabbit's nictitating membrane response: emergence of a conditioned response without alpha conditioning. *Psychobiology 18,* 312–320.

Schreurs, B.G., and Alkon, D.L. (1993). Rabbit cerebellar slice analysis of long-term depression and its role in classical conditioning. *Brain Research, 631,* 235–240.

Schreurs, B.G., Sanchez-Andres, J.V., and Alkon, D.L. (1991). Learning-specific differences in Purkinje-cell dendrites of lobule HVI (lobulus simplex): intracellular recording in a rabbit cerebellar slice. *Brain Research, 548,* 18–22.

Schreurs, B.G., Tomsic, D., Gusev, P.A., and Alkon, D.L. (1997a). Dendritic excitability microzones and occluded long-term depression after classical conditioning of the rabbit's nictitating membrane response. *Journal of Neurophysiology, 77,* 86–92.

Schreurs, B.G., McIntosh, A.R., Bahro, M., Herscovitch, P, Sunderland, T., and Molchan, S.E. (1997b). Lateralization and behavioral correlation of changes in regional cerebral blood flow with classical conditioning of the human eyeblink response. *Journal of Neurophysiology, 77,* 2153–2163.

Schreurs, B.G., Gusev, P.A., Tomsic, D., Alkon, D.L., and Shi, T. (1998). Intracellular correlates of acquisition and long-term memory of classical conditioning in Purkinje cell dendrites in slices of rabbit cerebellar lobule HVI. *Journal of Neuroscience, 18,* 5498–5507.

Skelton, R.W. (1988). Bilateral cerebellar lesions disrupt conditioned eyelid responses in unrestrained rats. *Behavioral Neuroscience, 102,* 586–590.

Solomon, P.R., Pomerleau, D., Bennet, L., James, J., and Morse, D.L. (1989). Acquisition of the classically conditioned eyeblink response in humans over the life span. *Psychology and Aging, 4,* 34–41.

Tang, Y.-P., Shimizu, E., Dube, G.R., Rampon, C., Kerchner, G.A., Zhou, M., Liu, G., and Tsien, J.Z. (1999). Genetic enhancement of learning and memory in mice. *Nature, 401,* 63–69.

Thompson, R.F. (1986). The neurobiology of learning and memory. *Science, 233,* 941–947.

Thompson, R.F., and Kim, J.J. (1996). Memory systems in the brain and localization of a memory. *Proceedings of the National Academy of Sciences of the United States of America, 93,* 13438–13444.
Tully, T. (1996). Discovery of genes involved with learning and memory: an experimental synthesis of Hirschian and Benzerian perspectives. *Proceedings of the National Academy of Sciences of the United States of America, 93,* 13460–13467.
Tully, T. (1997). Regulation of gene expression and its role in long-term memory and synaptic plasticity. *Proceedings of the National Academy of Sciences of the United States of America, 94,* 4239–4241.
Tully, T., and Quinn, W.G. (1985). Classical conditioning and retention in normal and mutant *Drosophila melanogaster. Journal of Comparative Physiology, 157,* 263–277.
Van der Zee, E.A., Luiten, P.G.M., and Disterhoft, J.F. (1997). Learning-induced alterations in hippocampal PKC-immunoreactivity: a review and hypothesis of its functional significance. *Progress in Neuro-Psychopharmacology and Biological Psychiatry, 21,* 531–572.
Vargo, M., and Hirsch, J. (1982). Central excitation in the fruit fly (Drosophila melanogaster). *Journal of Comparative and Physiological Psychology, 96,* 452–459.
Vaysse, G., and Medioni, J. (1976). Nouvelles experiences sur le conditionement et le pseudoconditionnement du reflexe tarsal chez la Drosophile (Drospohila melanogaster): effets de chocs electriques de faible intensite. *Comptes Rendus des Seances de la Societe de Biologie, 170,* 1299–1304.
Wallace, B., and Sperlich, D. (1988). Conditioning the behavior of *Drosophila melanogaster* by means of electric shocks. *Proceedings of the National Academy of Sciences of the United States of America, 85,* 2869–2872.
Woodruff-Pak, D.S. (1997). Classical conditioning. *International Review of Neurobiology, 41,* 341–366.
Woody, C.D. (1970). Conditioned eye blink: gross potential activity at coronal-pericruciate cortex of the cat. *Journal of Neurophysiology, 33,* 838–850.
Zamanillo, D., Sprengel, R., Hvalby, O., Jensen, V., Burnashev, N., Rozov, A., Kaiser, K.M.M., Kööster, H.J., Borchardt, T., Worley, P., Lüübke, J., Frotscher, M., Kelly, P.H., Sommer, B., Andersen, P., Seeburg, P.H., and Sakmann, B. (1999). Importance of AMPA receptors for hippocampal synaptic plasticity but not for spatial learning. *Science, 284,* 1805–1811.
Zawistkowski, S., and Hirsch, J. (1984). Conditioned discrimination in the blow fly, *Phormia regina:* controls and bidirectional selection. *Animal Learning & Behavior, 12,* 402–408.
Zucker, R.S. (1972). Crayfish escape behavior and central synapses. II. Neural circuit exciting lateral giant fiber. *Journal of Neurophysiology, 35,* 599–637.

3
Functional Utility and Neurobiology of Conditioned Autonomic Responses

RAY W. WINTERS, PHILIP M. MCCABE, AND NEIL SCHNEIDERMAN

Classical conditioning can be characterized as an experimenter-defined set of operations, involving a temporal contingency between two stimuli, that are independent of the subject's behavior. Viewed from this perspective, classical conditioning is unique among the various approaches to the study of behavior because it allows for a level of experimental control that is rarely encountered in the behavioral sciences. Furthermore, if the experimenter chooses the appropriate stimulus parameters, an analysis of the temporal relationships among the responses elicited and the stimuli used permit the externalization of otherwise internal events that may not be directly observed. The information gleaned from this type of analysis can be used to study the adaptive functions of learned associations (Hollis, 1984; Shettleworth, 1983; Culler, 1938; Palmerino et al., 1980; Dworkin, 1993) or the processes that underlie the associations formed during learning (Gormezano, 1972; Schneiderman, 1972, 1974; Moore, 1986; Thompson et al., 1983). In recent years, the "internal events" described by Black and Prokasy (1972) have been investigated directly using electrophysiological techniques to monitor neural activity during conditioning. In addition, modern neuroanatomic tracing techniques have allowed investigators to chart the central nervous system (CNS) pathways that provide the links between the stimuli and responses involved in conditioning (Weinberger & Diamond, 1987; Kapp et al., 1990; Thompson & Steinmetz, 1992; Davis, 1992; McCabe et al., 1992; Powell, 1994; Ledoux, 1995).

Although classical conditioning studies conducted in recent years have focused on somatomotor conditioned responses (e.g., nictitating membrane, behavioral freezing, potentiated startle), earlier investigators sought to understand the functional utility and mechanisms underlying conditioned autonomic responses. In fact, the seminal work in this field examined classically conditioned salivary and gastrointestinal activity (Pavlov, 1927; Kleitman & Crisler, 1927; Katzenelbogen et al., 1939). During the past 70 years, a variety of autonomically mediated responses have been classically conditioned, including heart rate (Schneiderman et al., 1966; deToledo & Black, 1966; Obrist & Webb, 1967; Cohen & Pitts, 1968, Fitzgerald et al., 1984; Powell et al., 1974; Kapp et al., 1979), blood pressure (Girden, 1942a; Yehle et al., 1967; Stebbins & Smith, 1964; Powell & Kazis, 1976; LeDoux et al., 1984; Dworkin & Dworkin, 1995), regional blood flow (Smith et al., 1980), pupillary responses (Girden, 1942b;

Gerall et al., 1957; Young, 1958; Oleson et al., 1972), galvanic skin response (Baxter, 1966; Holdstock & Schwarzbaum, 1965), and gastric motility (Ban & Shinoda, 1960).

The conditioned response (CR) that develops during conditioning involving the autonomic nervous system has been characterized in two ways. It has been considered to be a discrete response that has been elaborated from an unconditioned reflexive response to a highly specific unconditioned stimulus (Pavlov, 1910; Rescorla, 1967). A *discrete autonomic CR* can be viewed as a response that prepares the organism for an impending unconditioned stimulus that disrupts homeostasis (Pavlov, 1910; Dworkin, 1993). For example, a conditioned salivary response is a discrete autonomic CR that mitigates, in advance of the presentation of the unconditioned stimulus (US), the impact of an irritating acidic substance (the US) upon the oral mucosa. The autonomic CR has also been considered to be nonspecific in that it is not linked to any particular US (Weinberger et al., 1984a). A *nonspecific autonomic CR* is one of many conditioned responses elicited, and it only has functional utility within the context of the cluster of CRs that have developed during conditioning. According to this view, an aversive US, whether it be a shock to the leg or arm or an air puff to the eye, elicits a similar pattern of unconditioned responses (URs). The cluster of nonspecific autonomic CRs that develop during conditioning would be similar to these URs; the autonomic-mediated response pattern would include changes in heart rate, blood pressure, pupillary dilation, and a galvanic skin response. The hypothetical construct *conditioned fear* is used to refer to the change in the CNS that accounts for the cluster of autonomic-mediated responses that, as a result of conditioning, is elicited by the conditioned stimulus (CS). Conditioned fear is thought to promote the development of somatic CRs and learned behavioral changes that mitigate (or totally nullify) the impact of the US.

One aim of the present chapter is to characterize the adaptive functions of classically conditioned autonomic CRs. We first examine literature that addresses how unconditioned autonomic reflexes and *discrete autonomic CRs* facilitate adaptation to changes in the internal milieu and the external world. We also describe the adaptive role of conditioned fear and associated *nonspecific autonomic CRs* in the defense of homeostasis. Conditioning is a process that requires neurophysiological and neuroanatomic changes that allow for the enhancement of synaptic efficacy in the neural pathway that links a previously neutral stimulus to a conditioned autonomic response. The emergence of electrophysiological techniques in recent years has allowed for the direct observation of events in the CNS that occur during autonomic conditioning, and the utilization of modern neuroanatomic tracing techniques has allowed investigators to map the neural pathways that mediate CNS responses that occur during conditioning. A second aim of the present chapter, therefore, is to review research literature that has furthered our understanding of the CNS mechanisms that underlie autonomic conditioning, particularly the *nonspecific autonomic CRs* that develop during fear conditioning. We discuss the neurobiological basis of attentional mechanisms in the last part of the chapter.

The Functional Utility of Discrete Conditioned Autonomic Responses

A fundamental observation made by Pavlov (1910) was that a conditioned autonomic reflex, as it grows in strength, becomes increasingly similar to the unconditioned autonomic reflex from which it was elaborated. Developing a conditioned autonomic reflex was seen as a way to augment the effect of the unconditioned reflex. For example, placing an irritating acidic substance in the mouth will trigger an (unconditioned) reflexive salivary response. The saliva lubricates the mouth and diminishes the effect of the acidic substance upon the mucosa. The conditioned response to a neutral stimulus that precedes the unconditioned stimulus is also salivation. The latency of this CR decreases and its magnitude increases as conditioning progresses. Because this salivation (the CR) precedes the US, it improves the effectiveness of the unconditioned reflex. In fact, the enhanced effectiveness, in terms of protecting the mucosa, of this anticipatory response leads to a reduction in the magnitude of the unconditioned response.

Pavlov saw classical conditioning of autonomic responses as a way to study CNS processes involving the cerebral cortex, and most of his research efforts were dedicated to studying these processes rather than the adaptive functions of discrete conditioned autonomic responses. One of his students, Bykov, was among the first investigators to study the functional utility of conditioned autonomic reflexes. His research efforts were primarily dedicated to understanding the role of classical conditioning in the control of visceral functioning. Much of his work focused on the regulatory mechanisms involved in conditioned diuresis (Bykov, 1957, 1959). The UR in these studies was an increase in urine secretion in response to an intrarectal infusion of water. His observations regarding the temporal characteristics of the response during conditioning were similar to those made by Pavlov in that he found that the latency of the urine flow decreased after repeated exposure to the US. After several days, the urine flow began several minutes after the animal was placed in the experimental apparatus. Eventually, the urine flow occurred even before the presentation of the US: a conditioned autonomic reflex was developing and the CR was anticipatory in that it preceded the onset of the unconditioned stimulus. The CS in this situation was the laboratory setting.

Subsequently, Bykov used a tonal CS to gain better control of the CR. He also studied differential conditioning by using discriminative stimuli; the discriminative stimulus was the laboratory environment (the animals were tested in different rooms). Bykov (1959) also conducted studies that demonstrated that conditioned autonomic responses could be developed in the cardiovascular system. In these studies, the dependent measure was the increased cardiac output that occurs when an animal is required to run on a treadmill. A discrete tone was used as the CS for this experiment. Bykov found, as would be expected on the basis of his findings in the conditioned diuresis studies, that the latency of the increased cardiac output became shorter with repeated testing on the treadmill. Eventually, increases in cardiac output occurred well in advance of running. The conditioned autonomic response was smaller than the unconditioned response (40%–110% increase in cardiac output versus a 250%–400% increase), but it was adaptive in that it anticipated the increases in metabolic demands that are associated with running.

The observations made by Pavlov and by Bykov regarding the temporal relation between the CR and the US suggest that the autonomic conditioning process is a form of *predictive homeostasis*. Predictive homeostasis is distinguished from *reactive homeostasis* by the time at which the autonomic response mechanism is engaged. Response mechanisms that mediate reactive homeostasis are engaged *after* the occurrence of a regulatory challenge. For example, insulin secretion is increased in response to a rise in blood glucose level. Similarly, heart rate and blood flow to the muscles increase in response to the heightened metabolic requirements of exercise. The regulatory response usually remains engaged until the controlled variable is returned to a preset reference level (set point) or the regulatory challenge has been removed. In contrast to a reactive homeostatic mechanism, a predictive homeostatic mechanism is engaged *before* the onset of the regulatory challenge. It is a preemptive response that mitigates or nullifies the impact of the forthcoming stimulus which poses a challenge to homeostasis. Thus, based on the observations of Pavlov and Bykov, one possible adaptive function of the conditioned autonomic response in classical conditioning is that it promotes homeostasis by mitigating, in advance, the impact of the regulatory challenge (the US).

It is axiomatic that if conditioned autonomic responses are involved in predictive homeostasis, the conditioned response must be very similar to the unconditioned response. The degree of similarity is dependent on the complexity of the US. An electric shock to a muscle, for example, is far less selective with respect to stimulus and response systems than the acidic solutions used in Pavlov's salivary-conditioning studies. In general, research findings are consistent with the assertion that the UR and CR are similar, particularly when interoceptors detect the energy fluctuations generated by the US and the CS. A notable exception to this rule is heart rate conditioning in which shock is employed as a US (Schneiderman, 1974). The connection between cardiovascular activity and movement is important to the understanding of this exception to the rule. Obrist and colleagues (Obrist et al., 1974) conducted a series of studies that provide convincing evidence that heart rate is closely coupled with somatic activity. They found that the cardiovascular system and somatic musculature are so finely tuned that the smallest alteration in somatic activity will be accompanied by an equally precise alteration in heart rate. Cardiac and somatic activities only become uncoupled at extremely high levels of stress.

In view of this somatic–cardiac link, the most adaptive cardiac response during conditioning is dependent on the availability of a coping response (i.e., the amount of motor activity required). The cardiovascular system's role in the defense of homeostasis is to deliver nutrients and remove wastes from body tissue, and the heart's level of activity is dependent on the metabolic needs of these tissues. The UR to a shock US may be an increase in heart rate (or a biphasic response) because of increased motor activity (an unconditioned escape response), but in classical conditioning experiments in which the animal is restrained it would not be adaptive to develop increases in heart rate as a CR. In contrast, one would expect the CR and UR to be similar in experiments such as the treadmill studies discussed earlier (Bykov, 1957) because a conditioned increase in cardiac output would promote homeostasis by anticipating the impending increase in metabolic needs.

The results of studies assessing conditioned responses to drug injections also appear to be paradoxical with respect to the view that autonomic conditioning is a form of predicative homeostasis. In general, the CR to a drug is opposite to the pharmacological effect of the drug (Siegel, 1983). For example, epinephrine causes a decrease in gastric secretion, whereas the autonomic CR is an increase in gastric secretion (Guha et al., 1974). The CRs to almost two dozen drugs have been studied and in each case the direction of the CR is opposite to the pharmacological effect of the drug (Siegel, 1983). These findings provided the foundation for a model of drug tolerance that is based on the idea that learned responses to environmental cues (the CS) associated with drug ingestion are conditioned compensatory responses.

If the autonomic CR is to be viewed as a component of a predictive homeostatic process, it is essential to delineate the regulatory characteristics of the unconditioned reflex; otherwise, it is not possible determine if the CR is a replica of the UR. A question that naturally arises from studies involving conditioned autonomic responses to drugs is the following. Among the various physiological changes that occur as a result of drug ingestion, what are the *effective US* and the *effective UR?* For example, eating candy will eventually lead to a conditioned hypoglycemic response that precedes the arrival of the candy in the stomach. There are a number of physiological reactions to the ingestion of a piece of candy including salivary secretions, secretions in the stomach, an initial rise in blood glucose level, the secretion of insulin by the pancreas, and a subsequent lowering of blood glucose level. Which of these reactions is the US and which is the UR? The tradition has been to refer to the drug as the US and the effect of the chemical (drug) as the UR. Thus, in the candy example the increase in glucose would be identified as the UR. Viewed from this perspective, the hypoglycemic CR is opposite in polarity to the UR and hence appears to be a compensatory response. Viewed in terms of regulatory mechanisms and homeostasis, the *effective UR* is the insulin response (which is reflected by the hypoglycemic response) and the *effective US* is the increase in glucose. The rise in glucose is considered to be a "challenge" to the control system involved in the regulation of blood glucose level, and this system responds to this challenge by secreting insulin. In general, most unconditioned autonomic reflexes are homeostatic in that they diminish the impact of the US. Thus, the UR serves to limit the disruptive effect of the US on the physiological state of the organism. Viewed in this way, the *effective US* is always the physiological change that engages a homeostatic response—the *effective UR*—that serves to return the controlled variable (blood glucose in this example) to its reference level. The CR is also an increase in insulin and, like the UR, is reflected by the hypoglycemic response. The CR is a response to the sweet taste of candy, the CS; it precedes the absorption of glucose (Deutsch, 1974) and thus promotes homeostasis by anticipating a rise in blood glucose.

In our view, the reason why the autonomic CR appears to be a compensatory response to a drug is rooted in the manner in which the results of these experiments have been conceptualized (see Dworkin, 1993, for a more thorough analysis). More specifically, the pharmacological effect of the ingested drug has been considered to be the UR. However, if autonomic conditioning is viewed in terms of homeostatic mechanisms, the UR is always a corrective homeostasis-promoting response that serves to return a controlled variable to its reference level. The (effective) US is always a change in the physiological state of the organism that poses a regulatory challenge. If unconditioned reflexes are viewed as participants in reactive homeostasis and conditioned reflexes participants in predicative homeostasis, the CR and the (effective) UR will be similar.

The Relationship Between the CS–US Interval and the Latency of the CR

If the development of the autonomic CR promotes homeostasis by diminishing the impact of the US, variations in the CS–US interval should lead to variations in the latency of the CR. More specifically, if the CR counteracts the US by anticipating its onset, the CR should track the US when the CS–US interval is varied. Indeed, Pavlov (1927) described a phenomenon, which he referred to as "inhibition of delay," that indicates that CR–US tracking does occur. As conditioning progresses the magnitude of the CR increases and its latency normally decreases, although the duration of the CR may increase and overlap the US. When long CS–US intervals are used, however, the latency of the CR ceases to shorten early in the conditioning process so that it is withheld until the presentation of the US. Oleson et al., (1972) studied the relationship between variations in the CS–US interval and the time of onset of conditioned pupillary responses in cats who were paralyzed by a neuromuscular blockade. They found that, for a wide variety of CS–US intervals, the point of peak dilation occurred, on average, about 0.73 s before the onset of the US, except for very short CS–US intervals. Their findings, along with those from other laboratories (Pavlov, 1927; Rescorla, 1967; Williams, 1965), provide evidence for the view that the CR is participating in a predicative homeostatic process that serves to maintain the current physiological state of the organism by mitigating the impact of the US.

Autonomic Conditioning Viewed from a Biobehavioral Control Systems Perspective

Most organisms utilize regulatory mechanisms to maintain physiological variables at values that sustain life and promote health. The unconditioned autonomic reflex is an example of this type of mechanism. Described in terms of a linear control system model, the unconditioned autonomic reflex mitigates the impact of a regulatory challenge by engaging a negative feedback mechanism (Dworkin, 1993). The (effective) US for a physiological control system is the physiological change that poses a regulatory challenge by altering the physiological state of the organism. The (effective) UR is the response that serves to restore the original physiological state by opposing the effect of the US. In terms of a linear control system model (Dworkin, 1993), this process is referred to as a negative feedback mechanism because the UR limits the disruptive effect of the stimulus that produced it (the effective US). For example, an injection of the alpha-adrenergic agonist phenylephrine causes vasoconstriction and increased blood pressure, followed by a bradycardia that lowers cardiac output, thereby returning blood pressure to its original level. The rise in blood pressure in this example is the (effective) US, and the bradycardia is the (effective) UR. This unconditioned autonomic reflex invokes a negative feedback mechanism because the bradycardia diminishes the impact of the regulatory challenge (the increase in blood pressure); it accomplishes this by reducing cardiac output.

If the effectiveness of unconditioned autonomic reflexes are assessed by the quantitative methods used in a model of linear control systems, these reflexes often turn out to be relatively weak regulators of physiological variables because they exhibit regulatory lags in the negative feedback mechanism, particularly when the UR involves the neuroendocrine system (Dworkin, 1993). One adaptive function of a conditioned autonomic reflex is that it improves the regulatory properties of the unconditioned autonomic reflex; it improves the dynamic characteristics of the unconditioned reflex by strengthening the negative feedback mechanism. The CR is able to strengthen the negative feedback mechanism because it eliminates the regulatory lags associated with the unconditioned reflex and thus produces a greater reduction in the effective magnitude of the US. Because of the temporal relation it bears to the US, the CS provides a warning signal of an impending challenge to homeostasis. The CR mitigates the impact of the US by anticipating its occurrence and thus is a participant in a predictive homeostatic process. When analyzed by the quantitative techniques utilized in a linear control systems model, the CR is found to augment the dynamic properties of the unconditioned autonomic reflex by allowing it to operate at a lower gain and with less oscillation; that is, it improves the dynamic stability of the reflex (Dworkin, 1993).

The Adaptive Utility of Conditioned Fear and Nonspecific Autonomic Conditioned Responses

A number of investigators have examined multiple autonomic CRs, or simultaneously conditioned autonomic and somatomotor responses. The results of these studies provide evidence that the process of conditioning is dependent on the response system being studied. Among the CR-dependent observed differences in the conditioning process are variations in the rates of CR acquisition, acuity in discriminative responding, the shape of the interstimulus interval function, and the level of performance under delay versus trace procedures (Gantt, 1953; Mowrer, 1950, 1960; Schneiderman, 1972). These types of observations have led to the theoretical position that aversive conditioning occurs in stages such that the initial phase consists of conditioned fear, which is followed by the acquisition of discrete skeletal motor responses (Miller, 1948; Mowrer, 1947). CRs acquired during the initial "emotional" phase were said to be diffuse and preparatory, whereas CRs in the latter phase were labeled as precise and adaptive (Konorski, 1967; Weinberger, 1982).

Although Mowrer (1947) originally distinguished between visceral, or problem posing, responses during conditioned fear, and skeletal motor, or problem-solving, responses during the latter phase of avoidance learning, Weinberger (1982) has pointed out some CRs during fear conditioning are, in fact, skeletal motor responses (e.g., behavioral freezing, potentiated startle). Therefore, Weinberger distinguished between "nonspecific" (many of which are autonomic) CRs in the early phase versus "specific" CRs during the latter phase. By nonspecific he meant that it was not linked to any particular US (Weinberger, 1982) and that functional utility is only gained within the context of the cluster of CRs that had been elaborated from the UR. Thus, an aversive US, regardless of it specific characteristics (e.g., air puff versus electric shock) or location, elicits a similar pattern of URs. A particular UR, such as a decrease in heart rate, can be considered to be an element of a response pattern that facilitates learning by enhancing sensitivity to the environment (Sokolov, 1963). The constellation of nonspecific autonomic CRs that develop during conditioning would be similar to their corresponding URs. Conditioned fear (or anxiety) is the term used to refer to the CNS state that generates the cluster of nonspecific CRs. Conditioned fear is thought to develop from the same CNS structures that underlie an unconditioned fear response to an aversive US.

Motivational Properties of Conditioned Fear

Conditioned fear is considered to be a conditioned response that can be involved in predictive homeostasis because it motivates behavioral changes that lead to avoidance, in advance, of the presentation of an aversive stimulus that would disrupt homeostasis. Viewed from this perspective, conditioned fear is characterized as a state of the CNS associated with the anticipation of harm or vulnerability in response to a conditioned stimulus. A variety of coping responses are motivated by fear, including phasic inhibition of behavior (the behavioral freezing response), species-typical defensive behaviors (the fight/flight response), or active avoidance. As pointed out by Rosen and Schulkin (1998), the activation of the neuronal circuitry that forms the substrate of conditioned fear not only motivates behavioral changes, it alters the organism's orientation toward the environment by increasing nonspecific arousal and vigilance and by reallocating attentional resources. These CNS changes serve to improve the organism's ability to detect sensory stimuli that may be linked to danger and threat or which may guide coping responses such as instrumentally conditioned responses that allow the organism to avoid an aversive stimulus.

Alternatively, the autonomic CRs in aversive conditioning studies may be interpreted solely in terms of autonomic responses involved in predictive homeostasis (Schneiderman, 1972). The UR to an electric shock US in the restrained rabbit, for instance, involves an increase in arterial blood pressure, whereas the CR consists of bradycardia. Thus, the CR appears to mitigate the blood pressure increase elicited by the US, which becomes attenuated after several conditioning trials. In contrast, the UR to an aversive US in the unrestrained rat consists of an increase in arterial pressure, as does the CR. Thus, the blood pressure CR appears to be facilitating the metabolic needs of the organism (i.e., predictive homeostasis) by what Obrist and Webb (1967) referred to as the cardiac–somatic linkage. In conclusion, it appears that much of autonomic classical conditioning can be interpreted within the context of predictive homeostasis, although its relationship to such concepts as conditioned fear remains to be explored.

Conditioned Fear and the Development of Classically Conditioned Somatic CRs

In general, autonomic CRs develop more rapidly than most somatic CRs (Schneiderman, 1972; Weinberger, 1982), and in many cases this rapid development has adaptive utility. As a case in point, conditioned fear responses are thought to facilitate the development of somatic CRs that attenuate (or totally nullify) the impact of the US (Weinberger et al., 1984, a, b). The autonomic CRs associated with conditioned fear are thought to reflect a general change in behavioral state. As discussed earlier, conditioned fear not only leads to changes in activity in the autonomic nervous system but also alters the organism's orientation toward the environment in a way that enhances the reliability and efficiency of feature extraction from sensory input, thereby facilitating the detection of stimuli that are most relevant to the development of adaptive somatic CRs. For example, conditioned bradycardia responses to a tonal CS are observed well in advance of the development of eyeblink CRs to the same CS. The bradycardia CR is thought to be one of many autonomic CRs that emerge as a result of the CS-US pairings. The altered CNS state (i.e., conditioned fear) that underlies the constellation of autonomic CRs is thought to facilitate the development of the eyeblink response. Once this adaptive somatic CR develops, the previously conditioned bradycardia disappears (Jarrell et al., 1986a; Powell et al., 1990). This pattern of rapid conditioning of nonspecific autonomic CRs (and conditioned fear), followed by the development of specific adaptive somatic CRs (i.e., those that attenuate the impact of the US), has led Powell and colleagues (Powell et al., 1990) to argue for a two-stage classical conditioning model in experimental protocols characterized by early development of autonomic-mediated CRs.

CNS Circuitry Underlying Classical Fear Conditioning

Much of the recent research examining autonomic/behavioral conditioning has focused on the neurobiology of fear conditioning (Davis, 1992; Fanselow & Kim, 1994; Kapp et al., 1990; LeDoux, 1995). In the fear conditioning model, a neutral auditory stimulus is repeatedly paired with an aversive somatosensory US. Within relatively few pairings of the neutral stimulus and US, the organism begins to exhibit a variety of autonomic and behavioral conditioned responses (CRs) including changes in heart rate (Schneiderman et al., 1966; Cohen, 1969; Fitzgerald et al., 1984; Kapp et al., 1979; Powell et al., 1974; Supple & Leaton, 1990), blood pressure (LeDoux et al., 1984), pupillary responses (Oleson et al., 1975), behavioral suppression of operant responses (LeDoux et al., 1984; DeToledo & Black, 1966; Parrish, 1967; Swadlow et al., 1971), and potentiated startle responses (Davis et al., 1982). These response are thought to result from an altered state of the CNS that is the product of the conditioning process, that is, conditioned fear.

According to this view, rapidly acquired CRs occur together, and therefore it seems likely that they are generated by the same CNS mechanism. It is reasoned that components of the CNS circuitry that pertains to stimulus detection and emotional significance will overlap with some of the circuitry involved in the generation of nonspecific conditioned responses linked to fear. The learning of the emotional significance of the CS is thought to result from synaptic changes that take place on the sensory side of the circuitry. Consequently, the neural representation of the previously neutral stimulus (the CS) develops the capability of generating the CRs. Individual pathways mediating each fear-CR diverge and project separately to the different effector systems (e.g., heart rate (HR) versus pupillary responses versus potentiated startle; see Davis, 1992).

The Role of the Amygdala in Fear Conditioning and CR Pathways

The amygdala appears to be a critical structure in the development and expression of conditioned fear responses (Kapp et al., 1984; Davis, 1992; Fanselow & Kim, 1994; LeDoux, 1995). Disruption of amygdaloid function with either electrolytic or chemical lesions, or by pharmacological blockade, interferes with the acquisition and expression of autonomic and behavioral fear-conditioned responses (Blanchard & Blanchard, 1972; Fanselow & Kim, 1994; Kapp et al., 1979; Gentile et al., 1986; Hitchcock & Davis, 1986; Iwata et al., 1986; LeDoux et al., 1990a; McCabe et al., 1992). In addition, individual neurons within the amygdala have been shown to exhibit learning-related changes in electrophysiological activity during fear conditioning (Pascoe & Kapp, 1985; Quirk et al., 1995; McEchron et al., 1995).

The lateral nucleus of the amygdala appears to be the primary sensory interface of this structure (LeDoux et al., 1990a). The lateral nucleus receives CS and US inputs from the thalamus and cerebral cortex and, in turn, relays this information to other amygdaloid subnuclei. In particular, the central nucleus of the amygdala (ACe) functions as an output nucleus and may be critical for the integration and expression of fear CRs (Kapp et al., 1984; Davis, 1992; McCabe et al., 1992). The ACe receives direct projections from the lateral nucleus and indirect projections from the basolateral and basomedial subnuclei (Rogan & LeDoux, 1996). Cells in the ACe project to hypothalamic and brain stem regions that connect to the peripheral structures, such as the heart, that express autonomic, hormonal, and behavioral CRs observed as a result of fear conditioning (Davis, 1992). For example, ACe projections to the dorsal motor nucleus of the vagus nerve are involved in the mediation of heart rate CRs, projections to the nucleus reticularis pontis caudalis are involved in the potentiation of startle reflexes, and projections to the periaqueductal grey are involved in the behavioral freezing that is seen during conditioned fear (Davis, 1992).

The amygdala may be a critical site of cellular plasticity underlying fear conditioning. The cellular basis of associative learning is thought to involve changes in synaptic transmission at critical points within the conditioning circuit (Hebb, 1949). In the fear conditioning model, it is likely that convergence of CS and US inputs leads to changes in the synaptic efficacy of the CS inputs, such that following training, presentation of the CS allows these inputs to activate the neuronal circuitry involved in the expression of a CR. Importantly, it has been demonstrated that the amygdala receives CS and US information (Pascoe & Kapp, 1985; Quirk et al., 1995; McEchron et al., 1995), and it has been shown that amygdaloid neurons support experimentally induced long-term potentiation (LTP) (Clugnet & LeDoux, 1990). LTP has been proposed as a mechanism of synaptic plasticity and typically is dependent on glutamatergic N-methyl-D-aspartate (NMDA) receptors. Interestingly, it has been demonstrated that the CS inputs to the amygdala are glutamatergic (LeDoux & Farb, 1991), and that synaptic transmission through this pathway is blocked by NMDA and non-NMDA antagonists (Li et al., 1995). Intraamygdaloid administration of NMDA antagonists disrupts the acquisition of fear conditioning but does not block the expression of previously acquired fear conditioning (Campeau et al., 1992; Kim et al., 1993). Taken together, these findings suggests that the amygdala may be an important site of neuronal plasticity in the fear conditioning circuit.

Sensory Pathways in Fear Conditioning

Both Pavlov (1927) and Hebb (1949) believed that classical conditioning, which is regarded as a form of associative learning, requires the convergence of sensory information. More specifically, they believed that it is necessary for neurons conveying CS and US information to converge at some point, or points, for conditioning to take place (i.e., so that the CS will be able to substitute for the US in the generation of a response). If the amygdala is viewed as a potential site for associative plasticity in the fear conditioning paradigm, then CS and US information must project to amygdaloid neurons that are a part of the conditioning circuitry. Because the CS in the fear conditioning protocol is typically an acoustic stimulus, it has been hypothesized that structures within the CNS auditory pathway convey CS information to the amygdala.

Injections of retrograde neuroanatomic tracers into the amygdala have been found to label cell bodies of neurons in the medial subnucleus of the medial geniculate nucleus (mMG) (Veening, 1978; Ottersen & Ben-Ari, 1979; LeDoux et al., 1985; Jarrell et al., 1986a). The ventrolateral portion of the medial geniculate nucleus is an important thalamic structure in the auditory pathway (primary lemniscal pathway) that connects auditory receptors to projection areas in the auditory cortex. In contrast, mMG receives inputs from many sensory systems (auditory, visual, somatosensory) and is viewed as part of a "secondary lemniscal" system involved in other aspects of sensory processing such as polymodal associations (Weinberger & Diamond, 1987). Using neuroanatomic techniques, LeDoux and colleagues examined the anatomic connections between the auditory thalamus and the amygdala in the rat. They observed that the mMG and an additional adjacent region, the posterior intralaminar nucleus (PIN), convey acoustic information to the lateral nucleus of the amygdala (AL) but not directly to ACe (LeDoux et al., 1990b). This finding is consistent with a model in which the AL serves as the sensory interface for CS–US information, which is then conveyed to other amygdaloid nuclei such as ACe.

Both the acquisition and retention of a fear-conditioned response is prevented by electrolytic or chemical lesions of mMG/PIN (Jarrell et al., 1986a, b; McCabe et al., 1993; LeDoux et al., 1984; Jarrell et al., 1987). Electrophysiological recordings from single mMG/PIN neurons have revealed that these cells exhibit associative activity during the acquisition (McEchron et al., 1995), retention (Gabriel et al., 1975; 1976; McEchron et al., 1995; Supple & Kapp, 1989; Ryugo & Weinberger, 1978), and extinction (McEchron et al., 1995) phases of conditioning (Figure 3.1). Following CS onset, many mMG/PIN neurons fire at a shorter latency than amygdaloid neurons, providing evidence that CS information is processed in the thalamus before being relayed to the amygdala (McEchron et al., 1995). Furthermore, the latencies of these mMG/PIN neurons decrease during conditioning, suggesting that synaptic plasticity or rerouting of information occurs at, or before, this level of the conditioning circuitry. The results of studies by Edeline and Weinberger (1992) provide evidence that neurons in mMG exhibit sensory plasticity as a result of classical conditioning. Individual cells in the mMG showed associative retuning of auditory receptive fields such that the tuning curve of the mMG neuron shifts toward the CS frequency as a result of conditioning.

As mentioned previously, the cellular mechanism for CNS plasticity underlying associative learning may be a change in synaptic efficacy at critical points

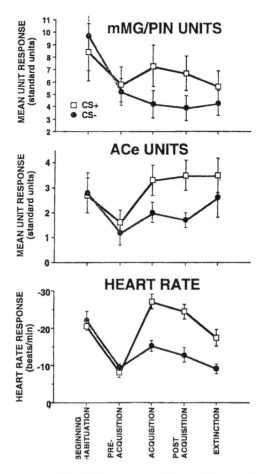

FIGURE 3.1. Magnitude of mMG/PIN (*upper;* $n = 19$) and ACe (*middle;* $n = 15$) unit responses to the CSs recorded simultaneously within the same animals across different phases of training. *Lower panel* shows heart rate responses of the rabbits ($n = 12$) from which these units were recorded. *Bars* indicate mean ± **SE**. *mMG/PIN,* medial subnucleus of medial geniculate nucleus/posterior intralaminary nucleus of thalamus; *ACE,* central nucleus of amygdala. (Reprinted from *Brain Research,* Vol. 682, McEchron et al., Simultaneous single unit recording in the medial nucleus of the medial geniculate and amygdaloid central nucleus throughout habituation, acquisition, and extinction of the rabbit's classically conditioned heart rate, pp. 157–166, Copyright 1995, with permission from Elsevier Science.)

within the conditioning circuit (Hebb, 1949). A study conducted in our laboratory tested the hypothesis that inputs to the mMG, conveying information about an auditory CS, altered mMG neural activity in a manner that supported changes in synaptic efficacy (McEchron et al., 1996a). Figure 3.2 shows the arrangement of the stimulating and recording electrodes in this experiment. We monitored the activity of single units in the mMG while stimulating either the brachium of the inferior colliculus (BIC), an auditory structure, or the superior colliculus (SC), a vi-

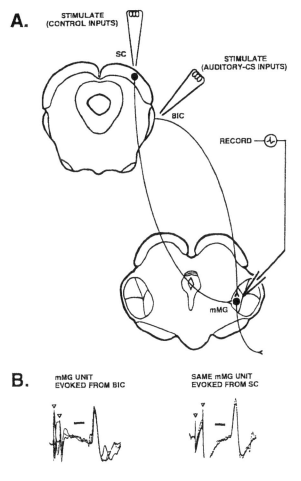

FIGURE 3.2. (**A**) Schematic representation of the sites of stimulation and the placement of the recording electrode in the area of the mMG immediately after training. Single mMG unit responses evoked by test stimulation of the BIC were used to examine conditioning-induced changes in the efficacy of synapses that transmit auditory CS information. The same mMG neurons were activated by test stimuli delivered to the SC to examine pathway-specific changes in unit responding. (**B**) Three waveforms of an mMG neuron 120 mV in amplitude (negative voltage upward) that was evoked by both BIC (*left*) and SC (*right*) test stimuli. *Open inverted triangles* indicate onset and offset of the biphasic test stimulus; *bars* represent 1 ms. For this neuron, the test stimulus-evoked unit latency from stimulus offset was 3.0 ms. *BIC*, brachium of inferior colliculus; *SC,* superior colliculus; *mMG,* medial subnucleus of medial geniculate nucleus. (From McEchron et al., 1996a, Copyright 1996, Society for Neuroscience.)

sual structure, in animals who had received one session of classical conditioning or pseudoconditioning. The animals in the classical conditioning group exhibited significant conditioned heart rate responses (bradycardia) to a tonal stimulus when compared to the pseudoconditioning controls.

Conditioning-induced changes in synaptic efficacy in the auditory neurons that transmit CS information were assessed by examining the responses of single mMG neurons to stimulation of the BIC. Changes in synaptic efficacy were assessed by measuring the latency, reliability, and spike frequency of mMG unit activity evoked by BIC test stimuli during the pretraining and posttraining phases of the experiment. We considered the possibility that observed changes in unit activity might have resulted from general changes in cellular excitability rather than classical conditioning. To control for general excitability effects, a test stimulating electrode was also placed in the SC; analyses were only performed on mMG neurons whose activity could be modulated by both BIC and SC stimulation. As shown in Figure 3.3, electrical stimulation of the BIC in HR-conditioned animals led to significant decreases in the latency of mMG units and significant increases in the reliability of the responses evoked (as measured by time-locking) and in spike frequency. The increase in synaptic strength did not occur in pseudoconditioned controls, suggesting that the changes in synaptic efficacy in the CS inputs were caused by associative mechanisms, that is, a previously neutral stimulus acquired the capacity to produce a CR. Thus, these data provide evidence of a specific site of neuronal plasticity within the portion of the fear conditioning circuit that conveys information about the CS.

In a recent experiment, we found that single neurons in mMG and in ACe exhibit short latency responses to presentation of the US alone (McEchron et al., 1995). It has been shown that auditory and spinal somatosensory projection fields overlap within mMG/PIN (LeDoux et al., 1987) and that neurons in this region respond to polymodal stimulation (Wepsic, 1966; Love & Scott, 1969). How does the US information get from the point of stimulation to the point of association? In the rabbit HR conditioning paradigm, which is used in our laboratory, the US is a corneal air puff that is conveyed to the CNS via the infraorbital branch of the trigeminal nerve (McEchron et al., 1996b). We observed that presentation of the corneal air puff US leads to expression of the c-Fos protein, a cellular marker of functional activity, in the ventral portion of the ipsilateral spinal trigeminal subnuclei caudalis and interpolaris. Interestingly, these spinal trigeminal subnuclei also project monosynaptically to mMG (McEchron et al., 1996b), thereby providing a potentially direct US input to the CS pathway. Although amygdaloid neurons respond electrophysiologically to somatosensory stimuli, there is little evidence that somatosensory inputs access the amygdala by direct projections. Instead, Turner and Herkenham (1991) proposed that somatosensory information is sent to mMG/PIN, which then conveys this information by its projections to the amygdala. According to this hypothesis, auditory CS and somatosensory US information converges and is integrated within mMG/PIN before being sent to the amygdala. In support of this idea, Cruikshank and colleagues (Cruikshank et al., 1992) showed that when paired with an acoustic CS, electrical stimulation of PIN can serve as an effective US that supports fear conditioning.

The results of trace conditioning studies also provide evidence that the mMG is the site of CS–US convergence during classical eyeblink conditioning (O'Connor et al., 1997). In classical trace conditioning, CR acquisition occurs despite the fact that the CS terminates before US onset. This type of conditioning requires the in-

IMMEDIATELY AFTER TRAINING

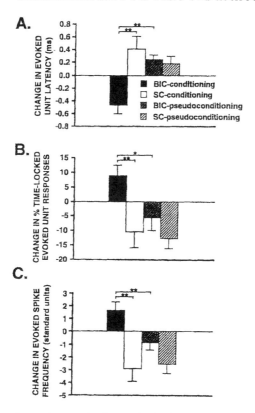

FIGURE 3.3. Mean change in test stimulus-evoked unit response latency (**A**), reliability (**B**) % time-locked unit responses), and spike frequency (**C**) immediately after training. Values obtained in the pretraining phase of stimulation were subtracted from values obtained in the posttraining phase of stimulation. In animals that received HR conditioning, neurons in the medial subnucleus of the mMG were activated by test stimuli delivered to both sites, the BIC (*black bar;* BIC conditioning) and the SC (*white bar;* SC conditioning). In animals that received HR pseudoconditioning, mMG neurons were activated by test stimuli delivered to BIC (*bar with thick diagonal lines,* BIC pseudoconditioning) or SC (*bar with thin diagonal lines,* SC pseudoconditioning). *Asterisks* indicate significant group comparisons (*$p < 0.05$; **$p < 0.01$). Comparison of the BIC conditioning group with control groups suggests that conditioning-related plasticity occurs at the synapses carrying auditory CS information. *Bars* indicate SE. *BIC,* brachium of inferior colliculus; *SC,* superior colliculus; *mMG,* medial subnucleus of medial geniculate nucleus. (From McEchron et al., 1996a, Copyright 1996, Society for Neuroscience.)

tegrity of a neural representation of the CS after it has been terminated, thereby allowing for an associative link with the US and the development of the capacity to generate a CR. The results of single cell studies (O'Connor et al., 1997) provide support for the view that mMG neurons are involved in the maintenance of the memory trace necessary for this type of conditioning and that these neurons play a role in CR generation and timing. Many of the mMG neurons were found to show activity that was sustained long after stimulus termination, and the modulation of the activity of these neurons reflected significant properties of the CS, both when it was present and following its termination. Single cell activity was also found to be related to the CR, particularly its temporal aspects. Moreover, many of the mMG units were modulated by the US. These findings suggested to O'Connor and colleagues that, in addition to being the site of CS–US convergence, the mMG may serve as an adaptive filter or selective attention gating mechanism, as suggested by Weinberger and colleagues (Diamond & Weinberger, 1989; Edeline & Weinber, 1992).

Conditioned Arousal and Central Cholinergic Modulation

It has been proposed that the classically conditioned fear response has a "learned arousal" component. Learned arousal is defined as a CNS state characterized by an enhanced readiness to detect stimuli during situations that pose a threat to the organism (Gallagher & Holland, 1994; Kapp et al., 1990; Silvestri & Kapp, 1998). According to this view, central cholinergic neurons, which are thought to be involved in Pavlovian conditioning, modulate the excitability of thalamic, limbic, and cortical neurons (Kapp et al., 1990; Silvestri & Kapp, 1998). Indeed, there is abundant evidence that application of acetylcholine or activation of basal forebrain cholinergic neurons can alter the excitability of cortical sensory neurons (Metherate et al., 1987; Metherate & Ashe, 1991) and can modulate learning-induced cortical plasticity (Weinberger & Diamond, 1987). Virtually all regions of the allo- and neocortex, including limbic structures, have been reported to receive cholinergic input from the basal forebrain. This input is believed to play an important role in cortical activation, attentional mechanisms, and the formation of memory traces (Page & Sofroniew, 1996). The thalamus also receives cholinergic input from the mesopontine cholinergic cell groups (Hallanger et al., 1987; Pare et al., 1990; Steraide et al., 1988), including a dense projection to the medial geniculate nucleus from the pedunculopontine tegmental nucleus (Ch5 region). Several electrophysiological studies have demonstrated that cholinergic input directly excites thalamocortical neurons (e.g., lateral geniculate nucleus, anterior thalamic nucleus) and augments sensory-evoked responses of these thalamic cells (Francesconi et al., 1988; McCormick & Prince, 1987; Pare et al., 1990), a function similar to that of basal forebrain cholinergic inputs to the cortex. In regards to classical conditioning, the cholinergic inputs to the thalamus could facilitate conditioning by modulating the excitability of mMG/PIN neurons.

Role of the Cerebral Cortex in Fear Conditioning

Although subcortical circuitry including the acoustic thalamus, amygdala, and brain stem regulatory nuclei is sufficient for the acquisition and expression of simple fear conditioning, the cerebral cortex also plays an important role in learned emotional responses. Auditory CS and somatosensory US information are conveyed to sensory cortical regions such as the primary and secondary auditory cortices (Anderson et al., 1980; Weinberger & Diamond, 1987; Weinberger et al., 1977), and the primary and secondary somatosensory cortices, respectively. Individual neurons in the auditory cortex exhibit associative changes in firing rate (Diamond & Weinberger, 1984; Weinberger & Diamond, 1987; Weinberger et al., 1984b, 1995) and retuning of auditory receptive fields (Diamond & Weinberger, 1989; Edeline et al., 1993; Edeline & Weinberger, 1993; Weinberger et al., 1993) as a result of conditioning. Lesions in the auditory cortex interfere with the acquisition (Teich et al., 1988), retention (Jarrell et al., 1987), and extinction (Teich et al., 1989) of conditioned heart rate responses. It has been proposed (Jarrell et al., 1987; Teich et al., 1989) that the auditory cortex, through descending connections with mMG, is involved in inhibitory responses seen in extinction training or differential conditioning (i.e., involving inhibition of the response to a nonreinforced CS).

Summary of the CNS Circuitry for Learned Emotional Responses

The development of a conditioned fear response to an auditory CS requires processing by several CNS structures, including neurons in the auditory pathway leading to the thalamus, the amygdala, the cerebral cortex, and the hypothalamic and brainstem nuclei that modulate visceral and somatic effector organs. Auditory CS information ascends via the primary and secondary lemniscal systems to cells in the medial geniculate, which then connect to the auditory cortex. A subset of medial geniculate neurons in mMG/PIN also receive somatosensory US information (spinothalamic or trigeminal), along with information about the CS. There is substantial evidence that mMG is the site of CS–US convergence in which acoustic and tactile information is associated during the conditioning process. This associative information is conveyed subcortically to the amygdala via cells in the mMG/PIN region and to the auditory cortex. The auditory cortex processes the associative information, as well as CS information alone, and through multisynaptic corticocortical connections, projects to the amygdala. In addition, the auditory cortex influences cells in mMG, in part through inhibitory connections that may play a role in extinction and differential conditioning. The amygdala integrates information from mMG and other sources and, through widespread descending projections to diencephalic and brain stem regulatory nuclei, modulates the constellation of autonomic and behavioral CRs associated with fear conditioning (e.g., bradycardia, blood pressure responses, behavioral inhibition, potentiated startle, pupillary responses).

Selective Attention Mechanisms

The term *selective attention* refers to central nervous system mechanisms that alter the level of processing assigned to a stimulus so as to change the probability that a particular response will occur. A number of terms are used by neuroscientists to refer to this process, including tuning (Oades, 1985), changing the signal-to-noise ratio (Aston-Jones, 1985), altering the salience of the representation of a stimulus (Depue & Zald, 1993), filtering (Swerdlow & Koob, 1987), tuning out irrelevant stimuli (Moore, 1979), gating (Hestenes, 1992), or changing the degree of activation of the representation of a stimulus (Gazzaniga et al., 1998). It is generally believed that some neural mechanisms involved in selective attention lead to an increase in the level of processing assigned to a stimulus, whereas others diminish the salience of the CNS representation of a stimulus. The CNS undoubtedly has a number of stimulus selection mechanisms, and there is considerable evidence for an executive attention mechanism, involving the anterior cingulate cortex, that coordinates activity across other attentional systems (Posner & Raichle, 1994). In this section of the chapter we discuss two putative selective attention mechanisms relevant to both classical and instrumental conditioning, one in which the mMG is a focal structure and one involving the hippocampus, the nucleus accumbens, and the ventral tegmental area.

The mMG as a Selective Attention Gating Mechanism

Weinberger and colleagues have proposed that the mMG, in addition to its associative function, is a key structure in a neural circuit that selectively filters acoustic stimuli based on their emotional significance to the organism. As discussed previously, Edeline and Weinberger (1992) observed that neurons in the mMG show retuning of auditory receptive fields during conditioning such that the peak of the tuning curve shifts toward the frequency of the CS. The results of single cell studies from the same laboratory revealed that neurons in the primary auditory cortex show a similar shift in the tuning curves (Bakin & Weinberger, 1990). These investigators believe that the mMG neurons are largely responsible for the changes in the cortical tuning curves. More specifically, the mMG is considered to be a focal structure in a circuit that modulates the salience of auditory stimuli according to their affective significance. Classically conditioned emotional responses in experiments with a single CS can occur in the absence of an auditory cortex (LeDoux et al., 1984), but this fact does not diminish the importance of these changes in the cortical tuning curves. As discussed next, in natural settings and in the laboratory, the CS in classical conditioning often becomes a signal for a behavioral change learned through instrumental conditioning, thereby enabling the animal to avoid an aversive stimulus (the US). Thus, for example, a previously neutral tonal stimulus, when paired with an aversive US such as foot shock, leads to classically conditioned fear, but the animal may learn through instrumental conditioning that it can avoid this shock by running down an alley. One possibility is that the representation of the CS in the auditory cortex participates in the learning involving active avoidance of the foot shock, although the auditory cortex is not the site of associative changes (Duvauchelle & Ettenberg, 1991; Hernandez & Hoebel, 1990). For instrumental learning to take place, the tonal CS must acquire emotional significance to the animal, and thus enhancing the salience of the representation of the CS in regions such as the auditory cortex has considerable adaptive significance.

The selective attention mechanism proposed by Weinberger and colleagues is composed of two systems, the primary lemniscal pathway in which information reaches the auditory cortex by direct projections from the ventrolateral portion of the medial geniculate nucleus, and the secondary lemniscal system in which neurons in the auditory cortex are modulated indirectly by mMG neurons. Information from the mMG is thought to reach the auditory cortex by the following path: mMG–lateral nucleus of the amygdala–central nucleus of the amygdala–nucleus basalis-primary auditory cortex. An auditory CS acquires affective significance after it is repeatedly paired with an aversive US such as a foot shock, and mMG neurons are thought to convey this information to the auditory cortex by the secondary lemniscal pathway. The cortical neurons that receive this information from the mMG also receive input from the primary lemniscal pathway (i.e., the ventrolateral MGN). The mMG input alters the level of processing assigned to the representations of stimuli in the auditory cortex. Specifically, it increases the salience of the representations of stimuli that correspond to the frequency of the CS and decreases the salience of others. The results of single cell studies conducted by Bakin and Weinberger (1996) provide evidence that the secretion of acetylcholine in the auditory cortex is particularly important to this dynamic filtering mechanism. These investigators presented a tonal stimulus of a particular frequency while concurrently stimulating the nucleus basalis electrically, and found that cortical neurons became more sensitive to that frequency. Moreover, the effect was eliminated by the cholinergic blockade atropine. Thus, it appears that the influence of mMG on stimulus salience is mediated by the secretion of acetylcholine by nucleus basalis neurons that innervate the auditory cortex.

Latent Inhibition as a Model for Selective Attention

Studies of the neural substrates of the phenomenon of latent inhibition have proven to be particularly beneficial in advancing our understanding of attentional mechanisms involved in classical conditioning. Latent inhibition (LI) refers to the retardation in conditioning that occurs when an animal is exposed to the to-be-conditioned stimulus before it is paired with the US. In its basic form, the latent inhibition experiment has two stages: an initial preexposure phase in which animals placed in the test apparatus are exposed repeatedly to a neutral stimulus such as a tone, but without consequence, and a second stage in which the same tonal stimulus is paired with a US such as a foot shock. A control group spends an equivalent amount of time in the test apparatus before conditioning trials but is not exposed to the to-be-conditioned stimulus. LI refers to the observation that the preexposed animals learn the CS–US association at a slower rate than the nonexposed control animals. LI is considered to be an index of an animal's ability to ignore or "tune out" irrelevant stimuli, and disruption of LI is thought to reflect deficits in selective attention. Thus, understanding the neural underpinnings of LI would serve as a means to further the understanding of CNS selective attention mechanisms.

The initial investigations of the neural basis of LI focused on the role of the hippocampus in the phenomenon. Several investigators used the LI experimental protocol in animals with damage to the hippocampus and found no evidence for retarded classical conditioning. Animals with extensive damage to the hippocampus behave as if they had never been exposed to the to-be-conditioned stimulus before the CS–US pairings (Solomon & Moore, 1975; McFarland et al., 1978). These findings are consistent with theories of hippocampal functioning that emphasize the role of this structure in selective attention. For example, both Kimble (1969) and Douglas and Pribram (1966) suggested that the function of the hippocampus is to exclude from attention stimuli that have no significant consequences with respect to reinforcement. More recently, Pribram (1986) suggested that the hippocampus is actively involved in learning to ignore stimuli that do not lead to reinforcement. In contrast to the views espoused by these investigators, Solomon (1979) contends that the hippocampus is involved in the "tuning out" of any stimuli that are irrelevant to the current task, not just nonreinforced stimuli. According to this view, an irrelevant stimulus is defined as any stimulus that would lead to behaviors that are insufficient, inappropriate, or maladaptive (Moore, 1979).

Gray (1982) proposed that the hippocampus functions as a compactor that interrupts ongoing behavior when there is a mismatch between predicted and expected events. Following the same line of thinking, Schmajuk and Moore (1985, 1988) developed an attentional-associative model to account for the effect of hippocampal lesions on classical conditioning protocols such as the one used in LI experiments. According to this model, the hippocampus computes a stimulus associability (alpha) value based on the extent of mismatch between actual and predicted events arising from the occurrence of a CS. The rate of learning factor alpha in their mathematical model is used to quantify the salience or associability of a CS, that is, the amount of attention afforded the CS. An alpha value for a CS is based on its associability on previous occasions and the extent to which predictions of the US are made by all CSs present at a given point in time. The hippocampus uses this information to determine the level of processing to be assigned to the CS. When the hippocampus is lesioned, the associability of a CS cannot be accurately determined; neither its associability on previous occasions nor the predictions of the US by other CSs can influence the salience of the CS.

When the Schmajuk and Moore (1985, 1988) model is applied to the LI, it can be seen that animals that have been previously exposed to the CS in the absence of the US will have an associability value for the CS that is quite low during the initial classical conditioning trials as a result of the preexposure trials. This low value of alpha increases as conditioning proceeds, and the animal's response becomes less dependent on its preexposure to the CS. Hippocampal lesions disrupt

LI because previous CS associability values do not contribute to the currently computed value. Stated differently, hippocampal lesions prevent alpha from decreasing when the CS is presented without a US.

Weiner and colleagues have used findings from their studies of the effects of pharmacological treatments on LI (Weiner et al., 1981, 1988, 1997) as a basis for a neural model of this phenomenon. Their "switching model" of LI (Weiner, 1990; Weiner & Feldon, 1997) uses the ideas of Schmajuk and Moore (1985, 1988) regarding hippocampal functioning as a cornerstone for their theoretical views. According to the switching model, the hippocampus uses CS associability values as a basis for a signal it sends to the nucleus accumbens (NAC). After the hippocampus detects a mismatch between old and new predictions about the relation between the CS and the US, it computes the associability value for the CS by averaging past and present CS associabilities. This computed associability value would be low if the CS did not predict the US in the past. It would also be low if there were one or more CS in the present environment that reliably predicted the US in the past. It would be high if it previously was a good predictor of the US, particularly if other CSs in the current stimulus array were not reliable predictors of the US in the past. The computed associability value determines the level of processing assigned to the CS. If it reaches a value that requires that the CS receive further processing, a signal is sent to the NAC that would increase the chances the NAC will engage a different behavioral program. If the CS associability value is low, the hippocampus sends a signal to the NAC that increases the probability that the animal will resume a previously established behavior rather than use alternative behavioral strategies. In general, the higher the associability value the more likely the animal is to switch to an alternative behavior pattern. In agreement with both Gray (1982) and Gabriel et al., (1986), Weiner (1990) asserts that the animal inhibits ongoing behavior when the hippocampus detects a mismatch between a predicted and actual event. In Weiner's model, CS associability is computed during this period of behavioral inhibition.

Another key aspect of the switching model are dopamine-secreting neurons of the ventral tegmental area (VTA). A number of studies have demonstrated that the enhancement of dopamine (DA) transmission by low doses of the DA agonist amphetamine (Weiner et al., 1987) disrupts LI, whereas blockade of DA transmission by neuroleptics, such as haloperidol (Weiner & Feldon, 1987) or sulpiride (Feldon & Weiner, 1991), facilitates the development of LI. Several lines of evidence indicate that these effects are mediated by dopamine-secreting neurons of the ventral tegmental area (VTA DA) that project to the NAC (see Weiner, 1990, and Weiner & Feldon, 1997, for a review of the evidence). VTA DA neurons are activated by four types of stimuli: an unexpected reward (Mirenowicz & Schultz, 1994), novel stimuli (Wickelgren, 1997), unconditioned aversive stimuli (Salamone et al., 1997; Besson & Louillot, 1995; Saulskaya & Marsden, 1995; Young et al., 1993), and conditioned incentive stimuli (Robbins & Everitt, 1996; Besson & Louillot, 1995; Salamone et al., 1997; Saulskaya & Marsden, 1995; Young et al., 1993). A conditioned incentive stimulus is a classically conditioned stimulus, that is, a CS, that predicts reward or the relief of punishment; more details about this type of CS are provided later. According to the switching model, increased DA secretion by VTA DA neurons that innervate the NAC facilitates switching to a new behavioral program. Thus, enhancement of DA activity by low doses of amphetamine disrupts LI because it promotes a rapid switch of responding according to the current CS–US contingency, whereas the blockage of DA transmission by neuroleptics enhances the control by the "CS–no event contingency" established during the preexposure period. Stated differently, the decrease in DA activity produced by neuroleptics reduces the capacity of the animal to switch responding according to the changed contingency of reinforcement. It is important to point out that altering DA activity does not affect an animal's ability to learn to ignore irrelevant stimuli during the preexposure trials, but it does affect the subsequent expression of that learning during the conditioning trials. Presumably, VTA DA neurons are activated in the conditioning phase because these neurons respond to an unexpected reward.

According to the switching model, neurons of the medial raphe modulate the NAC switching mechanism. This assertion is based primarily on the observation that electrolytic lesions of serotonin-secreting neurons of the medial raphe lead to the disruption of LI (Solomon et al., 1980; Asin et al., 1980). This effect may occur because these lesions lead to the depletion of serotonin in the hippocampus (Solomon et al., 1980) or because these lesions have been found to lead to diminished restraint of VTA DA activity by serotonin at the NAC.

Role of Classically Conditioned Emotional Responses in the Nucleus Accumbens (NAC) Switching Mechanism

Latent inhibition has been observed in the experimental protocols used in both instrumental and classical conditioning (see reviews by Lubow & Gewirtz, 1995; Weiner, 1990; Weiner & Feldon, 1997), so if the attentional-associative model of Schmajuk and Moore (1985,1988) and the switching model (Weiner, 1990; Weiner & Feldon, 1997) are valid, the CNS representations of the conditioned stimuli used in these two procedures should be equivalent, at least in terms of their ability to activate key structures in these models such as the hippocampus, NAC, and the VTA. Indeed, as described in detail here, there is good reason to believe that stimuli that were neutral before instrumental learning acquire their ability to control behavior (i.e., become conditioned stimuli) by classical conditioning that takes place during instrumental learning.

Function of the Nucleus Accumbens

The NAC is considered to be a particularly important modulator of affective behavior because it serves as an interface between limbic structures, such as the amygdala and hippocampus, that process affective information, and structures such as the basal ganglia which are involved in the generation of motor activity (Morgenson & Nielsen, 1984; Swerdlow & Koob, 1987; Kalivas et al., 1993; Everitt & Robbins, 1992). Although research on the NAC has focused on its role in the expression of behaviors learned through instrumental conditioning, there is ample evidence that this structure is capable of modulating an unconditioned reflex, specifically the acoustic startle reflex (Sorenson & Swerdlow, 1982; Humby et al., 1996; Yamada et al., 1998) and an integrated unconditioned cardiorespiratory/behavioral pattern, the defense reaction (Brutus et al., 1989; al Maskati & Zbro zyna, 1989). Similarly, the role of VTA DA projections to the NAC in instrumentally learned behaviors has been the subject of an extensive body of research literature, yet there are also findings that provide evidence that projections from VTA mediate the effects of DA on acoustic startle (Ellison et al., 1978; Swerdlow et al., 1990, 1992).

It is generally believed that the NAC is a final common element in the acquisition (Mark et al., 1989; Taylor & Robbins, 1986) and expression (Kalivas et al., 1993; Salamone et al., 1991) of behaviors learned through instrumental conditioning, including those in which the US is a reward and those in which the animal learns to successfully avoid an aversive US (i.e., active avoidance). The NAC and VTA DA neurons are considered to be key elements in the CNS circuitry that underlies appetitive motivational states. The term appetitive motivation is a hypothetical construct used to refer to a state of the CNS that exists when environmental signals trigger goal-directed behaviors, the goal being either a US that has reward value or the safety signals present when an organism successfully avoids an aversive US. Kalivas et al., (1993) refer to the neural circuitry that forms the substrate for this state as the motive circuit. The neural system underlying appetitive motivation has also been referred to as the Behavioral Approach System or simply BAS (Gray, 1979), the Behavioral Activation System, also called BAS (Fowles, 1980), and the Behavioral Facilitation System, or BFS (Depue & Zald, 1993). There is a strong emphasis on the role of VTA DA neurons for investigators who use the terms BAS or BFS, rather than motive circuit, to refer to the CNS circuitry underlying appetitive motivational states. There is general agreement, however, that these neural systems lead to behavioral changes that bring an animal in contact with broad classes of unconditioned and conditioned stimuli.

Classical Conditioning That Takes Place During Instrumental Learning

Although the hypothetical construct *appetitive motivation* is typically used in reference to behaviors acquired through operant (instrumental) conditioning, the conditioned stimuli that activate many of the elements of the neural circuitry associated with this state of the CNS are acquired through classical conditioning (Seward, 1952; Spence, 1956; Deutsch, 1964; Gray, 1975). During the process of instrumental conditioning, a number of neutral stimuli occur in close temporal contiguity with the US and after frequent pairings become CSs. For example, when an animal learns to press a lever to receive a food pellet (the US), a number of neutral stimuli located in the test chamber become classically conditioned CSs. In the language of instrumental learning theory, the US is referred to as a primary reinforcer and the CS as either a conditioned incentive stimulus, a conditioned reinforcer, or a secondary reinforcer. A primary positive reinforcer is an appetitive US that increases the probability of a response that preceded the occurrence of the US. A primary negative reinforcer is an aversive US that increases the probability of behaviors that prevent the animal from experiencing the aversive stimulus, that is, an avoidance response, or simply active avoidance. During instrumental conditioning, the CS not only predicts the occurrence of the US, it also acquires reinforcing properties. The evidence that these CSs acquire reinforcing properties is based on demonstrations that they can function as a reward in the learning of a new response (Wolfe, 1936; Cowles, 1937).

When a CS acquires reinforcing properties, it becomes a conditioned incentive stimulus (also referred to as secondary reinforcer or conditioned reinforcer). Subsequently, it is these conditioned incentive stimuli (CSs) that activate CNS structures involved in appetitive motivation and goal-directed behavior and the active avoidance of aversive USs. This property of the CS is extremely important because the attainment of most goals requires a sequence of behaviors that are spatially and temporally removed from the US (the primary reinforcer). Goal attainment necessitates a series of subgoals utilizing secondary reinforcers to enable a chain of approach behaviors (Seward, 1952; Spence, 1956; Deutsch, 1964; Gray, 1975). The establishment of this behavioral chain is made possible by the development of classically conditioned associations between initially neutral stimuli (cues) and the primary reinforcer (US), the cue now becoming a conditioned incentive stimulus for the next behavior in the chain. This is followed by the formation of other classically conditioned associations between new cues and those already established as secondary reinforcers. Conditioned incentive stimuli can be developed from neutral stimuli linked to a positive reinforcer, or by cues occurring in close proximity to the termination of an aversive US. Thus, conditioned incentive stimuli learned through classical conditioning can lead to approach or avoidance behaviors.

Role of the Amygdala in the Nucleus Accumbens Switching Mechanism

In view of the important role of the amygdala in the assignment of emotional significance to CSs in classical conditioning and conditioned incentive stimuli in instrumental conditioning (Aggleton & Mishkin, 1986; LeDoux, 1994, for reviews), it seems reasonable to speculate that information about these conditioned stimuli is conveyed to the structures that form the corpus of the proposed selective attention mechanism, the hippocampus and the NAC. Indeed, there is evidence for functional connections between the amygdala and the hippocampus (Cahill & McGaugh, 1998; Packard & Teather, 1998), and there are dense neuroanatomic connections between the amygdala and the NAC (Heimer & Wilson, 1975; Nauta & Domesick, 1984). Single unit studies provide convincing evidence that NAC neurons are responsive to both primary and conditioned reinforcers (Apicella et al., 1991; Henriksen & Giacchino, 1993) and CSs that predict aversive stimuli (Young et al., 1993). There is also evidence that a primary source of this information is the amygdala (Kalivas et al., 1993; Everitt & Robbins, 1992).

According to Everitt and Robbins (1992), conditioned incentive stimuli affect behavior by two independent processes. First, information from the amygdala concerning the affective significance of a conditioned incentive stimulus (i.e., the relation a discrete CS bears to an appetitive or aversive US) is conveyed to the NAC. This association is learned by classical conditioning. Second, the VTA DA neurons connecting to the NAC amplify the impact of this stimulus. As discussed previously, VTA neurons respond to conditioned incentive stimuli (Schultz et al., 1993). The switching model adds a third factor, the input to the NAC from the hippocampus. Viewed in terms of selective attention mechanisms, the input to the NAC from the amygdala simply conveys information, learned through classical conditioning, about the relationship between the CS and a US. The level of processing assigned to the CS is dependent on the hippocampal mechanisms that alter the salience of the CS based on the organism's conditioning history with the CS and other CSs in the current stimulus array. After determining the level of processing to be assigned to the CS, it sends a signal to the NAC that either promotes switching to a new behavior, if the associability value is high, or inhibits switching if the associability value is low.

It has been proposed that VTA DA neurons modulate the salience of the US or the CS, depending on the conditioning history of the organism with these stimuli (Shultz et al., 1993). Single cell recording studies conducted by Shultz and colleagues (Ljungberg et al., 1992; Romo & Shultz, 1990; Shultz et al., 1993) revealed that during the early phases of instrumental conditioning VTA DA neurons respond to the US (primary positive reinforcer), but once the CS–US connection has been established and the animal learns the behavior that will lead to the primary positive reinforcer, VTA DA neurons cease to respond to the US. Instead, these neurons respond to CSs in the environment that predict the US; the stimuli that were initially neutral have become conditioned incentive stimuli. The secretion of dopamine in response to the US in the initial phase of instrumental conditioning appears to be important to the associative changes that take place; that is, the associative link in memory between the CS and the response that leads to the primary positive reinforcer (Stein & Belluzzi, 1989; Duvauchelle & Ettenberg, 1991; Hernandez & Hoebel, 1990).

Increased VTA DA activity promotes response switching by the NAC, so during the initial conditioning trials in the LI protocol the VTA DA input to the NAC promotes switching because the CS–US contingency is new. That is, the VTA DA neuronal activity is increased by the US; subsequently, it is driven by the CS.

Thus, deficits in the ability to "tune out" irrelevant stimuli may result from a VTA with high reactivity or the inability of the hippocampus to accurately calculate the CS associability value.

Application of the Attentional-Associative Model and
the Switching Model to Psychopathology

One of the most salient characteristics of several types of psychopathologies is the inability to ignore irrelevant stimuli and maintain a response set. Accordingly, the ideas of Schmajuk and Moore (1985, 1988) and Weiner (Weiner, 1990; Weiner & Feldon, 1997) can be used to account for the attention deficits observed in schizophrenics (see review by Weiner & Feldon, 1997) and manic episodes experienced in bipolar patients (Johnson et al., in press). Briefly, the ability to sustain goal-directed behavior is proposed to be dependent on the hippocampus–NAC–VTA attentional mechanism because imbalances in the interactions of these structures affects how distractible the individual is. It is hypothesized that an imbalance is caused by high VTA DA activity in both schizophrenia and during a manic episode. VTA DA-mediated switching to a different behavioral programs or cognitive sets may be particularly beneficial to goal-directed behaviors when the contingencies of reinforcement have changed and new behavioral strategies must be generated to reach the goal. However, high levels of VTA DA activity may impair the ability to maintain a current cognitive or response set and thereby render the schizophrenic or manic patient too dependent on current context and immediate situational demands. Thus, these patients would show enhanced behavioral and cognitive switching characterized by attention to irrelevant stimuli.

Summary and Conclusions

When viewed in terms of homeostasis, both discrete and nonspecific autonomic CRs are seen as homeostasis-promoting responses that prepare an organism for an impending unconditioned aversive stimulus. From this perspective, the unconditioned autonomic reflex is considered to be a reactive homeostatic mechanism and discrete URs as compensatory regulatory responses. A quantitative assessment of the effectiveness of unconditioned autonomic reflexes reveals that they are relatively weak regulators of physiological variables, primarily because of regulatory lags involving negative feedback. The conditioned autonomic reflex is considered to be a predictive homeostatic mechanism because the discrete CR that develops mitigates (or totally nullifies), in advance, the impact of the regulatory challenge (the US), thereby improving the regulatory properties of the unconditioned reflex.

Autonomic conditioning involving a nonspecific CR is also considered to be a predictive homeostatic process. A nonspecific CR is one of many conditioned autonomic responses elicited by the CS, and its functional utility is gained within the

3. Functional Utility and Neurobiology of Conditioned Autonomic Responses   73

context of the cluster of CRs that have been elaborated from URs. The constellation of nonspecific CRs that develop during autonomic conditioning are thought to be produced by an altered CNS state that results from the conditioning process. The altered CNS state is referred to as conditioned fear, and it not only leads to changes in autonomic activity, it is also thought to enhance the reliability and efficiency of feature extraction in sensory modalities. The change in behavioral state associated with conditioned fear serves two purposes. It facilitates the development of somatic CRs that mitigate the impact of the US (e.g., the development of an eyeblink CR that attenuates the impact of an air puff to the cornea), or the development of a instrumentally learned response that allows the organism to actively avoid the US.

Conditioned fear is considered to be an altered CNS state that mediates nonspecific autonomic CRs. An unconditioned aversive stimulus elicits a variety of autonomic responses, and the amygdala is a structure that is necessary for the acquisition and expression of the various nonspecific CRs that are elaborated from these URs. Viewed in this way, individual pathways that mediate the various nonspecific CRs diverge from the amygdala and project to separate effector systems such as those controlling heart rate or pupil size. Information about the CS and US converge at neurons in the mMG/PIN, which in turn project to the amygdala. Electrophysiological changes in individual neurons in the amygdala have been shown to parallel the changes that occur during the development of CRs to an aversive stimulus. Moreover, there is evidence that the neuronal plasticity necessary for the conditioning can occur at either the amygdala or the mMG. There is also evidence that is consistent with the view that conditioned changes in arousal parallel the rapidly acquired autonomic CRs.

Selective attention mechanisms are also important to the development and expression of conditioned autonomic responses. There is evidence that the mMG, in addition to its associative function, is a key structure in a neural circuit that selectively filters acoustic stimuli based on their emotional significance to the organism. Specifically, it increases the level of processing assigned to the representations of stimuli in the auditory cortex that correspond to the frequency of the CS and decreases the salience of other auditory stimuli.

The hippocampus, the ventral tegmental area, and the nucleus accumbens appear to be components of a neural circuit that alter the salience of stimuli involved in both classical and instrumental conditioning. According to the attentional-associative model, the hippocampus computes an associability value for a CS based on the extent to which the CS and other stimuli present were able to predict the US in the past. The associability value is a measure of the level of processing to be assigned to the CS. The switching model adopts the attentional-associative model view of hippocampal functioning and asserts that the hippocampus uses the associability value as a basis for a signal that it sends to the nucleus accumbens. If it reaches a value that requires that the CS receive further processing, a signal is sent to the nucleus accumbens which increases the probability that this structure will switch to a different behavioral program. If the associability value is low, the hippocampus sends a signal to the nucleus accumbens that increases the probability the animal will resume previously established behavior, rather than alternative behavioral strategies. According to the switching model, projections to the nucleus accumbens from the dopamine-secreting neurons of the ventral tegmental area promote switching to a new behavioral program. When these models are applied to the latent inhibition experimental protocol, it can be seen that signals from the hippocampus promote responding to the CS–no event contingency established during the preexposure period when the CS was presented in the absence of the US. In contrast, enhanced ventral tegmental area activity promotes rapid switching of responding according to the current CS–US contingency.

References

Aggleston, J.P., and Mishkin, M. (1986). The amygdala: sensory gateway to the emotions. In R. Plutchik and H. Kellerman (Eds.), *Emotion: Theory, research and experience* (Vol. 3) Biological foundations of emotions (pp. 281–299). New York: Academic Press.

al Maskati, H.A., and Zbro zyna, A.W. (1989). Stimulation in prefrontal cortex inhibits cardiovascular and motor components of the defence reaction in rats. *Journal of Autonomic Nervous System, 28*(2), 117–125.

Anderson, R.A., Knight, P.I., and Merzenich, M.M. (1980). The thalamocortical and corticothalamic connections of AI, AII, and the anterior auditory field (AAF) in the cat: evidence for two largely segregated systems of connections. *Journal of Comparative Neurology, 194*, 663–701.

Apicella, P., Ljungberg, T., Scarnati, E., and Schultz, W. (1991). Responses to reward in monkey dorsal and ventral striatum. *Experimental Brain Research, 85*, 491–511.

Asin, K.E., Wirtshafter, D., and Kent, E.W. (1980). The effects of electrolytic median raphe lesions on two measures of latent inhibition. *Behavioral and Neural Biology, 28*, 408–417.

Aston-Jones, G.A. (1985). The locus ceoruleus: behavioral functions of locus coeruleus derived from cellular attributes. *Physiological Psychology, 13*(3), 118–126.

Bakin, J.S., and Weinberger, N.M. (1990). Classical conditioning induces CS-specific receptive field plasticity in the auditory cortex of the guinea pig. *Brain Research, 536*(1–2), 271–286.

Bakin, J.S., and Weinberger, N.M. (1996). Induction of a physiological memory in the cerebral cortex by stimulation of the nucleus basalis. *Proceedings of the National Academy of Science of the United States, 93*(20), 10546–10547.

Ban, T., and Shinoda, H. (1960). Experimental studies on the relation between the hypothalamus and conditioned reflex. II. On the conditioned response in EEG and gastric motility. *Medical Journal of Osaka University, 11*, 85–93.

Baxter, R. (1966). Diminution and recovery of the UCR in delayed and trace classical GSR conditioning. *Journal of Experimental Psychology, 71*, 447–451.

Besson, C., and Louillot, A. (1995). Asymmetrical involvement of mesolimbic dopamine neurons in addective perception. *Neuroscience, 68*, 963–968.

Black, A.H., and Prokasy, W.F. (1972). Introduction. In A.H. Black and W.F. Prokasy (Eds.), *Classical conditioning. II: Current research and theory* (pp. xi–xii). New York: Appleton-Century-Crofts.

Blanchard, D.C., and Blanchard, R.J. (1972). Inate and conditioned reactions to threat in rats with amygdaloid lesions. *Journal of Comparative and Physiological Psychology, 81*, 281–290.

Brutus, M., Zuabi, S., and Siegel, A. (1989). Microinjections of D-Ala2-Met5-enkephalinamide placed into nucleus accumbens suppress hypothalamically elicited hissing in the cat. *Experimental Neurology, 104*(1), 55–61.
Bykov, K.M. (1957). *The cerebral cortex and the internal organs.* (Translated from the Russian and edited by W. Horsley Gantt.) New York: Chemical Publishing.
Bykov, K.M. (1959). *The cerebral cortex and the internal organs.* (Translated from the Russian and edited by R. Hodes and A. Kolbey.) Moscow: Foreign Languages Publishing House.
Cahill, L., and McGaugh, J.L. (1998). Mechanisms of emotional arousal and lasting declarative memory. *Trends in Neuroscience, 21*, 294–299.
Campeau, S., Miserendino, M.J., and Davis, M. (1992). Intra-amygdala infusion of the *N*-methyl-D-aspartate receptor antagonist AP5 blocks acquisition but not expression of fear-potentiated startle to an auditory conditioned stimulus. *Behavioral Neuroscience, 106*, 569–574.
Clugnet, M.C., and LeDoux, J.E. (1990). Synaptic plasticity in fear conditioning circuits: induction of LTP in the lateral nucleus of the amygdala by stimulation of the medial geniculate body. *Journal of Neuroscience, 10*, 2812–2824.
Cohen, D.H. (1969). Development of a vertebrate experimental model for cellular neurophysiologic studies of learning. *Conditioned Reflexes, 4*, 61–80.
Cohen, D.H., and Pitts, L.H. (1968). Vagal and sympathetic components of conditioned cardioacceleration in the pigeon. *Brain Research, 9*, 15–31.
Cowles, J.T. (1937). Food-tokens as incentive for learning by chimpanzees. *Comparative Psychology Monographs, 14*(17).
Cruikshank, S.D., Edeline, J.-M., and Weinberger, N.M. (1992). Stimulation at a site of auditory-somatosensory convergence in the medial geniculate nucleus is an effective unconditioned stimulus for fear conditioning. *Behavioral Neuroscience, 106*, 471–483.
Culler, E.A. (1938). Recent advances in some concepts of conditioning. *Psychological Review, 45*, 134–153.
Davis, M. (1992). The role of the amygdala in fear and anxiety. *Annual Review of Neuroscience, 15*, 353–375.
Davis, M., Gendelman, D.S., Tischler, M.D., and Gendelman, P.M. (1982). A primary acoustic startle circuit: lesion and stimulation studies. *Journal of Neuroscience, 6*, 791–805.
Depue, R.A., and Zald, D.H. (1993). Biological and environmental processes in nonpsychotic psychopathology: a neurobehavioral perspective. In C.G. Costello (Ed.), *Basic issues in psychopathology* (pp. 127–237). New York: Guilford Press.
DeToledo, L., and Black, A.H. (1966). Heart rate: change during conditioned suppression in rats. *Science, 152*, 1404–1460.
Deutsch, J.A. (1964). *The structural basis of behaviour.* Cambridge: Cambridge University Press.
Deutsch, R. (1974). A mechanism for saccharin-induced sensitivity to insulin in the rat. *Journal of Comparative and Physiological Psychology, 86*, 350–358.
Diamond, D.M., and Weinberger, N.M. (1984). Physiological plasticity of single neurons in auditory cortex of the cat during acquisition of the pupillary conditioned responses: II. Secondary field (AII). *Behavioral Neuroscience, 98*, 189–210.
Diamond, D.M., and Weinberger, N.M. (1989). Role of context in the expression of learning-induced plasticity of single neurons in auditory cortex. *Behavioral Neuroscience, 103*, 471–494.
Douglas, R.J., and Pribram, K.H. (1966). Learning and limbic lesions. *Neuropsychologia, 4*, 197–219.

Duvauchelle, C.L., and Ettenberg, A. (1991). Haloperidol attenuates conditioned place preferences produced by electrical stimulation of the medial prefrontal cortex. *Pharmacology, Biochemistry and Behavior, 38*(3), 645–650.

Dworkin, B.R. (1993). *Learning and physiological regulation.* Chicago: University of Chicago Press.

Dworkin, B.R., and Dworkin, S. (1995). Learning of physiological responses: II. Classical conditioning of the baroreflex. *Behavioral Neuroscience, 109,* 1119–1136.

Edeline, J.-M., and Weinberger, N.M. (1992). Associative retuning in the thalamic source of input to the amygdala and auditory cortex: receptive field plasticity in the medial division of the medial geniculate body. *Behavioral Neuroscience, 106,* 81–105.

Edeline, J.-M., and Weinberger, N.M. (1993). Receptive field plasticity in the auditory cortex during frequency discrimination training; selective retuning independent of task difficulty. *Behavioral Neuroscience, 107,* 82–103.

Edeline, J.-M., Pham, P., and Weinberger, N.M. (1993). Rapid development of learning-induced receptive field plasticity in the auditory cortex. *Behavioral Neuroscience, 107,* 539–551.

Ellison, G., Eison, M., and Huberman, H. (1978). Stages of constant amphetamine intoxication: delayed appearance of normal social behaviors in rat colonies. *Psychopharmacology, 56,* 293–299.

Everitt, B.J., and Robbins T.W. (1992). Amygdala-ventral striatal interactions and reward-related processs. *The amygdala: neurobiological aspects of emotion, memory, and mental dysfunction* (pp. 401–429). New York: Wiley-Liss, Inc.

Faneslow, M.S., and Kim, J.J. (1994). Acquisition of contextual Pavlovian fear conditioning is blocked by application of an NMDA receptor antagonist D,L-2-amino-5-phosphonovaleric acid to the basolateral amygdala. *Behavioral Neuroscience, 108,* 210–212.

Feldon, J., and Weiner, I. (1991). The latent inhibition model of schizophrenic attention disorder: haloperidol and sulpiride enhance rats' ability to ignore irrelevant stimuli. *Biological Psychiatry, 29,* 635–646.

Fitzgerald, R.D., Hatton, D.C., Foutz, S., Gilden, E., and Martinsen, D. (1984). Effects of drug-induced changes in resting blood pressure on classically conditioned heart rate and blood pressure in restrained rats. *Behavioral Neuroscience, 98,* 829–839.

Fowles, D.C. (1980). The three arousal model: implications of Gray's two-factory learning theory for heart rate, electrodermal activity, and psychopathy. *Psychophysiology, 17*(2), 87–104.

Francesconi, W., Müller, C.M., and Singer, W. (1988). Cholinergic mechanisms in the reticular control of transmission in the cat lateral geniculate nucleus. *Journal of Neurophysiology, 59*(6), 1690–1718.

Gabriel, M., Saltwick, S.E., and Miller, J.D. (1975). Conditioning and reversal of short-latency multiple-unit responses in the rabbit medial geniculate nucleus. *Science, 189,* 1108–1109.

Gabriel, M., Miller, J.D., and Saltwick, S.E. (1976). Multiple-unit activity of the rabbit medial geniculate nucleus in conditioning, extinction, and reversal. *Physiological Psychology, 4,* 124–134.

Gabriel, M., Sparenborg, S.P., and Stolar, N. (1986). An executive function of the hippocampus: pathway selection for thalamic neuronal significance code. In R.L. Isaacson and K.H. Pribram (Eds.) *The hippocampus* (Vol. 4) (pp. 1–40). New York: Plenum Press.

Gallagher, M., and Holland, P.C. (1994). The amygdala complex: multiple roles in associative learning and attention. *Proceedings of the National Academy of Sciences of the United States of America, 91,* 11771–11776.

Gantt, W.H. (1953). Principles of nervous breakdown—schizokinesis and autokinesis. *Annals of the New York Academy of Sciences, 56,* 143–163.
Gazzaniga, U.S., Ivry, B., and Mangun, E.R. (1998). *Cognitive neuroscience: the biology of the mind.* New York: W.W. Norton.
Gentile, C.G., Jarrell, T.W., Teich, A.H., McCabe, P.M., and Schneiderman, N. (1986). The role of amygdaloid central nucleus in differential Pavlovian conditioning of bradycardia in rabbits. *Behavioral Brain Research, 20,* 263–273.
Gerall, A.A., Sampson, P.B., and Boslov, G.L. (1957). Classical conditioning of human pupillary dilation. *Journal of Experimental Psychology, 54,* 457–474.
Girden, E. (1942a). The dissociation of blood pressure conditioned responses under curare and erythroidine. *Journal of Experimental Psychology, 31,* 219–231.
Girden, E. (1942b). The dissociation pupillary conditioned reflexes under curare and erythroidine. *Journal of Experimental Psychology, 31,* 322–332.
Gormezano, I. (1972). Investigations of defense and reward conditioning in the rabbit. In A.H. Black and W.F. Prokasy (Eds.), *Classical conditioning. II: Current research and theory* (pp. 151–181). New York: Appleton-Century-Crofts.
Gray, J.A., (1975). *Elements of a two-process theory of learning.* London: Academic.
Gray, J.A., (1979). A neuropsychological theory of anxiety. In C.E. Isard (Ed.), *Emotions and personality.* New York: Plenum Press.
Gray, J.A., (1982). *The neuropsychology of anxiety: an inquiry into the functions of the septo-hippocampal system.* New York: Oxford University Press.
Guha, D., Dutta, S.N., and Pradhan, S.N. (1974). Conditioning of gastric secretion by epinephrine in rats. *Proceedings of the Society for Experimental Biology and Medicine, 147,* 817–819.
Hallanger, A.E., Levey, A.I., Lee, H.J., Rye, D.G., and Wainer, B.H. (1987). The origins of cholinergic and other subcortical afferents to the thalamus in the rat. *Journal of Comprehensive Neurology, 262,* 105–124.
Hebb, D.O. (1949). *The organization of behavior.* New York: Wiley.
Heimer, L., and Wilson, R.D. (1975). The subcortal projections of the allocortex: similarities in the neural associations of the hippocampus, the pyriform cortex and the neocortex. In M. Santini (Ed.), *Golgi centennial symposium* (pp. 177–192). New York: Raven Press.
Henriksen, S., and Giacchino, J. (1993). Functional characteristics of nucleus accumbens neurons: evidence obtained from in vivo electrophysical recordings. In P. Kalivas and C. Barnes (Eds.), *Limbic motor circuits and neuropsychiatry.* Boca Raton, FL: CRC Press.
Hernandez, L., and Hoebel, B.G. (1990). Feeding can enhance dopamine turnover in the prefrontal cortex. *Brain Research Bulletin, 25*(6), 975–979.
Hestenes, D. (1992). A neural network theory of manic-depressive illness. In D.S. Levine and S.J. Leven (Eds.), *Motivation, emotion, and goal direction in neural networks* (pp. 209–257). Hillsdale, NJ: Lawrence Erlbaum Associates.
Hitchcock, J.M., and Davis, M. (1986). Lesions of the amygdala, but not of the cerebellum of red nucleus, block conditioned fear as measured with the potentiated startle paradigm. *Behavioral Neuroscience, 100,* 11–22.
Holdstock, T.L., and Schwarzbaum, J.S. (1965). Classical conditioning of heart rate and galvanic skin response in the rat. *Psychophysiology, 2,* 25–38.
Hollis, K.L. (1984). The biological function of Pavlovian conditioning: the best defense is a good offense. *Journal of Experimental Psychology: Animal Learning and Behavior, 10,* 413–425.
Humby, T., Wilkinson, L.S., Robbins, T.W., and Geyer, M.A. (1996). Prepulses inhibit startle-induced redcutions of extracellular dopamine in the nucleus accumbens of rat. *Journal of Neuroscience, 16*(5), 2149–2156.

Iwata, J., LeDoux, J.E., Meely, M.P., Arneric, S., and Reis, D. J. (1986). Intrinsic neurons in the amygdaloid field projected to by the medial geniculate body mediate emotional responses conditioned to acoustic stimuli. *Brain Research, 383,* 195–214.

Jarrell, T.W., Gentile, C.G., McCabe, P.M., and Schneiderman, N. (1986a). The role of medial geniculate nucleus in differential Pavlovian conditioning of bradycardia in rabbits. *Brain Research, 374,* 126–136.

Jarrell, T.W., Gentile, C.G., Romanski, L.M., McCabe, P.M., and Schneiderman, N. (1987). Involvement of cortical and thalamic auditory regions in retention of differential bradycardiac conditioning stimuli in rabbits. *Brain Research, 412,* 285–294.

Jarrell, T.W., Romanski, L.M., Gentile, C.G., McCabe, P.M., and Schneiderman, N. (1986b). Ibotenic acid lesions in the medical geniculate region prevent the acquisition of differential Pavlovian conditioning of bradycardia to acoustic stimuli in rabbits. *Brain Research, 382,* 199–203.

Johnson, S.L., Sandrow, D., Meyer, B., Wnters, R., Miller, I., Keitner, G., and Solomon, D. (in press). Increases in manic symptoms following life events involving goal-attainment. *Journal of Abnormal Psychology.*

Kalivas, P., Churchill, L., and Klitenick, M. (1993). The circuitry mediating the translating of motivational stimuli into adaptive motor responses. In P. Kalivas and C. Barnes (Eds.), *Limbic motor circuits and neuropsychiatry* (pp. 310–354). Boca Raton, FL: CRC Press.

Kapp, B.S., Frysinger, R.C., Gallagher, M., and Haselton, J.R. (1979). Amygdala central nucleus lesions. Effect on heart rate conditioning in the rabbit. *Physiology and Behavior, 23,* 1109–1117.

Kapp, B.S., Pascoe, J.P., and Bixler, M.A. (1984). The amygdala: a neuroanatomical systems approach to its contribution to aversive conditioning. In L.R. Squire (Ed.), *Neuropsychology of memory* (pp. 473–488). New York: Guilford Press.

Kapp, B.S., Wilson, A., Pascoe, J.P., Supple, W.F., and Whalen, P.J. (1990). A neuroanatomical systems analysis of conditioned bradycardia in the rabbit. In M. Gabriel and J. Moore (Eds.), *Neurocomputation and learning: foundations of adaptive networks.* New York: Bradford Books.

Katzenelbogen, S., Loucks, R.B., and Gantt, W.H. (1939). An attempt to condition gastric secretion to histamine. *American Journal of Physiology, 128,* 10–12.

Kim, J.J., Rison, R.A., and Fanselow, M.S. (1993). Effects of amygdala, hippocampus, and periaqueductal gray lesions on short- and long-term contextual fear. *Behavioral Neuroscience, 107,* 1093–1098.

Kimble, D.P. (1969). Possible inhibitory functions of the hippocampus. *Neuropsychologia, 7,* 235–244.

Kleitman, N., and Crisler, G.A. (1927). A quantitative study of a salivary conditioned reflex. *American Journal of Physiology, 79,* 571–614.

Konorski, J. (1967). *Integrative activity of the brain: an interdisciplinary approach.* Chicago: University of Chicago Press.

LeDoux, J.E. (1994). Cognitive emotional interactions in the brain. In P. Ekman and R.J. Davidson (Eds.), *The nature of emotions* (pp. 216–223). Oxford University Press.

LeDoux, J.E. (1995). Emotion: clues from the brain. *Annual Review of Psychology, 46,* 209–235.

LeDoux, J.E., and Farb, C.R. (1991). Neurons of the acoustic thalamus that project to the amygdala contain glutamate. *Neuroscience Letters, 134*(1), 145–149.

LeDoux, J.E., Sakaguchi, A., and Reis, D.J. (1984). Subcortical efferent projections of the medial geniculate nucleus mediate emotional responses conditioned to acoustic stimuli. *Journal of Neuroscience, 4,* 683–698.

LeDoux, J.E., Chicchetti, P., Xagoraris, A., and Romanski, L. (1990a). The lateral amygdaloid nucleus: sensory interface of the amygdala in fear conditioning. *Journal of Neuroscience, 10,* 1062–1069.
LeDoux, J.E., Farb, C., and Ruggiero, D.A. (1990b). Topographic organization of neurons in the acoustic thalamus that project to the amygdala. *Journal of Neuroscience, 10,* 1043–1054.
LeDoux, J.E., Ruggiero, D.A., Forest, R., Stornetta, R., and Reis, D.J. (1987). Topographic organization of convergent projections to the thalamus from the inferior colliculus and spinal cord in the rat. *Journal of Comparative Neurology, 264,* 123–146.
LeDoux, J.R., Ruggiero, D.A., and Reis, D.J. (1985). Projections of the subcortical forebrain from anatomically defined regions of the medial geniculate body in the rat. *Journal of Comparative Neurology, 242,* 182–213.
Li, X.F., Phillips, R., and LeDoux, J.E. (1995). NMDA and non-NMDA receptors contribute to synaptic transmission between the medial geniculate body and the lateral nucleus of the amygdala. *Experimental Brain Research, 105,* 87–100.
Ljungberg, T., Apicella, P., and Schultz, W. (1992). Responses of monkey dopamine neurons during learning of behavioral reactions. *Journal of Neurophysiology, 67,* 145–163.
Love, J.A., and Scott, J.W. (1969). Some response characteristics of cells of the magnocellular division of the medial geniculate body of the cat. *Canadian Journal of Physiology and Pharmacology, 47,* 881–888.
Lubow, R.E., and Gewirtz, J.C. (1995). Latent inhibition in humans: data, theory, and implications for schizophrenia. *Psychological Bulletin, 117*(1), 87–103.
Mark, G.P., Blander, D.S., Hernandez, L., and Hoebel, B.G. (1989). Effects of salt intake, rehydration and conditioned taste aversion (CTA) development on dopamine output in the rat nucleus accumbens. *Appetite, 12,* 224.
McCabe, P.M., Schneiderman, N., Jarrell, T.W., Gentile, C.G., Teich, A.H., Winters, R.W., and Liskowsky, D. (1992). Central pathways involved in classical differential conditioning of heart rate responses in rabbits. In I. Gormezano and E.A. Wasserman (Eds.), *Learning and memory: the behavioral and biological substrates* (pp. 321–346). Hillsdale, NJ: Erlbaum.
McCabe, P.M., McEchron, M.D., Green, E.J., and Schneiderman, N. (1993). Effects of electrolytic and ibotenic acid lesions of the medial nucleus of the medial geniculate nucleus on single tone heart rate conditioning. *Brain Research, 619,* 291–298.
McCormick, D.A., and Prince, D.A. (1987). Actions of acetylcholine in the guinea-pig and cat medial and lateral geniculate nuclei, in vitro. *Journal of Physiology (London), 392,* 147–165.
McEchron, M.D., McCabe, P.M., Green, E.J., Llabre, M.M., and Schneiderman, N. (1992). Air puff versus shock unconditioned stimuli in rabbit heart rate conditioning. *Physiology and Behavior, 51,* 195–199.
McEchron, M.D., McCabe, P.M., Green, E.J., Llabre, M.M., and Schneiderman, N. (1995). Simultaneous single unit recording in the medial nucleus of the medial geniculate and amygdaloid central nucleus throughout habituation, acquisition, and extinction of the rabbit's classically conditioned heart rate. *Brain Research, 682,* 157–166.
McEchron, M.D., Green, E.J., Winters, R.W., Nolen, T.G., Schneiderman, N., and McCabe, P.M., (1996a). Changes of synaptic efficacy in the medial geniculate nucleus as a result of auditory classical conditioning. *Journal of Neuroscience, 16,* 1273–1283.
McEchron, M.D., McCabe, P.M., Green, E.J., Hitchcock, J.M., and Schneiderman, N. (1996b). Immunohistochemical expression of the c-Fos protein in the spinal trigeminal nucleus following presentation of a corneal airpuff stimulus. *Brain Research, 710,* 112–120.

McFarland, D.J.,Kostas, J., and Drew, W.G. (1978). Dorsal hippocampal lesions: effects of preconditioning CS exposure on flavor aversion. *Behavioral Biology, 22*(3), 397–404.

Metherate, R., and Ashe, J.H. (1991). Basal forebrain stimulation modifies auditory cortex responsiveness by an action at muscarinic receptors. *Brain Research, 559,* 163–167.

Metherate, R., Tremblay, N., and Dykes, R.W. (1987). Acetylcholine permits long-term enhancement of neuronal responsiveness in cat primary somatosensory cortex. *Neuroscience, 22,* 75–81.

Miller, N.E. (1948). Studies of fear as an acquirable drive: I. Fear as motivation and fear-reduction as reinforcement in the learning of new responses. *Journal of Experimental Psychology, 38,* 89–101.

Mirenowicz, J., and Schultz, W. (1994). Improtance of unpredictability for reward responses in primate dopamine neurons. *Journal of Neurophisology, 72,* 1024–1027.

Mogenson, G.L., and Nielsen, M.A. (1984). A study of the contribution of hippocampal-accumbens subpallidal projections to locomotor activity. *Behavioral and Neural Biology, 42,* 38–51.

Moore, J.W. (1979). Information processing in space-time by the hippocampus. *Physiological Psychology, 7*(3), 224–232.

Moore, J.W. (1986). Two model systems. In D.L. Alkon and C.D. Woody (Eds.), *Neural mechanisms of conditioning* (pp. 209–219). New York: Plenum Press.

Mowrer, O.H. (1947). On the dual nature of learning: a reinterpretation of "conditioning" and "problem-solving." *Harvard Educational Review, 17,* 102–150.

Mowrer, O.H. (1950). *Learning theory and personality dynamics.* New York: Ronald Press.

Mowrer, O.H. (1960). *Learning and theory and behavior.* New York: Wiley.

Nauta, W.J.H. and Domesick, V.B. (1984). Afferent and efferent relationships of the basal ganglia. *CIBA Foundation Symposium, 107,* 3–29.

Oades, R.D. (1985). The role of nonadrenaline in tuning and dopamine in switching between signals in the CNS. *Neuroscience and Biobehavioral Reviews, 9,* 261–282.

Obrist, P.A., and Webb, R.A. (1967). Heart rate during conditioning in dogs: relationship to somatic-motor activity. *Psychophysiology, 4,* 734.

Obrist, P.A., Howard, J.L., Lawler, J.E., Galosy, R.A., Meyers, A., and Gaebelein, C.J. (1974). The cardiac–somatic interaction. In P.A. Obrist, A.H. Black, J. Brener, and L.V. DiCara (Eds.), *Cardiovascular psychophysiology* (pp. 136–162). Chicago: Aldine.

O'Connor, K.N., Allison, T.L., Rosenfield, M.E., and Moore, J.W. (1997). Neural activity in the medial geniculate nucleus during auditory trace conditioning. *Experimental Brain Research, 113,* 534–556.

Oleson, T.D., Westenberg, I.S., and Weinberger, N.M. (1972). Characteristics of the pupillary dilation response during Pavlovian conditioning in paralyzed cats. *Behavioral Biology, 7,* 829–840.

Oleson, T.D., Ashe, J.H., and Weinberger, N.M. (1975). Modification of auditory and somatosensory system activity during pupillary conditioning in the paralyzed cat. *Journal of Neurophysiology, 38,* 1114–1139.

Ottersen, O.P., and Ben-Ari, Y. (1979). Afferent connections to the amygdaloid complex of the rat and cat. I. Projections of the thalamus. *Journal of Comparative Neurology, 187,* 401–424.

Packard, M.G., and Teather, L.A. (1998). Amygdala modulation of multiple memory systems: hippocampus and caudate-putamen. *Neurobiology, Learning and Memory, 69*(2), 163–203.

Page, K.J., and Sofroniew, M.V. (1996). The ascending basal forebrain cholinergic system. *Progress in Brain Research, 107,* 513–522.

Palmerino, C.C., Rusiniak, K.W., and Garcia, J. (1980). Flavor-illness aversions: the peculiar roles of odor and taste in memory for poison. *Science, 208,* 753–755.

Pare, D., Steraide, M., Deschenes, M., and Bouhassira, D. (1990). Prolonged enhancement of anterior thalamic synaptic responsiveness by stimulation of a brain-stem cholinergic group. *Journal of Neuroscience, 10,* 20–33.

Parrish, J. (1967). Classical discrimination conditioning of heart rate and bar-press suppression in the rat. *Psychonomic Science, 9,* 267–268.

Pascoe, J.D., and Kapp, B.S. (1985). Electrophysiological characteristics of amygdaloid central nucleus neurons during differential Pavlovian conditioned heart rate responding in the rabbit. *Behavioral Brain Research, 16,* 117–133.

Pavlov, I.P. (1910). *The work of the digestive glands* (2nd English ed.). (Translated from Russian by W.H. Thompson.) London: Charles Griffin.

Pavlov, I.P. (1927). *Conditioned reflexes: an investigation of the physiological activity of the cerebral cortex.* (Translated from Russian and edited by G.V. Anrep.) New York: Dover.

Posner, M.I., and Raichle, M.E. (1994). *Images of the brain.* New York: W.H. Freeman.

Powell, D.A. (1994). Rapid associative learning: conditioned bradycardia and its central nervous system substrates. *Integrative Physiological and Behavioral Science, 29,* 109–133.

Powell, D.A., and Kazis, E. (1976). Blood pressure and heart rate changes accompanying classical eyeblink conditioning in the rabbit (oryctolagus cuniculus). *Psychophysiology, 13,* 441–448.

Powell, D.A., Buchanan, S.L., and Gibbs, C.M. (1990). Role of the prefrontal-thalamic axis in classical conditioning. *Progress in Brain Research, 85,* 433–465.

Powell, D.A., Lipkin, M., and Milligan, W.L. (1974). Concomitant changes in classically conditioned heart rate and corneoretinal potential discrimination in the rabbit (oryctolagus cuniculus). *Learning and Motivation, 5,* 532–547.

Pribram, K.H. (1986). The hippocampal system and recombinant processing. In R. Isaacson and K.H. Pribram (Eds.) *The hippocampus* (Vol. 4, pp. 627–656). New York: Plenum Press.

Quirk, G.J., Repa, C., and LeDoux, J.E. (1995). Fear conditioning enhances short-latency auditory responses of lateral amygdala neurons; parallel recordings in the freely behaving rat. *Neuron, 15,* 1029–1039.

Rescorla, R.A. (1967). Inhibition of delay in Pavlovian fear conditioning. *Journal of Comparative and Physiological Psychology, 64,* 114–120.

Robbins, T.W., and Everitt, B.J. (1996). Neurobehavioural mechanisms of reward and motivation. *Current Opinions in Neurobiology, 6*(2), 228–236.

Rogan, M.T., and LeDoux, J.E. (1996). Emotion: systems, cells, synaptic plasticity. *Cell, 85,* 469–475.

Romo, R., and Schultz, W. (1990). Dopamine neurons of the monkey midbrain: contingencies of responses to active touch during self-initiated arm movement. *Journal of Neurophysiology, 63,* 592–606.

Rosen, R.B., & Schulkin, J. (1998). From normal fear to pathological anxiety. *Psychological Review, 105,* 325–350.

Ryugo, D.K., and Weinberger, N.M. (1978). Differential plasticity of morphologically distinct neuron populations in the medial geniculate body of the cat during classical conditioning. *Behavioral Biology, 22,* 275–301.

Salamone, J.D., Steinpreis, R.E., McCullough, L.D., Smith, P., Grebel, D., and Mahan, K. (1991). Haloperidol and nucleus accumbens dopamine depletion suppress lever pressing for food but increase free food consumption in a novel food choice procedure. *Psychopharmacology* (Berlin), 104, 515–251.

Salamone, J.D., Cousins, M.S., and Snyder, B.J. (1997). Behavioral functions of nucleus accumbens dopamine: empirical and conceptual problems with the anhedonia hypothesis. *Neuroscience and Biobehavioral Reviews, 21*(3), 341–359.

Saulskaya, N., and Marsden, C.A. (1995). Conditional depamine release: dependence on N–mythyl–D–aspartate receptors. *Neuroscience, 67*, 57–63.

Schmajuk, N., and Moore, J.W. (1985). Real-time attentional models for classical conditioning and the hippocampus. *Physiological Psychology, 13*, 278–290.

Schmajuk, N.A., and Moore, J.W. (1988). The hippocampus and the classically conditioned nicitating membrane response: a real-time attentional-associative model. *Psychobiology, 16*, 20–35.

Schneiderman, N. (1972). Response system divergencies in aversive classical conditioning. In A.H. Black and W.F. Prokasy (Eds.), *Classical conditioning. II: Current research and theory* (pp. 341–376). New York: Appleton-Century-Crofts.

Schneiderman, N. (1974). The relationship between learned and unlearned cardiovascular responses. In P.A. Obrist, A.H. Black, J. Brener, and L.V. DiCara (Eds.), *Cardiovascular psychophysiology* (pp. 190–210). Chicago: Aldine.

Schneiderman, N., Smith, M.S., Smith, A.C., and Gormezano, I. (1966). Heart rate classical conditioning in rabbits. *Psychonomic Science, 6*, 241–242.

Schultz, W., Apicella, P., and Ljungberg, T. (1993). Repsonses of monkey dopamine neurons to reward and conditioned stimuli during successive steps of learning a delayed response task. *Journal of Neuroscience, 13*, 900–913.

Seward, J.P. (1952). Introduction to a theory of motivation in learning. *Psychological Review, 59*, 405–413.

Shettleworth, S.J. (1983). Function and mechanism in learning. In M. D. Zeiler and P. Harzen (Eds.), *Advances in analysis of behaviour*, Vol. 3. Chichester, England: Wiley.

Siegel, S. (1983). Classical conditioning, drug tolerance, and drug dependence. In Y. Israel, F.B. Glaser, H. Kalant, R.E. Popham, W. Schmidt, and R.G. Smart (Eds.), *Research advances in alcohol and drug problems*, Vol. 7 (pp. 207–243). New York: Plenum Press.

Silvestri, A.J., and Kapp, B.S. (1998). Amygdaloid modulation of mesopontine peribrachial neuronal activity: implications for arousal. *Behavioral Neuroscience, 122*, 571–588.

Smith, O.A., Astley, C.A., DeVito, J.L., Stein, J.M., and Walsh, K.E. (1980). Functional analysis of hypothalamic control of the cardiovascular response accompanying emotional behavior. *Federal Proceedings, 39*, 2487–2494.

Solomon, P.R. (1979). Temporal versus spatial information processing theories of the hippocampal function. *Psychological Bulletin 86*(6), 1272–1279.

Solomon, P.R., and Moore, J.W. (1975). Latent inhibition and stimulus generalization of the classically conditioned nictitating membrane response in rabbits (Oryctolagus cuniculus) following forsal hippocampal ablation. *Journal of Comprehensive Physiological Psychology, 89*(10), 1192–1203.

Solomon, P., Nichols, G.L., Kiernan, J.M., Kamer, R.S., and Kaplan, L.J. (1980). Differential effects of lesions in medial and dorsal raphe of the rat: latent inhibition and septohippocampal serotonin levels. *Journal of Comparative Physiological Psychology, 94*, 145–154.

Sorenson, C., and Swerdlow, N. (1982). The effect of tail pinch on the acoustic startle response in rats. *Brain Research, 247,* 105–113.

Spence, K.W. (1956). *Behavior theory and conditioning.* New Haven: Yale University Press.

Stein, L., and Belluzzi, J.D. (1989). Cellular investigations of behavioral reinforcement. *Neuroscience and Biobehavioral Reviews, 13,* 69–80.

Sokolov, E.N. (1963). Perception and the conditioned reflex (pp. 5–19, 49–53, 295–303). New York: Pergamon Press.

Steraide, M., Pare, D., Parent, A., and Smith, Y. (1988). Projections of cholinergic and non-cholinergic neurons of the brainstem core to relay and associational thalamic nuclei in the cat and macaque monkey. *Neuroscience, 25,* 47–67.

Stebbins, W.C., and Smith, O.A. (1964). Cardiovascular concomitants of the conditioned emotional response in the monkey. *Science, 144,* 881–882.

Supple, W.F., and Kapp, B.A. (1989). Response characteristics of neurons in the medial component of the medial geniculate nucleus during Pavlovian differential fear conditioning in rabbits. *Behavioral Neuroscience, 103,* 1276–1286.

Supple, W.F., and Leaton, R.N. (1990). Lesions of the cerebellar vermis and cerebellar hemispheres: effects on heart rate conditioning in rats. *Behavioral Neuroscience, 104,* 934–947.

Swadlow, H.A., Hosking, K.E., and Schneiderman, N. (1971). Differential heart rate conditioning and lever lift suppression in restrained rabbits. *Physiology and Behavior, 7,* 257–260.

Swerdlow, N., and Koob, G. (1987). Dopamine, schizophrenia, mania, and depression: toward a unified hypothesis of cortico-striato-pallido-thalamic function. *Behavior and Brain Sciences, 10,* 197–248.

Swerdlow, N.R., Braff, D.L., Masten, V.L., and Geyer, M.A. (1990). Schizophrenic-like sensorimotor gating abnormalities in rats following dopamine infusion into the nucleus accumbens. *Psychopharmacology* (Berlin), *101*(3), 414–420.

Swerdlow, N.R., Caine, S.B., and Geyer, M.A. (1992). Regionally selective effects of intracerebral dopamine infusion on sensorimotor gating of the startle reflex in rats. *Psychopharmacology* (Berlin), *108*(102), 189–195.

Taylor, J.R., and Robbins, T.W. (1986). 6-Hydroxydopamine lesions of the nucleus accumbens, but not of the caudate nucleus, attenuate enhanced responding with reward-related stimuli produced by intra-accumbens d-amphetamine. *Psychopharmacology* (Berlin), *90*(3), 390–397.

Teich, A.H., McCabe, P.M., Gentile, C.G., Winters, R.W., Liskowsky, D.R., and Schneiderman, L. (1988). Role of auditory cortex in the acquisition of differential heart rate conditioning. *Physiology and Behavior, 44,* 405–412.

Teich, A.H., McCabe, P.M., Gentile, C.G., Schneiderman, L., Winters, R.W., Liskowsky, D.R., and Schneiderman, L. (1989). Auditory cortex lesions prevent the extinction of Pavlovian differential heart rate conditioning to tonal stimuli in rabbits. *Brain Research, 480,* 210–218.

Thompson, R.F., and Steinmetz, J.E. (1992). The essential memory trace circuit for a basic form of associative learning. In I. Gormezano and E.A. Wasserman (Eds.), *Learning and memory: the behavioral and biological substrates* (pp. 369–386). Hillsdale, NJ: Erlbaum.

Thompson, R.F., Berger, T.W., and Madden, J., IV (1983). Cellular processes of learning and memory in the mammalian CNS. *Annual Review of Neuroscience, 6,* 447–491.

Turner, B. H., and Herkenham, M. (1991). Thalamoamygdaloid projections in the rat: a test of the amygdala's role in sensory processing. *Journal of Comparative Neurology, 313,* 295–325.

Veening, J.G. (1978). Subcortical afferents of the amygdaloid complex in the rat: an HRP study. *Neuroscience Letters, 8,* 197–202.

Weinberger, N.M. (1982). Sensory plasticity and learning: the magnocellular medial geniculate nucleus of the auditory system. In C.D. Woody (Ed.), *Conditioning: representation of involved neural functions* (pp. 697–718). New York: Plenum Press.

Weinberger, N.M. and Diamond, D.M. (1987). Physiological plasticity in auditory cortex: rapid induction by learning. *Progressions in Neurobiology, 29,* 1–55.

Weinberger, N.M., Diamond, D.M., and McKenna, T.M. (1984a). Initial events in conditioning: plasticity in the pupillomotor and auditory systems. In G. Lynch, J.L. McGaugh, and N.M. Weinberger (Eds.), *Neurobiology of learning and memory* (pp. 197–227). New York: Guilford Press.

Weinberger, N.M., Hopkins, W., and Diamond, D.M. (1984b). Physiological plasticity of single neurons in auditory cortex of the cat during acquisition of the pupillary conditioned response: I. Primary field (AI). *Behavioral Neuroscience, 98,* 171–188.

Weinberger, N.M., Javid, R., and Lepan, B. (1993). Long term retention of learning-induced receptive field plasticity in the auditory cortex. *Proceedings of the National Academy of Sciences of the United States of America, 90,* 2394–2398.

Weinberger, N.M., Javid, R., and Lepan, B. (1995). Heterosynaptic long-term facilitation of sensory-evoked responses in the auditory cortex by stimulation of the magnocellular medial geniculate in guinea pigs. *Behavioral Neuroscience, 109,* 10–17.

Weiner, I. (1990). Neural substrates of latent inhibition: the switching model. *Psychological Bulletin, 108*(3), 442–461.

Weiner, I., and Feldon, J. (1987). Facilitation of latent inhibition by haloperidol. *Psychopharmacology, 91,* 248–253.

Weiner, I., and Feldon, J. (1997). The switching model of latent inhibition: an update of neural substrates. *Behavioural Brain Research, 88,* 11–25.

Weiner, I., Lubow, R.E., and Feldon, J. (1981). Chronic amphetamine and latent inhibition. *Behavior Brain Research, 2,* 285–286.

Weiner, I., Israeli-Telerant, A., and Feldon, J. (1987). Latent inhibition is not affected by acute or chronic administration of 6 mg-kg dl-amphetamine. *Psychopharmacology, 91,* 345–351.

Weiner, I., Lubow, R.E., and Feldon, J. (1988). Disruption of latent inhibition by acute administration of low doses of amphetamine. *Pharmacology and Biochemistry Behavior, 30,* 871–878.

Weiner, I., Shadach, E., Barkai, R., and Feldon, J. (1997). Haloperidol- and clozapine-induced enhancement of latent inhibition with extended conditioning implications for the mechanism of action of neuroleptic drugs. *Neuropsychopharmacology, 16,* 42–50.

Wepsic, J.G. (1966). Multimodal sensory activation cells in the magnocellular medial geniculate nucleus. *Experimental Neurology, 15,* 299–318.

Wickelgren, I. (1997). Getting the brain's attention. *Science, 278,* 35–37.

Williams, D.R. (1965). Classical conditioning and incentive motivation. In W.F. Prokasy (Ed.), *Classical conditioning* (pp. 340–357). New York: Appleton-Century-Crofts.

Winer, J.A., Diamond, I.T., and Raczowski, D. (1977). Subdivisions of the auditory cortex of the cat: the retrograde transport of horseradish peroxidase to the medial geniculate body and posterior thalamic nuclei. *Journal of Comparative Neurology, 176,* 387–418.

Wolfe, J.B. (1936). Effectiveness of token-rewards for chimpanzees. *Comparative Psychology Monographs, 12,* No. 60.

Yamada, S., Harano, M., and Tanaka, M. (1998). Dopamine autoreceptors in rat nucleus accumbens modulate prepulse inhibition of acoustic startle. *Pharmacology, Biochemistry and Behavior, 60*(4), 803–808.

Yehle, A.L., Dauth, G., and Schneiderman, N. (1967). Correlates of heart-rate classical conditioning in curarized rabbits. *Journal of Comparative and Physiological Psychology, 64,* 93–104.

Young, A.M.J., Joseph, M.H., and Gray, J.A. (1993). Latent inhibition of conditioned dopamine release in the nucleus accumbens. *Neuroscience, 54,* 5–9.

Young, F.A. (1958). Studies of pupillary conditioning. *Journal of Experimental Psychology, 55,* 97–110.

4
The Functional Anatomy of Skeletal Conditioning

GERMUND HESSLOW AND CHRISTOPHER H. YEO

On Skeletal Muscle Reflexes

The nervous system mediates responses to a wide variety of stimuli. For example, a painful stimulus to the sole of the foot will elicit a rapid contraction of flexor muscles in the leg to withdraw the foot. Bright light reaching the retina elicits the pupillary reflex that serves to constrict the pupil and reduce the amount of light passing through it. Irritation of the lining of the nose can cause a sneeze. All reflexes are elicited by specific stimuli and all are relatively consistent and stereotyped in their appearance.

Reflexes are controlled by sets of neurons that form reflex arcs, which always have three identifiable components: first, the afferent limb, which consists of a sensory receptor and nerve that feeds the sensory information into the nervous system; second, a reflex center that receives and integrates the sensory information; and third, the efferent limb of the reflex, which consists of motoneurons and nerves and the target muscle or gland that produces the reflex response. Figure 4.1 illustrates the general principles of a reflex arc. It is generally agreed that reflexes are inborn, so their basic neural circuitry is predetermined. As we shall see, however, there are mechanisms that allow reflexes to be modified and patterned because they can change throughout development and in adulthood.

The anatomy and physiology of several skeletal muscle responses have been well characterized. Figure 4.2 shows some of the simpler components of the leg flexion withdrawal reflex to a strong cutaneous stimulus. The spinal cord reflex center organizes a flexion of the leg by activating flexor muscle contraction at the same time as enabling extensor muscle relaxation. Further interneuron connections pass to motoneurons controlling synergist muscles for ankle and hip flexion, so that other joints within the limb are appropriately controlled as the foot moves upward. Other connections pass to the contralateral side of the spinal cord to activate upper leg extensors and inhibit flexors. This pattern of reflex activity on the contralateral side is opposite to that on the ipsilateral side and provides postural stabilization while the affected limb is withdrawn. Muscle synergies such as these are common in skeletal muscle reflexes and, as is shown, must be considered in the analysis of neural mechanisms of reflex modification during learning.

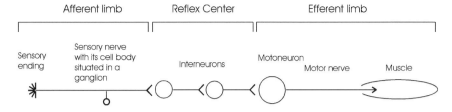

FIGURE 4.1. The reflex arc.

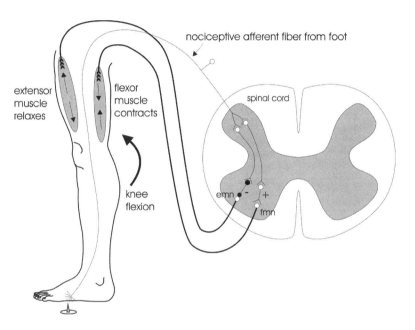

FIGURE 4.2. The leg flexion withdrawal reflex. A sharp, painful stimulus to the sole of the foot activates cutaneous nociceptors and their Aδ afferent fibers to the spinal cord. The Aδ fibers, with their cell bodies in a dorsal root ganglion just outside the spinal cord, terminate upon neurons in the dorsal horn of the spinal cord. These dorsal horn neurons project to a further set of interneurons within the ventral horn of the spinal cord. Note that the same sensory input activates both excitatory (+) and inhibitory (-) interneurons that project to the upper leg flexor (*fmn*) and extensor muscle motoneurons (*emn*), respectively. Thus, the sensory input from the sole of the foot activates flexor muscles and extensor muscles in the upper leg resulting in a knee flexion and withdrawal of the leg.

On Conditioned Skeletal Muscle Reflexes and Their Suitability for Neurobiological Analysis of Learning

Several skeletal muscle reflexes have been used in conditioning studies. In dogs, leg flexion responses to electrical stimulation were used in early investigations (Bekhterev, 1932; Brogden, 1939) and, in humans, finger withdrawal to electrical stimulation has been conditioned (Watson, 1916). In both these examples, an aversive electrical stimulus activates the reflex but skeletal muscle responses to appetitive stimuli can also be conditioned. In rabbits, the jaw-opening response to delivery of saccharin solution or water inside the mouth is a useful model for analysis (Smith et al., 1966).

Limb flexion responses are controlled by complex patterns of activity in spinal interneurons. In the case of leg flexion, spinal interneurons regulate agonist and antagonist muscle activities across the knee joint and also across hip and ankle joints on the side of limb withdrawal. Additionally, they control muscles in the trunk and contralateral limb to provide postural preparation and compensation as the affected limb is withdrawn. Descending control through the vestibulospinal, reticulospinal, and corticospinal tracts modulates these reflex patterns and allows voluntary control over many aspects of limb movement. When neuroscientists began to turn their attention toward an analysis of the neural mechanisms underlying these types of conditioning, it was immediately clear that the complexity of the spinal and supraspinal interactions would prove to be a problem. One simplifying approach was to isolate the spinal cord from descending influences by transecting it a high level and to use discrete electrical stimulation of different afferents as conditioned stimulus (CS) and unconditioned stimulus (US) (Patterson et al., 1973). In this preparation, learning takes the form of an enhancement of a preexisting response to the CS. Although undoubtedly associative in nature, one feature of the learning in this reduced spinal preparation is that the CS produces a UR-like response before conditioning begins. A key characteristic of conditioning in the intact subject is that, under normal circumstances, the CS need not initially produce a response similar to the unconditional response (UR) for subsequent conditioning to take place (see Schreurs, 1989 for a review).

An alternative approach to the selection of a conditioning preparation suitable for analysis has been to use a reflex behavior with a much simpler efferent limb. The mammalian eyeblink response is one such reflex. In cats, Woody and his colleagues (Woody, 1970) have paired an auditory click CS with a glabellar (forehead) tap US to produce short-latency eyeblink CRs. Direct recording of the motor output to the eyelids from the facial nucleus motoneurons (see following) and in parts of the motor cortex revealed that this short-latency form of eyeblink conditioning produces a combination of associative and nonassociative changes (Woody, 1982).

In some animals, the eyes are additionally protected by a reflex closure of the nictitating membrane (NM). Gormezano et al. (1962) demonstrated that, under appropriate conditions, conditioning of the NM response is very purely associative. Using a tone CS and an air puff to the cornea as a US, conditioned NM responses develop at latencies appropriate to the interstimulus interval (ISI) and are essentially free from alpha responses, sensitization, and pseudoconditioning effects (Gormezano, 1996; Gormezano & Moore, 1969; Gormezano et al., 1983). For these reasons, NMR conditioning provides an ideal model for the analysis of neural mechanisms underlying associative learning (Thompson, 1976).

The Eyelid Blink and Nictitating Membrane Response in Rabbits

Upper and lower eyelids, known together as the external eyelids, are present in all mammals but some species, including rabbits, also have a NM. The NM is sometimes known as the third eyelid, although it is anatomically and functionally quite distinct from the external eyelids. Any tactile stimulation of a variety of face areas close to the eye (including the eyelids and eyelashes, the conjunctiva, and the surface of the cornea) will cause a reflex, protective closure of the external eyelids and a horizontal sweep of the NM across the corneal surface in a nasal to temporal direction (see inset, Figure 4.3). Both responses depend on simple reflex arcs that have cranial nerve afferent and efferent limbs linked by brain stem interneurons. A detailed understanding of the brain stem circuitry controlling the eyeblink/NM unconditioned reflex is a necessary preliminary to our understanding of the circuitry and neural mechanisms that mediate the conditioned response.

The Afferent Limb of the Eyeblink and Nictitating Membrane Response

All primary sensory afferents from the face run in branches of the trigeminal nerve. The supraorbital nerve supplies the upper eyelid and face above the eye and the infraorbital nerve supplies the lower eyelid and much of the face and head below the eye. These primary afferent fibers project to nuclei of the trigeminal system in the brain stem and to some of the higher segments of the spinal cord. In rabbits, as in other species, the brain stem trigeminal nuclei are subdivided on the basis of cytoarchitectonic criteria into the nucleus principalis (Vp) and the nucleus spinalis (Vsp). From rostral to caudal, the nucleus spinalis is further subdivided into pars oralis (Vo), pars interpolaris (Vi), and pars caudalis (Vc). Physiological studies in other species show that there is a sensory representation of the entire head in each trigeminal subdivision with the periorbital structures located ventrally in the Vp, Vo, and Vi, and caudally in Vc.

van Ham and Yeo (1996a) traced sensory inputs from the cornea and periorbital areas in rabbits and showed that they have different patterns of termination in the brain stem and spinal cord. Sensory afferents in the periorbital skin and conjunctiva distribute most heavily to pars caudalis of the spinal trigeminal nucleus (Vc) and to the dorsal horn of spinal segment C1 (dhC1) with much weaker projections to spinal segment C2, rostral Vc, and adjacent reticular formation and to the lateral part of pars interpolaris of the spinal trigeminal nucleus (Vi). There are weak inputs from the periorbital skin to ventral pars oralis of the spinal trigeminal nucleus (Vo) and to the principal trigeminal nucleus (Vp). Corneal afferents distribute most densely in the ventral part of Vi and more weakly to caudal Vc and the adjacent dhC1. There are sparse projections to the ventral and dorsal part of Vp and

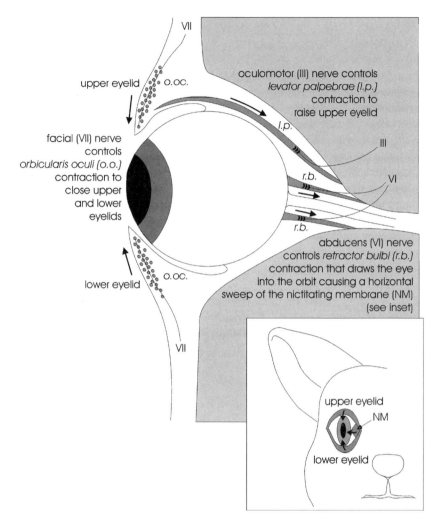

FIGURE 4.3. The rabbit eyeblink and nictating membrane response (NMR): muscles and innervation.

to the ventral part of Vo. All the inputs to the caudal trigeminal and upper spinal areas strictly observe laminar patterns of termination (van Ham & Yeo, 1996a).

In summary, there are strong corneal and periocular afferent inputs to Vc and level C1 of the spinal cord and a strong input to Vi from the cornea only. There are weaker inputs to other subdivisions of the trigeminal system and dorsal segments of the spinal cord. An earlier idea, that the eyeblink reflex is mediated by relatively restricted circuitry, was based on a finding that the afferent limb of the blink reflex terminated in a small group of neurons in Vo (Harvey et al., 1984). The more sensitive anatomic tracing techniques used in the studies reviewed here reveal that the afferent limb of the blink reflex is more complex and so the eyeblink reflex center is likely to be more distributed than was originally envisaged.

The Efferent Limb of the Eyeblink and Nictitating Membrane Response

The efferent limbs of the external eyelid blink and of the nictitating membrane response (NMR) are distinct and different (see Figure 4.3). Closure of the external eyelids is produced by contraction of the *orbicularis oculi* (o.o.) muscle that forms a flattened ring around the eye. The o.o. is controlled by a subset of motoneurons in the dorsal part of the facial (VIIth) nucleus of the brain stem via efferents that travel in the VIIth nerve. The upper eyelid can be raised by contraction of the small, *levator palpebrae* (l.p.) muscle below the o.o. that is innervated by fibers from the contralateral oculomotor (IIIrd cranial nerve) nucleus. Thus, the l.p. muscle is an antagonist of the o.o. muscle, and the position of the upper eyelid is determined by the balance of IIIrd and VIIth nerve activity supplying these two muscles.

The NM has no direct muscle attachment to produce its movement. The NMR is a consequence of retraction of the eyeball into the orbit by the extraocular muscles, primarily the *retractor bulbi* (r.b.) muscle. This muscle has four slips, arranged around the exit of the optic nerve from the back of the eye, which contract to draw the eyeball into the bony orbit. This retraction displaces some of the soft tissues behind the eye to press on the cartilaginous NM and so it sweeps across the eye. When the r.b. muscle relaxes, elastic properties of the tissues restore the NM to a rest position in the nasal canthus of the eye. The r.b. muscle is innervated by a subset of abducens (VIth cranial nerve) fibers that arise from a small, accessory abducens nucleus (AccVI) in the brain stem (Cegavske et al., 1976; Disterhoft et al., 1985).

The Premotor "Blink" Area: A Reflex Center for NM and Eyeblink Responses

Analysis of the pathways linking the afferent and efferent limbs of the NM and eyeblink reflexes has been guided by accurate measurement of the time course of the muscle responses. If the blink reflex is activated with a very brief electrical stimulus, then an electromyogram (EMG) of the eyelid reflex response reveals a short- (6–7 ms) latency, R1 component and a longer-latency (>15 ms), R2 component. Its latency indicates that the R2 component is polysynaptic, but the short minimum latency of the R1 component indicates that at least part of it must be disynaptic, that is, three neurons connected across two synapses. These three neurons must be (1) the sensory afferent, with its cell body located in the trigeminal

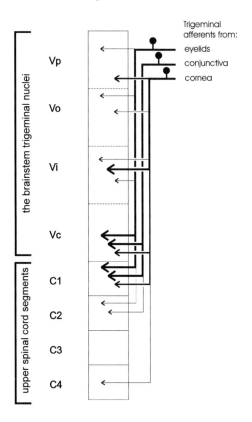

FIGURE 4.4. The afferent limb of the rabbit eyeblink/NM response. *C1–C4*, spinal cord cervical segments 1–4; *Vp*, principal trigeminal nucleus; *Vo*, spinal trigeminal nucleus, pars oralis; *Vi*, spinal trigeminal nucleus, pars interpolaris; *Vc*, spinal trigeminal nucleus, pars caudalis.

ganglion and synapsing upon (2) a trigeminal nucleus (or upper spinal cord) neuron projecting to, and synapsing upon (3) an orbicularis oculi or retractor bulbi motoneuron in the VII or AccVI nucleus, respectively (Figure 4.4).

A further property of the NMR and eyeblink system has helped in the identification of its neural substrate. Because a single, tactile stimulation of the face usually elicits an eyelid blink *and* the NMR, it is likely that they share a common, premotor component. van Ham and Yeo (1996b) looked for neurons within the trigeminal nuclei that receive periocular/corneal afferents and that have a common projection to the ipsilateral o.o. and AccVI motoneurons. A cluster of neurons in Vp, rostral Vo, and the adjacent reticular areas satisfied these criteria. Through connections identified as Path 1 (Figure 4.5), they were seen to project to o.o. and AccVI motoneurons and, in addition, they projected to a region immediately lateral and dorsolateral to the contralateral oculomotor nucleus where

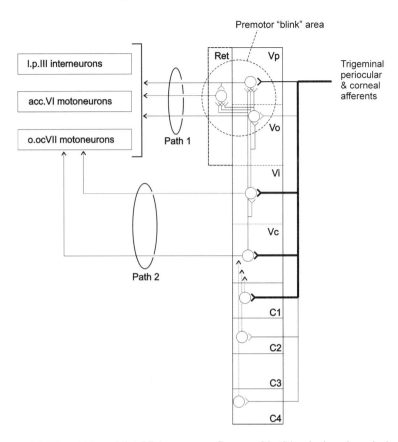

FIGURE 4.5. The rabbit eyeblink/NM response reflex arc. *C1–C4*, spinal cord cervical segments 1–4; *Vp*, principal trigeminal nucleus; *Vo*, spinal trigeminal nucleus, pars oralis; *Vi*, spinal trigeminal nucleus, pars interpolaris; *Vc*, spinal trigeminal nucleus, pars caudalis; *Ret*, reticular region; *l.p.III interneurons*, interneurons projecting to levator palpebrae motoneurons in the oculomotor (III) nucleus; *acc.VI motoneurons*, motoneurons in the accessory VI nucleus supplying the retractor bulbi muscle which retracts the eyeball and produces the NM response; *o.ocVII motoneurons*, motoneurons in the facial (VII) nucleus that supply the orbicularis oculi muscle which produces eyelid closure.

there are inhibitory neurons projecting back across the midline to the levator palpebrae motoneurons. It is suggested that neurons in Vp/Vo form a premotor "blink" area that can simultaneously activate the orbicularis oculi and retractor bulbi muscle to produce a concerted external eyelid blink and NM response and, simultaneously, inhibit the levator palpebrae muscle that is the upper eyelid antagonist of lid closure. This premotor "blink" area receives moderately strong inputs from the periocular skin and a weaker input from the cornea suggesting that it can mediate the disynaptic R1 response. There is a similar "blink" area in Vo in cats, although it lacks the projection to the l.p. interneuron area (Holstege et al., 1986a, b).

The premotor "blink" area also receives a rich set of direct and multisynaptic projections, previously identified in guinea pigs (Pellegrini et al., 1995) from neurons more caudally in Vi, Vc, and upper spinal cord that themselves receive strong afferent periocular inputs. We suggest that these polysynaptic pathways with output through the "blink" area and Path 1 mediate the R2 component of the eyeblink response.

The more caudal trigeminal regions also have a direct output to the facial (VIIth) nucleus that distributes widely to o.o. motoneurons and to motoneurons controlling many other facial muscles through connections we have identified as Path 2 (see Figure 4.5). Significantly, these caudal trigeminal regions do not project directly to AccVI or to III, so they cannot mediate a concerted blink response. Instead, they may provide a more general, disynaptic (including the eyelid R1) preparatory activation of many facial muscles to somatosensory or nociceptive stimulation. These projections are very similar to those identified arising from the Vi/Vc border in guinea pigs (Pellegrini et al., 1995).

The identification of the blink area that coordinates eyelid and NM responses is an important step in the analysis of mechanisms supporting eyeblink conditioning because it is a likely point of convergence for activity relating to conditioned and unconditioned responses. In particular, there are important projections from the red nucleus (RN), an important recipient of cerebellar output, to the blink area (Holstege & Tan, 1988). As we describe next, it is now beyond doubt that the cerebellum plays a critical role in the regulation of both conditioned and unconditioned eyeblink/NM responses.

Neural Substrates of Skeletal Muscle Conditioning: Early Studies

The early history of attempts to identify the neural substrates of conditioning does not represent a steady line of progress where each step builds on previous findings and in turn serves as the starting point for further investigation. On the contrary, many potentially important findings became historical blind alleys, not because they could not have been fruitfully pursued but because they were forgotten or ignored.

Pavlov thought that there was one basic associative mechanism that was located in the cerebral cortex, and some of the early attempts to find the neural mechanisms of conditioning seemed consistent with his assumption. For instance, in the early 1930s, Loucks (1933, 1935) managed to condition animals to a CS that consisted of direct 'faradic' stimulation of the cerebral cortex, a finding that was later replicated by several investigators. However, the implication that the cerebral cortex is exclusively involved in conditioning was challenged by Doty and collaborators (Doty et al., 1956; Doty, 1969), who reported that stimulation of almost any part of the brain could serve as a CS. Although Doty did not succeed in establishing conditioning to a cerebellar stimulation CS, others have subsequently done so (Donhoffer, 1966).

Lesion studies revealed that the cerebral cortex is not necessary for delay conditioning. There had been early reports of conditioning in decorticate animals (Culler & Mettler, 1934) and a clear demonstration, with histological verification of the lesion, of conditioned leg flexion in a decorticate dog (Bromiley, 1948). A more systematic analysis of the role of the cortex was later provided by Oakley and Russell (1972, 1975), who demonstrated that acquisition of NMR conditioning in rabbits is relatively normal after complete bilateral removal of the neocortex. In some instances, conditioning in the decorticate subjects actually proceeded more rapidly than normal. Such experiments clearly demonstrated that subcortical structures alone can support conditioning, but they did not rule out that the neocortex normally plays an important role in intact animals. This possibility was tested in a later experiment, and it was shown that conditioned NM responses acquired in intact rabbits were retained after decortication (Oakley & Russell, 1977). It has also been reported that neocortical lesions do not affect conditioned inhibition in rabbit NMR conditioning (Moore et al., 1980).

Further lesion studies also revealed that other forebrain structures, widely recognized as important in more complex forms of learning, are not essential for delay conditioning of the rabbit NMR. Lesions of the hippocampus (which was mainly spared in the experiments by Oakley and Russell) failed to prevent conditioning (Schmaltz & Theios, 1972). So too did lesions of the amygdala (Kemble et al., 1972). Norman et al. (1974) demonstrated normal conditioning of eyelid responses, including extinction and discrimination learning, in cats in which the ventral posteromedial nucleus of the thalamus and all telencephalic structures such as the neocortex, hippocampus, septum, basal ganglia, and amygdala had been removed. Similarly, Mauk and Thompson (1987) showed complete retention of rabbit NMR conditioning after decerebration rostral to the red nucleus.

These lesion experiments unequivocally showed that delay conditioning of NMR or eyeblink responses can be supported entirely by lower midbrain or hindbrain neural circuits. Findings such as these have been reported many times since, and the decerebrate is now recognized as an important preparation for the analysis of hindbrain mechanisms in conditioning. There are several electrophysiological analyses of eyeblink conditioning in cats and ferrets with decerebrations just rostral to the superior colliculus and the red nucleus (Hesslow, 1994a; Ivarsson et al., 1997).

Fortunately, lesion experiments of conditioning have not only yielded negative results. Several early lesion studies indicated that the cerebellum might play a necessary role. Popov (1929) studied leg-flexion conditioning in dogs using a flashing light CS and electrical stimulation to the distal hindlimb as the US. The US produced hindlimb flexion, and a hindlimb flexion CR developed with CS pairings. After a complete removal of the cerebellum, the conditioned flexion response had been abolished and replaced by a weak extension. Later Russian reports (Fanardjian, 1961) indicated that cerebellar ablations impaired but did not abolish conditioning, consistent with the view that the cerebellum might regulate the performance of the CR but that the memory trace is located elsewhere. Sadly, this work, which was mainly published in Russian, went largely unnoticed in the west (but see Welsh & Harvey, 1992 for a recent review of this work).

More recently, it was shown that lesions of the red nucleus abolished conditioned leg flexion responses to a tone CS in cats (Smith, 1970). The red nucleus receives major inputs from the cerebral cortex and from the cerebellum. Because it was already known that the cerebral cortex is not essential for this conditioning, a good case could have been made in the early 1970s that conditioning is critically dependent on the cerebellum. The experiment that established the importance of the cerebellum came 10 years later.

In 1982, Richard Thompson and colleagues reported a striking and exciting new finding. A unilateral lesion of the cerebellum, including the cortex and the underlying dentate and interpositus nuclei, abolished NMR conditioning that had previously been acquired and prevented relearning of the response (McCormick et al., 1982a). The abolition of CRs was ipsilateral to the lesion, and the unconditional responses to corneal air puff were normal. When the corneal air puff US was transferred to the other side, contralateral to the lesion, conditioning proceeded rapidly. Here was a clear demonstration that the cerebellum is critically involved in NMR/eyeblink conditioning. In another study, Desmond and Moore (1982) impaired NMR conditioning by interrupting cerebellar outflow in the superior cerebellar peduncle. These findings opened a new chapter in the analysis of neural mechanisms of conditioning and posed two fundamental questions. Are memories for NMR/eyeblink conditioning stored within the cerebellum and what are the underlying mechanisms? After nearly two decades of intensive research, these questions are beginning to find answers. In the following pages, we review what is understood about these cerebellar memories. To begin, we must examine the cerebellar anatomy and physiology that underpin current ideas about cerebellar function.

An Overview of the Anatomy and Physiology of the Cerebellum

Basic Anatomy and Connections of the Cerebellum

The mammalian cerebellum lies above the hindbrain fourth ventricle and consists of a highly folded (foliated) sheet of cortex lying above a set of cerebellar nuclei (sometimes known as the cerebellar "deep" nuclei). The cerebellar cortex receives inputs from a variety of sources and projects outputs ventrally to the cerebellar nuclei below. The cerebellar nuclei project on to the thalamus, red nucleus, and brain stem regions to influence a variety of motor and sensory gating systems.

The cortical folia, and the fissures that separate them, run in the transverse plane as they cross the midline but they diverge into complex patterns within the lateral cerebellar hemispheres. Two prominent fissures divide the cerebellum into three major lobes. The primary fissure divides the most rostral, anterior lobe from the more caudal, posterior lobe, and the posterolateral fissure divides the posterior lobe from the flocculonodular lobe. The complexity of the cerebellar folia has attracted many schemes of nomenclature, but the most useful of these is that of Larsell, who identified a consistent pattern of 10 lobules (always identified by Roman numerals I–X). The anterior lobe consists of lobules I–V, the posterior lobe contains lobules VI–IX and the flocculonodular lobe is lobule X. This lobular pattern can be seen in all mammalian species, including humans (Larsell, 1970; Larsell & Jansen, 1972). In the midline vermis, the lobular structure is clear and Larsell's 10 lobules are easily recognized, but it is often difficult to trace the continuities of the folial chain more laterally (see Voogd & Glickstein, 1998 for a review).

All cerebellar afferent and efferent connections pass through one of the three cerebellar peduncles found on each side. The inferior cerebellar peduncle mainly supplies afferent input from the spinal cord and brain stem, the middle cerebellar peduncle contains afferents from the pons, and the superior cerebellar peduncle contains most of the efferent fibers from the cerebellum that run forward to midbrain and thalamus (see Brodal, 1981, 1992) for more detailed accounts of basic anatomy and connections).

There are two major types of cerebellar afferent, which differ considerably in their anatomy and physiological properties. The cerebellar mossy fibers are extremely numerous and arise from many sources, including the spinal cord, brain stem reticular nuclei, and the pontine nuclei. The middle cerebellar peduncle exclusively contains mossy fiber inputs from the pontine nuclei, and other mossy fibers mainly travel in the inferior cerebellar peduncle. In sharp contrast, climbing fiber afferents originate exclusively in the inferior olive and travel only in the inferior cerebellar peduncle. Within the cerebellum, all climbing fibers and many mossy fibers give collateral inputs to one of the deep cerebellar nuclei. Mossy and climbing fibers then travel up to the cerebellar cortex, where they terminate in entirely different ways upon their targets.

The Neuronal Architecture of the Cerebellum

The cerebellar cortex contains five major types of neuron. Across the whole cerebellar cortex, the cell bodies and processes of these neurons are arranged in a regular, three-dimensional lattice that has been likened to a crystalline structure (Figure 4.6). This arrangement allows for a very large number of highly ordered synaptic interactions across the entire cerebellar cortex. The great regularity of this arrangement of neurons and their connections means that any one region of cerebellar cortex is anatomically almost indistinguishable from any other. Regionally, the cortex differs only in the variety and type of sensory inputs that it samples and in the output pathways that it influences. This anatomy is consistent with the view that the cerebellar cortex applies a common principle of information processing to a diverse set of inputs and outputs.

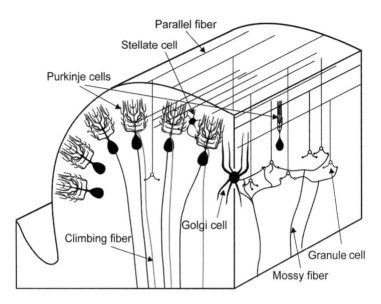

FIGURE 4.6. Diagram illustrating the three-dimensional arrangement of the cerebellar cortex. Notice that the flattened dendritic trees of the Purkinje cells lie perpendicular to the folia whereas the parallel fibers run along the folia.

The only cortical output neurons are the Purkinje cells whose soma lie in a monolayer between the molecular and granular layers. Purkinje cells are characterized by very large dendritic trees that branch richly in a flattened plane to form fan-shaped arbors aligned perpendicularly to the axis of the folium. Purkinje cells are the final targets of the two major afferent inputs to the cortex. They receive climbing fiber inputs directly and mossy fiber inputs indirectly via granule cells.

Ascending mossy fibers terminate upon the cortical granule cells, a huge population of neurons that accounts for more than half the neurons in the entire brain. Mossy fibers branch to contact many granule cell "rosettes" (specialized dendritic regions), each of which samples from several mossy fibers. Thus, there is both convergence and divergence of information through mossy fiber–granule cell synapses, with greater divergence than convergence. Mossy fiber–granule cell synapses have, therefore, been viewed as both mixing and expanding information coding to the cerebellar cortex (Marr, 1969; Albus, 1971; Gilbert, 1974, 1975). The granule cell ascending axon then travels to the molecular (surface) layer of the cortex where it forms a T-branch and sends out two parallel fibers that run for several millimeters along the long axis of the folium. There are synaptic contacts on the Purkinje cells from the ascending shaft of the granule cells and from parallel fibers as they pass through their dendritic trees of many Purkinje cells. A single Purkinje cell can receive synaptic inputs from many thousands of different granule cells. The organization of climbing fiber inputs to the cortex is in complete contrast. Each Purkinje cell receives input from only a single climbing fiber.

However, each climbing fiber makes multiple synaptic contacts upon its target Purkinje cell, ensuring a powerful drive upon the Purkinje cell. Each climbing fiber can form as many as 10 branches, each of which contacts a single Purkinje cell.

Within the cerebellar cortex, there are three types of GABA-ergic inhibitory interneurons. The stellate and basket cells receive their input from the parallel fibers and send their inhibitory outputs to the dendrites and the lower part of the soma of the Purkinje cells, respectively. The Golgi cells also receive input from parallel fibers and send inhibitory output back to the granule cell rosettes. The result of this arrangement is a negative feedback loop that allows Golgi cells to gate the mossy/parallel fiber input to the cerebellar cortex. Climbing fibers from the inferior olive also synapse upon Golgi cells, but they do not drive them to fire; instead, their firing rate is decreased. It is not known whether climbing fibers reduce Golgi cell activity directly or indirectly via inhibitory interneurons.

Output from the cerebellar cortical Purkinje cells is also inhibitory on their targets in the cerebellar and vestibular nuclei. So, of the five main types of neurons in the cerebellar cortex, only the glutaminergic granule cells have excitatory outputs upon their target neurons.

Afferent Systems to the Cerebellum

Climbing fibers and mossy fibers, the two major cerebellar afferent systems, are anatomically quite distinct. The single climbing fiber input to each Purkinje cell contrasts sharply with the hugely convergent input of many thousands of mossy/parallel fibers. This remarkable degree of convergence is quite unique in the brain, and it makes the cerebellar cortex a theoretically attractive structure for associative learning. A key feature of many models of cerebellar learning is that the mossy/parallel fiber inputs interact in a special way with the climbing fiber inputs. These models are supported by the distinctively different physiology of the climbing fiber and mossy/parallel fiber inputs.

The climbing fiber input from the inferior olive has two characteristic features. First, it normally fires at very low frequencies, usually around 1 Hz, and has a maximal rate of about 10 Hz, but only for very brief periods. The second unusual feature of the climbing fiber input is that it produces a massive depolarization of the Purkinje cell. This effect occurs partly because each firing of an olivary cell consists of one to four spikes in very close succession, but mainly results from the extensive synaptic contacts between climbing fiber and Purkinje cell. This depolarization generates in the dendrites a prolonged inward calcium current, which may last for several hundred milliseconds, and in the cell body an action potential followed by one to five smaller, secondary spikes. This response of the Purkinje cell to climbing fiber input is often called a complex spike to distinguish it from the normal action potentials, the 'simple spikes' generated by the parallel fiber input (Figure 4.7). The complex spike is often followed by a 15- to 30-ms pause in simple spike firing and was termed an inactivation response by Granit and

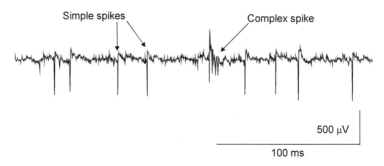

FIGURE 4.7. Simple and complex spikes. Extracellular recording from a Purkinje cell shows the characteristic response to an impulse in the climbing fiber, the 'complex spike', and the normal action potentials, 'simple spikes'.

Phillips (1956). Purkinje cells fire simple spikes at a rate that is highly variable between 50 and 100 Hz. This normal activity of the Purkinje cell is to a large extent determined by the parallel fiber input, but Purkinje cells also have an internal spike-generating mechanism and will fire at high rates even when inputs are totally absent.

Descriptions of cerebellar climbing fiber function are still incomplete and controversial (Simpson et al., 1996). Partly because the climbing fibers fire at such low rates, it has been suggested that their main function may be to control long-term excitability of the Purkinje cells. There is evidence both for a general control of excitability and for a synapse-specific modulation. Lesions of the olive, or cooling of the climbing fibers, cause a rise in tonic firing of Purkinje cells and a virtual shutdown of cerebellar output (Colin et al., 1980; Montarolo et al., 1982). Conversely, if climbing fiber frequency is increased beyond about 3 Hz by direct stimulation, the spontaneous activity of the Purkinje cells can be progressively reduced until it is completely suppressed (Rawson & Tilokskulchai, 1981; Andersson & Hesslow, 1987a, b). A synapse-specific form of long-term depression (LTD) has also been demonstrated. Climbing fiber inputs that occur in temporal conjunction with particular parallel fiber inputs induce LTD of the synaptic efficacy of the active parallel fiber synapses. This form of plasticity underpins several models of cerebellar learning. Cerebellar LTD is discussed in greater detail at the cellular level in Chapter 3 of this volume.

There are several other views of climbing fiber function, including suggestions that they select or regulate ongoing activity in Purkinje cells. Because neurons in the inferior olive are electrotonically coupled by gap junctions (Sotelo et al., 1974; Llinas et al., 1974) and their membrane properties give them intrinsic oscillatory behavior (Llinas & Yarom, 1981), it has been suggested that climbing fibers may synchronously activate groups of Purkinje cells to elicit a coordinated and appropriately timed motor output (Llinas & Welsh, 1993).

There are also serotonergic and noradrenergic afferents to the cerebellar cortex. These inputs are highly divergent and therefore probably have a low information content. They are usually thought to have a modulatory role, as elsewhere in the brain.

Cerebellar Cortex: Zones and Microzones

In the cortex, climbing fiber connections form a pattern of strips that run orthogonally through the transverse cortical lobules (Andersson et al., 1987; Voogd, 1992). In the midline, these strips or "zones" are parasagittal, but they deviate into more complex shapes in the hemispheres where the axes of the folia diverge. Each zone receives climbing fibers from a distinct part of the olive and projects to a specific target cerebellar (or vestibular) nucleus. In Voogd's original scheme, there were four zones on each side; zone A is located medially in the vermis and zones B, C, and D are progressively more lateral. Zones A, B, C, and D project to the fastigial, vestibular, interpositus, and dentate nuclei, respectively.

Further analysis has extended Voogd's original scheme. The zones have been studied in greatest detail in lobules IV and V of the anterior lobe of the cat, lobules that are particularly important for hindlimb and forelimb control. The C zones, in particular, have been shown to be quite complex. They can be subdivided into C1, C2, and C3, and there are additional Cx and Y zones. One cell group in the dorsal accessory olive projects to the lateral C3 and the Y zones, another group to medial C3 and C1, and all three cortical zones project to the anterior interpositus nucleus. In contrast, the C2 zone receives its climbing fiber input from the medial accessory olive and projects to the posterior interpositus nucleus. Where they have been analyzed, other lobules show a basic conformity with the lobule IV/V scheme, but there can be variations from it (see Figure 4.8 for a full scheme).

Each cerebellar cortical zone may be subdivided further into microzones, oriented in narrow strips transverse to the folia and parallel with the dendritic trees of the Purkinje cells. Each microzone consists of a functionally homogenous group of Purkinje cells that controls a single muscle or muscle group. Each Purkinje cell in a microzone has an olivary input with a specific somatosensory receptive field related to the activity of that muscle group. The principles of this arrangement are illustrated in the schematic diagram of the C3 zone through lobules IV and IV (Figure 4.8). It can be seen to contain orderly sets of microzones that control muscle groups within lower and upper body regions.

The Olivo-Cortico-Nuclear Module

As we have seen, Purkinje cells of the cerebellar cortex receive climbing fiber inputs from specific regions of the inferior olive and their efferent axons terminate in specific areas of the cerebellar nuclei. As well as containing excitatory neurons projecting in to the motor system, the cerebellar nuclei also contain neurons that project back into the inferior olive. This nucleo-olivary projection was assumed to be excitatory until it was discovered that electrical stimulation of the superior cerebellar peduncle, where the nucleo-olivary fibers pass, caused a strong suppression of the olivary neurons (Hesslow, 1986). Consistent with this physiology, olive-projecting neurons in the interpositus nucleus contain the inhibitory neurotransmitter GABA (Nelson & Mugnaini, 1989). Further studies have confirmed the inhibitory nature of the nucleo-olivary projection (Andersson & Hesslow, 1987a, b) and that the inhibition is dependent on GABA (Andersson et al., 1988).

The inhibitory projection from the cerebellar nuclei to the inferior olive also seems to respect the microzonal organization. A group of neurons in the interpositus nucleus inhibit those olivary cells that project to the Purkinje cells con-

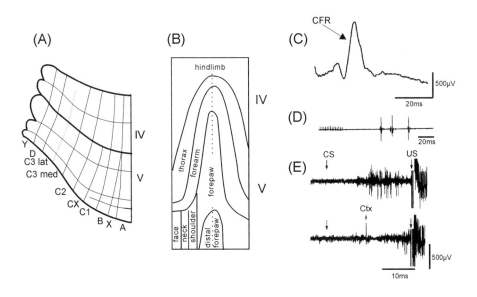

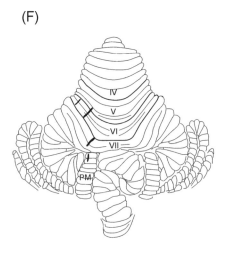

trolling the same interpositus neurons, but they do not inhibit olivary cells projecting to adjacent microzones controlling antagonistic muscles (Andersson & Hesslow, 1987a). This entire olivo-cortico-nucleo-olivary loop is now recognized as a cerebellar processing module.

Eyeblink Control Regions of the Cerebellum

From the brief description of cerebellar anatomy and physiology given here, it is clear that the microzonal organization of the cerebellum determines which cerebellar regions control different muscle groups. In later parts of this chapter, we consider how the cerebellum contributes to the regulation and conditioning of eyeblink responses. Therefore, we need to know which parts of the cerebellum regulate the face musculature and eyeblink muscles in particular.

The muscles of the face and neck are mainly controlled from microzones in lobules HVI and HVII (the hemispheral parts of lobules VI and VII of the posterior lobe). Although most face microzones have not been mapped in detail, eyeblink control areas in cats have been fully mapped and characterized using two defining characteristics. First, eyeblink microzones have climbing fiber inputs that can be activated by periocular stimulation; (see Figure 4.8C) and, second, electrical stimulation of these areas of the cerebellar cortex causes a localized inhibition of eyeblink-controlling neurons in the anterior interpositus nucleus (AIP), followed by a rebound excitation, and indirectly elicits a long-latency blink (Hesslow, 1994b) (see Figure 4.8D). An additional finding, discussed in more detail later, is that stimulation of these eyeblink control areas also completely suppresses an ongoing conditioned eyeblink response (Hesslow, 1994a) (see Figure

◄─────────────────────────────────────

FIGURE 4.8. Cerebellar zonation. (**A**) Cerebellar cortical zones in lobules IV and V of the cat cerebellar cortex as defined by their different olivary inputs. Zone A is medial and zone Y is lateral. (**B**) Detailed somatotopic organization of climbing fiber input to the C3 zone in lobules IV and V of the cat. The lateral and medial parts (which have different climbing fiber input) are separated by a *dashed line*. Only a very small part of lobule V contains a face representation. (**C**) Surface recording from lobule VI of the C3 zone of cat. Averaged field potential shows characteristic mossy and climbing fiber responses (*CFR*) to periocular stimulation. (**D**) EMG recording from a cat upper eyelid shows a long-latency blink response following a 40-ms train of stimuli to the site in cerebellar cortex from which the record in **C** was obtained. (**E**) Superimposed EMG records from the eyelid of a cat showing conditioned responses to a forelimb *CS* (*upper trace*). Lower trace shows the suppression of the CR by a brief electrical stimulus to the cerebellar cortex (Ctx) in the C3 eyeblink control area. (**F**) Diagram of the cerebellum of the cat with the cerebellar surface folded out in one plane (after Larsell 1970). Four eyeblink control areas, indicated in *black,* were found in the C1, C3, and Y zones of lobule HVI and HVII and one in the C3 zone in the paramedian lobe. These areas in the cat all receive their climbing fiber input from the dorsal accessory olive (DAO) and project to the anterior niterpositus nucleus (*AIP*), in good agreement with the findings in our lesion and anatomy studies in rabbits.

4.8E). The locations of the cat eyeblink control areas are shown in Figure 4.8F. Of these, the medial area in lobule HVI and the paramedian area are in the C3 zone. The area in lobule HVII is within the C1 zone, but the zonal properties of the lateral area in HVI have yet to be clarified. It may be in a lateral part of C3 or the Y zone.

There are other areas receiving climbing fiber input activated by periocular stimulation in the B and C2 zones, but they are unlikely to be involved in eyeblink control. The B zone projects to the lateral vestibular nucleus. The cells in the C2 zone, which receive climbing fiber input from the face, also receive bilateral input from hindlimbs and forelimbs as well as auditory input. Furthermore, electrical stimulation of these areas does not produce eyeblink responses. So, when interpreting electrophysiological recording studies of eyeblink/NMR conditioning, it must be borne in mind that climbing fiber responses activated by a periocular US are not sufficient to identify those Purkinje cells that control eyeblinks and that may, therefore, be critically involved in conditioning.

Theories of Cerebellar Function

Since the earliest studies (by Flourens, 1824 and Luciani, 1915), it has been known that cerebellar damage disturbs not only voluntary movements but also postural, proprioceptive, cutaneomuscular, and ocular reflexes (Bloedel & Bracha, 1995; Welsh & Harvey, 1992). As we have seen, the cerebellum receives and organizes an extensive range of afferent inputs and distributes its outputs selectively to different efferent targets. However, the anatomic and physiological similarity of every cerebellar cortical region indicates a consistent type of information processing within the olivo-cortico-nuclear modules distributed across the whole cerebellum.

Many models suggest that this common cerebellar processing principle involves a mechanism for motor learning (Albus, 1971; Gilbert, 1975; Ito, 1982; Marr, 1969). All these models suggest that the mossy fiber inputs to the Purkinje cell convey information about the context within which a movement is made, and the climbing fiber input to the Purkinje cell instructs a change in efficacy of currently active, context-encoding parallel fiber/Purkinje cell synapses. So, when the context occurs again, the changed synaptic strengths alter the probability of firing of the Purkinje cell such that a correct movement is made. The first of these theories (Marr, 1969) described how new voluntary movements might be controlled initially by the cerebral cortex which, at the same time, instructed olivary activity. So, with repeated practice, the new movement in its correct context would be learned by the cerebellum. Later theories by Albus (1971) and Ito (1972) freed the olive from a dependence upon cerebral cortical instruction and suggested that the olive coded an error signal that allowed the cerebellum to learn independently. Gilbert showed how a specific pattern of Purkinje cell activity might be learned and how the noradrenergic input to the cerebellum might contribute to reinforcing the learning (Gilbert, 1974, 1975). These later theories are applicable both to the learning of voluntary movements and also to the calibration and maintenance of existing reflexes, and they have been widely adapted and extended (see Houk et al., 1996; Smith, 1996; Thach, 1996, for recent reviews).

Cerebellar lesions do not prevent all movement; they cause movements to be inaccurate, clumsy, and uncoordinated with poor estimation of the necessary forces, metrics, and timing. The theories of cerebellar function just outlined suggest that memories for movements and learned calibrations of reflexes are stored within the cerebellum. Could the movement deficits seen after cerebellar lesions be due to a loss of these learned routines and calibrations?

The discovery (McCormick et al., 1982a) that cerebellar lesions disrupt NMR conditioning contributed very importantly to the cerebellar learning debate. These lesions appeared to leave the unconditioned response unchanged but the conditioned response, recently learned through an associative mechanism, was lost. These studies, and many that followed it, have contributed importantly to our understanding of cerebellar function and associative learning mechanisms.

Cerebellar Lesions and Eyeblink/NMR Conditioning

Lesions of the Cerebellar Nuclei Abolish NMR Conditioning

After the initial finding that a large cerebellar lesion abolished NMR conditioning (McCormick et al., 1982a), a series of experiments from several research groups used smaller lesions to refine our understanding of the essential cerebellar circuitry. A lesion confined to the cerebellar nuclei in the dentate/interpositus nucleus region completely abolished conditioned responses previously established to a tone CS and prevented their reappearance with continued training (Clark et al., 1984; McCormick & Thompson, 1984a). More discrete lesions restricted to regions in each of the individual cerebellar nuclei (dentate, fastigial, anterior interpositus, and posterior interpositus) revealed that it is only the *anterior interpositus nucleus* (AIP) that is critical for NMR conditioning both for an auditory and for a visual CS (Yeo et al., 1985a). The critical region within the interpositus nucleus was further characterized using a fiber-sparing lesion produced by the excitotoxin, kainic acid (Lavond et al., 1985).

The simplest interpretation of these lesion studies is that the plasticity essential for NMR conditioning may be within the cerebellar nuclei (see Lavond et al., 1987). This suggestion requires that the relatively simple cerebellar nuclear circuitry is sufficient to generate accurately timed CRs to auditory and visual CSs and to any other potential CS. However, another possibility is that plasticity essential for NMR conditioning is within the cerebellar cortex. Because the cerebellar nuclei are the output targets of the cerebellar cortex, nuclear lesions block the transmission of information from the cerebellar cortex to the motor system and would, therefore, block the expression of cortical learning. This view is consistent with several earlier theories of cerebellar cortical function that had recognized its enormous processing capacity and suggested extensive pattern recognition capabilities (Albus, 1971; Marr, 1969; Ito, 1972) and the capacity to learn movements by specifying sequences of Purkinje cell activity over time (Gilbert, 1975). In comparison to the relatively simple nuclear circuitry, cerebellar cortical structure would allow for a much richer repertoire of stimulus discriminations and for the accurate control of conditioned response topography that is the hallmark of conditioning across a range of ISIs. However, it has proven to be a difficult problem to identify the relative contributions of the cerebellar cortex and deep nuclei to eyeblink/NMR conditioning.

Cerebellar Cortex Lesions and NMR Conditioning

There is general agreement that lesions of the cerebellar nuclei produce severe effects upon the expression of CRs, although opinions differ as to whether these are learning or performance deficits (see following). However, there has been much less agreement on the effects of cortical lesions upon CRs. Some studies indicated lesions of the cerebellar cortex only altered the amplitude-time course of conditioned responses but did not abolish them (McCormick & Thompson, 1984a, b; Woodruff-Pak et al., 1985; Lavond et al., 1987). In contrast, other studies indicated that conditioned responses were always affected if the lesion included the lateral parts of lobule HVI and that when the deeper regions of lobule HVI were removed, conditioned responses were abolished (Yeo et al., 1984, 1985b) over five postoperative training sessions.

These apparently varying effects of cerebellar cortical lesions led to much debate and uncertainty over the role of the cerebellar cortex in NMR conditioning. Clearly, if cortical lesions do not produce severe disruption of CRs, then memory storage cannot be within the cortical circuitry. However, several issues must be carefully considered when interpreting the cortical lesion studies. First, the cerebellar cortex is mapped according to the muscle groups it controls, the receptive field properties of its climbing fiber inputs, and, to some extent, by the modality of its mossy fiber inputs. The effectiveness of a cortical lesion must be judged, therefore, according to the extent to which it involves the critical control areas. Second, there is considerable anatomic convergence from the cortex upon the cerebellar nuclei. For this reason alone, a discrete lesion within the cerebellar nuclei is much more likely to affect all essential circuitry than a discrete lesion of the cerebellar cortex. Third, there is a crucial difference between the general effects of cortical and nuclear lesions. Nuclear lesions depress motor excitability but cortical lesions enhance it. Finally, cerebellar cortical lesion effects may depend upon the strength of conditioning established before the lesion and the extent of the postlesion testing. We consider these four issues in detail.

On Cerebellar Cortical Lesions and the Location of Eyeblink Control Areas

At the time of the first rabbit cerebellar cortical lesion studies, eyeblink control areas of the cerebellar cortex had not been mapped (Hesslow, 1994a, b), so there was no clear view whether the critical cerebellar cortical areas had been covered by the lesions. As we reviewed earlier, the cerebellum is organized into olivo-cortico-nuclear zones. The earlier finding that the anterior interpositus (AIP) is critical for NMR conditioning identifies those cortical areas that might contribute. Cortical zones C1, C3, and Y have been clearly identified in anatomic and physiological studies of other species to project to AIP and to receive olivary inputs from the dorsal accessory olive (DAO). These zones extend through most of the cerebellar folia just lateral to the vermis. So, if the cerebellar cortex is important for eyeblink/NMR conditioning, it follows that the critical areas must lie within the C1, C3, and Y cortical zones. Cats have a main C3 eyeblink area in lobule HVI (with a small rostral extension into lobule V), a smaller area in HVI that is either lateral C3 or Y, and two additional eyeblink control areas in C1 or C3 regions of HVII and the paramedian lobe. Our original studies (Yeo et al., 1984, 1985b) indicated that lesions of the entire lobule HVI, including the depths of the lobule, abolished CRs to light and white noise CSs over five daily training sessions. If the distribution of eyeblink control areas is similar in cats and rabbits, then these lesions would have affected the two HVI control regions consistent with the loss of CRs. It is less clear whether cortical lesions in the other early studies would have included these control regions in any individual subject.

In a further study, Yeo and Hardiman (1992) reexamined cortical lesion effects and found that extended postoperative training could lead to some recovery of function after lesions of HVI alone, but lesions that additionally included parts of HVII and HVIII (parts of ansiform lobe and paramedian lobe) and lobules IV/V produced more sustained impairments. Assuming similarities with the location of eyeblink control areas in the cat, these lesions would have included three or perhaps all four eyeblink control areas, and they implicate the eyeblink control areas outside HVI as subsidiary circuitry in the learning process. The effective cortical lesions do not necessarily invade critical white matter surrounding the cerebellar nuclei because fiber-sparing, kainic acid lesions of the cerebellar cortex also produce effective losses of CRs (Hardiman & Yeo, 1992).

On Cerebellar Cortical Convergence to the Cerebellar Nuclei

There is considerable anatomic convergence from cerebellar Purkinje cells upon the cerebellar nuclei. For example, the AIP receives inputs from the nonadjacent C1, C3, and Y cortical strips that run through several lobules. Furthermore, the organization of these convergent inputs is orderly so as to produce motor maps of the body within the cerebellar nuclei (Thach, 1996), and within these body maps will be face control areas. A discrete lesion of the cerebellar nuclei is, therefore, more likely to affect all the eyeblink control circuitry than a discrete lesion of the cerebellar cortex, which would have to cover several cortical territories.

On Cerebellar Lesions and Excitability Changes

The cerebellum has a relatively steady, or tonic, output that provides a background level of drive upon the motor system. Upon these tonic outputs, more rapidly changing, or phasic, outputs are superimposed to create or modulate active movements. In the context of eyeblink conditioning, baseline levels of blink excitability will be influenced by tonic drive upon the motor circuitry and conditioned responses will be mediated by phasic commands.

The pathway from the cerebellar nuclei to the red nucleus and brain stem eyeblink control circuitry is excitatory. Lesions of the cerebellar nuclei remove tonic, excitatory drive upon the eyeblink circuitry and potentially depress the expression of CRs. In contrast, the cerebellar cortex is inhibitory upon the cerebellar nuclei so lesions of the cortex disinhibit the eyeblink reflex circuitry and, if the circuitry controlling the CR is distributed across several cortical microzones, lesions of only some of the areas involved might even increase the probability of CR expression. The different changes in general excitability caused by nuclear and cortical lesions increases the probability that the former will be more effective in impairing eyeblink CRs.

In a series of studies of eyeblink conditioning in rabbits, Mauk and colleagues have described some novel effects of cerebellar cortical lesions that relate to their disinhibitory effects. They showed that cortical lesions abolish long-latency, adaptive CRs and produce short-latency, CS-driven responses in previously trained subjects (Perrett et al., 1993; Perrett & Mauk, 1995). Because these responses are only seen subsequent to associative CS–US pairings they may be considered CRs, but they also have unusual properties in that they cannot be extinguished and their timing is not adapted to the CS–US interval. It has been suggested that these short-latency responses are mediated by plasticity in mossy fiber to cerebellar nuclei synapses. This plasticity is unmasked by the disinhibitory effects of the cortical lesion, and eyeblink conditioning may normally involve plasticity within the cortex and its cerebellar nuclear targets (Mauk, 1997; Garcia et al., 1999). This is an attractive proposition that will require further clarification. First, it is not clear why the cortical lesions effective in unmasking these short-latency responses are in regions of the anterior lobe outside those known to control eyeblink responses. Second, it is not yet certain that the short-latency responses are mediated by nuclear plasticity. The disinhibitory effects of the cerebellar cortical lesions could unmask excitability changes elsewhere in the eyeblink control circuitry. The motor cortex, which has more potent control of the eyeblink than the NMR, is known to be engaged in learning short-latency eyeblink responses in the cat (Brons & Woody, 1980) and may contribute to these unmasked responses.

On the Strength of Conditioning and Its Resistance to Unilateral and Bilateral Lesions

The sensitivity of conditioned responses to cerebellar cortical lesions might relate to the amount of preoperative conditioning. If subjects are highly overtrained before receiving large unilateral cortical lesions in HVI and HVII, impairments of CR frequency can be overcome with a few additional sessions of retraining (Harvey et al., 1993). This finding could be taken as evidence for storage of memories outside the cortex once learning is very strongly instated. However, further analysis has shown that *bilateral* cortical lesions in highly overtrained subjects produce major impairments of CR frequency that do not improve even with continued postoperative training (Gruart & Yeo, 1995), although low response levels remain. Importantly, the few residual CRs seen after such lesions are highly variable in their latency to peak, and their accurate timing does not recover. That bilateral lesions impair conditioning more completely than unilateral lesions is consistent with electrophysiological identification of eyeblink control from ipsilateral and contralateral cerebellar cortex via decussating output pathways (Ivarsson & Hesslow, 1993).

Lesions of Cerebellar Efferent Pathways

The cerebellar nuclei have different efferent targets. The fastigial nucleus projects mainly to vestibular and reticular brain stem regions. In contrast, the interpositus and dentate nuclei have major forward projections, via the superior cerebellar peduncle, to the contralateral red nucleus and ventrolateral thalamus. The red nucleus projection is to the source neurons of the contralaterally descending rubrobulbar and rubrospinal tracts. The ventrolateral thalamus target neurons project into the motor cortex to influence the corticobulbar and corticospinal tracts. Lesions of the superior cerebellar peduncle reveal that the dentate or interpositus nucleus efferents are essential for the expression of NM CRs (McCormick et al., 1982b; Rosenfield & Moore, 1983). Lesions of the red nucleus and of the rubrobulbar tract, which carries efferent fibers from the red nucleus down to the premotor blink areas of the brainstem, identify the cerebello-rubro-bulbar pathways as essential for the expression of NMR conditioning (Rosenfield & Moore, 1983; Rosenfield et al., 1985). These findings are consistent with an earlier decerebration study that indicates that thalamic and other forebrain circuitry is not essential for conditioning (Mauk & Thompson, 1987).

Cerebellar Lesions Define Circuitry Essential for Conditioning

If the many cerebellar lesion studies are considered with reference to the critical issues described here, it is clear that both the cerebellar nuclei and cortex play essential roles in the development and production of normal conditioned responses. Lesion studies identify circuitry necessary for the expression of conditioned responses, and they serve as guides for anatomic and physiological studies that begin to characterize the specific roles of the cerebellar cortex and nuclei in conditioning.

CS and US Information Converges in Lobule HVI

Lesion studies revealed that cortical lobule HVI is particularly important for the production of conditioned eyeblink/NM responses with appropriate amplitude and timing. Neuroanatomic tracing of the projections to lobule HVI has revealed a striking convergence of information relating to the CS and US (Yeo et al., 1985c). The inferior olive projection to HVI is from the medial parts of the rostral dorsal accessory olive (DAO) and medial parts of the rostral principal olive (PO). These olivary regions receive input from the trigeminal nucleus (Berkeley & Hand, 1978), especially from pars interpolaris (Vi) (van Ham & Yeo, 1992) and so provide face somatosensory information to HVI (Gellman et al., 1983; Miles & Wiesendanger, 1975a, b). Additionally, lobule HVI receives rich mossy fiber inputs from several divisions of the trigeminal system (Yeo et al., 1985c; van Ham & Yeo, 1992; Rosenfield & Moore, 1995) directly as mossy fiber inputs. Thus, both climbing fiber and mossy fiber inputs can provide lobule HVI with information related to a periorbital shock or corneal airpuff US.

In addition to the face somatosensory inputs, there is a powerful input to lobule HVI from dorsolateral and lateral divisions of caudal parts of the pontine nuclei, which themselves are targets for visual information from the superior colliculus (Holstege & Collewijn, 1982; Wells et al., 1989) and for auditory information from the inferior colliculi (Kawamura, 1975; Burne et al., 1981). These pontocerebellar pathways can provide information about auditory and visual CSs to HVI during NMR conditioning.

A Cerebellar Cortical Conditioning (CCC) Model

In an early anatomic study, the convergence of US-related information through the climbing fiber input with CS-related information through mossy fiber inputs was shown to be consistent with a simple implementation for NMR conditioning of the Marr–Albus models of cerebellar learning (Yeo et al., 1985d). We now describe here the assumptions within this simple model and outline some of its properties.

Model Assumptions

Learning Occurs in Eyeblink Control Purkinje Cells

The first assumption is that learning occurs in a set of eyeblink control Purkinje cells. In Marr's original formulation, the fundamental associative unit of the model was a single Purkinje cell. We now know, however, that Purkinje cells are organized into sagittally oriented microzones containing hundreds of Purkinje cells. Each microzone has similar climbing fiber inputs but a range of different parallel fiber inputs. Movements can be controlled by more than one microzone. There are at least four microzones in the C1 and C3 zones (possibly also in other zones), which control eyelid and NM movement via the anterior interpositus nucleus and that receive climbing fiber input from the cornea. Each Purkinje cell within these microzones receives climbing fiber input derived from the cornea and the periorbital area, and we propose that learning occurs across all of them.

Mossy/Parallel Fibers Transmit the CS and Learning Involves Change at Parallel Fiber–Purkinje Cell Synapses

The second assumption is that mossy/parallel fibers transmit the CS. Learning involves changes in the efficacy of sets of parallel fiber synapses on groups of eyeblink control Purkinje cells. A large proportion of mossy fibers originates in the pontine nuclei, which receives direct sensory input from all sensory modalities and from sensory and motor areas in the cerebral cortex. Auditory, visual, or somatosensory CSs are represented by these inputs to the cerebellum. Because each Purkinje cell has a very large number of such synapses, about 80,000 in cats (see Ito, 1984 for a discussion of the numerical aspects of cerebellar circuitry in different species), the eyeblink control groups can learn to respond to, and discriminate between, a large number of mossy/parallel fiber input patterns.

Parallel Fiber–Purkinje Cell Synaptic Change Is "Instructed" by Climbing Fiber Inputs Related to the US

The third assumption, as in previous theories (Marr, 1969; Albus, 1971), is that climbing fibers provide "teaching" signals to the Purkinje cells. Here, they are driven by the US and they enable a change in the synaptic weights of parallel fiber to Purkinje cell synapses encoding the CS. Only CS-activated parallel fibers synapses that are active in close temporal contiguity with US-driven climbing fiber inputs undergo changes in efficacy. Because the sensory receptive fields of climbing fibers to the eyeblink control areas closely match areas of the face activated by the US, stimuli that elicit blinks will only instruct synaptic efficacy changes on eyeblink control Purkinje cells.

Although it may seem straightforward to assign to the climbing fibers the role of transmitting the US information, this is not a trivial assumption. Mossy fibers also transmit strong input from the periorbital area to Purkinje cells in the relevant microzones, and there are no *anatomic* reasons for preferring the climbing fibers as the US pathway.

There Is CS Processing at Mossy Fiber–Granule Cell Synapses

A crucial feature of the Marr–Albus theory of the cerebellum is the information processing that occurs at the interface between mossy fibers and granule cells. It has been calculated that there are about 10^{10} to 10^{11} granule cells in the human cerebellum (that is, roughly as many as there are neurons in the rest of the brain), which means that there must be a vast convergence as well as divergence at these synapses. Each mossy fiber influences, via the granule cells, thousands of Purkinje cells. A single Purkinje cell receives information from 80,000 granule cells, each of which is innervated by five mossy fibers. These numbers apply to single Purkinje cells, and they increase if all the Purkinje cells in a microzone are considered. The transmission from mossy fibers to granule cells is modulated by inhibition from Golgi cells, which are excited by mossy fibers and parallel fibers and inhibit transmission in the granule cells; this tends to reduce variation in the total granule cell activity. Marr (1969) showed how Golgi cell control of granule cell transmission powerfully enhances the pattern-recognizing capacities of Purkinje cells. (The details of these mechanisms are beyond the scope of this review. For a clear exposition, see Tyrell & Willshaw, 1992.)

Lacking in this formulation of the model is a cellular learning mechanism. Such mechanisms are the subject of Chapter 2 of this volume and are not covered here.

Cortical and Nuclear Plasticity

Because both mossy and climbing fibers send collaterals to the deep cerebellar nuclei, it might be considered arbitrary to suggest a cortical rather than a nuclear learning model. There are strong theoretical arguments for proposing the cerebellar cortex as the main site of plasticity, however. Put simply, the cerebellar cortex has features that make it ideally suited for associative learning. First, as noted earlier, it has the requisite convergence of sensory input. Many features of the mossy fiber input to the deep nuclei are unknown, but it seems highly unlikely that the degree of sensory convergence is anything like that of the cerebellar cortex. Second, the expansion of the mossy fiber input to the granule cells and the ability of the Golgi cells to regulate the transmission at these synapses greatly increases the pattern recognition capacity of the cortical network (Tyrell & Willshaw, 1992).

It should be noted, however, that one consequence of learning in the CCC model is that, if more and more parallel fiber to Purkine cell synapses become depressed as more and more motor sequences are learned, the overall output from cerebellar cortex would decline. A decline in tonic output from Purkinje cells would lead to disinhibition of nuclear neurons and increased levels of tone in the motor system. A counterbalancing plasticity at cerebellar nuclei neurons, either at mossy fiber synapses or by decreased overall neuronal exciatbility, could maintain a balanced output to the motor system.

In another example of cerebellar-dependent motor learning, the modification of gain in the vestibulo-ocular reflex (VOR), there is evidence of plasticity within the cerebellar cortical control area, the flocculus (Ito, 1982, 1998), and in the brain stem vestibular targets of the floccular Purkinje cells (Miles & Lisberger, 1981; du Lac et al., 1995). This evidence has led to the idea that there is learning-related plasticity within both the cerebellar cortex and in its cortical target neurons for the VOR. Earlier, we reviewed some lesion experiments that may offer support for nuclear plasticity in eyeblink conditioning evidence relevant to these two-site models (Perrett et al., 1993; Perrett & Mauk, 1995). By analogy with the VOR, it has now been proposed that there is cortical and nuclear plasticity supporting eyeblink/NMR conditioning (Raymond et al., 1996; Mauk & Donegan, 1997; Mauk, 1997; Medina & Mauk, 1999). The main feature of these models is that mossy fiber to nuclear synapses are potentiated to provide CS-activated drive to the eyeblink circuitry whereas the temporal form of the CR is sculpted on this output by inhibitory modulation from the Purkinje cell. The timing of this inhibitory sculpting is learned within the cerebellar cortex from the CS–US association. These models, therefore, point to critical plasticity in the cerebellar cortex and nuclei.

Commentary

The earlier theorists did not directly suggest that the cerebellum might be the site of learning in classical conditioning, but they came close. The type of learning postulated by Marr to occur in the cerebellum fits a description of classical conditioning of motor responses. For instance, he suggested that "visual cues and information about mood and so forth can form enough of a context actually to initiate an action" (Marr, 1969, p. 468), but he did not explicitly connect this to classical conditioning. Marr did make a suggestion about "learned conditional reflexes," but by this he meant ordinary reflexes that could be suppressed or facilitated depending on the context. The context would not elicit a response, according to this scheme, but rather "enable" a US to elicit a UR. Albus (1971) explicitly adopted the terminology of Pavlovian conditioning, although he used it metaphorically in a description of the cellular events underlying learning. Thus, he suggested that "the inactivation response pause in Purkinje cell spike rate is an unconditioned response (UR) in a classical learning sense caused by the unconditioned stimulus (US) of a climbing fiber burst. It is further hypothesised that the mossy fiber activity pattern ongoing at the time of the climbing fiber burst is the conditioned stimulus (CS). If this is true, the effect of learning should be that eventually the particular mossy fiber pattern (CS) should elicit a pause (CR) in Purkinje cell activity similar to the inactivation response (UR) that previously had been elicited only by the climbing fiber burst (US)" (Albus, 1971, p. 44).

Cerebellum and Conditioning: Permanent and Reversible Lesions Define Necessity and Function

Cerebellar Lesions: Support for Cerebellar Learning?

We have described how lesions of the AIP or appropriate regions of the cerebellar cortex impair conditioned NM/eyeblink responses, reviewed anatomic evidence for convergence of CS- and US-related information in the cerebellum, and incorporated these findings within a simple implementation of the Marr–Albus learning model applied to NM/eyeblink conditioning. Before developing the ideas in this model further, it is necessary to consider carefully our central assumption: that cerebellar lesions disturb circuitry involved in learning and storing CS–US associations. Is it possible that cerebellar lesions do not damage memories or prevent learning? Do such lesions only prevent the expression of learning that develops elsewhere in the brain? In other words, do cerebellar lesions produce performance, rather than learning, deficits?

In Figure 4.9, we show a simplified view of some of the cerebellar and brain stem circuitry important in eyeblink conditioning. Additionally, there is an extracerebellar location where the "performance hypothesis" suggests that conditioning might occur. The exact location of this putative site is not specified in the performance hypothesis, but it may be considered as some brain stem region where CS and US information converge. There is an obvious question. If conditioning is mediated by this extracerebellar circuitry, why do cerebellar lesions impair CRs? We have previously described how lesions can disrupt tonic, regulatory output from the cerebellum. The performance hypothesis suggests that loss of this tonic outflow to the premotor and motor components of the blink reflex reduces excitability in the premotor and motor elements of the reflex. So, a neural instruction for the CR, produced from within the extracerebellar circuit, now cannot sufficiently raise excitability in the depressed reflex output stages and so the motoneurons do not fire. What evidence supports this hypothesis?

Welsh and Harvey (1989) carefully analyzed the effects of interpositus nucleus lesions on both the CR *and* the UR. As expected, CRs were greatly impaired but there were also subtle changes in the UR. When tested with unpaired presentations of a low-intensity air puff US, it was seen that URs were slightly weaker after IP lesions. UR amplitude was slightly, although not significantly, lower and the UR rise times were significantly longer. As the CR and UR share a common motor pathway and the ability to express URs is clearly impaired, this is evidence for a general "performance" deficit that may underlie the loss of CR expression. Although effects upon the UR were rather weak compared with the total loss of CRs, this is because CRs to a tone CS are inherently weaker responses than URs driven by periocular stimuli. Welsh and Harvey (1989) suggested that theirs was evidence against a cerebellar learning mechanism. However, other evidence must be considered before reaching this singular conclusion.

Manipulations of the AIP produce variable effects on the UR. One further study using pharmacological inactivation of AIP (Bracha et al., 1994) supported the previous observation of a UR deficit but another lesion study refuted them (Steinmetz et al., 1992). It seems that depression of UR amplitude is not inevitably associated with CR losses. Indeed, cortical lesion studies provide clear evidence to the contrary. Because the cerebellar cortex is inhibitory upon the cerebellar nuclei, cerebellar cortical lesions would be expected to produce increases in reflex excitability, an opposite action to that produced by AIP lesions, and this is the case. Lesions of lobule HVI greatly *increase* UR amplitudes while, at the same time, abolishing or depressing CR amplitudes (Yeo & Hardiman, 1992; Hardiman

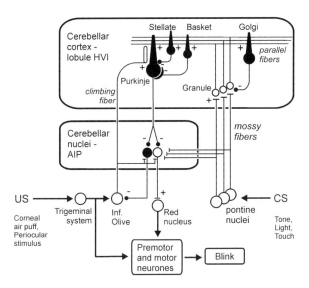

FIGURE 4.9. Cerebellar cortical conditioning model. A simplified view of the crucial elements in the cerebellar cortical conditioning (CCC) model in which many elements such as interneurons and collaterals of cerebellar afferents are omitted for clarity. The unconditioned reflex pathway is from the periorbital area or the cornea via the trigeminal nucleus to the premotor and motor neurons producing the blink. The US pathway is from the trigeminal nucleus to the inferior olive and via climbing fibers to the Purkinje cells of eyeblink microzones of the cerebellar cortex. The CS can be one of a variety of sensory stimuli (tone, light, touch, etc.). CS information is transmitted through mossy fibers from an appropriate precerebellar nucleus (e.g., pontine nuclei for a tone *CS*) to granule cells and then through parallel fibers to the Purkinje cells. The inhibitory output from the Purkinje cells is via the anterior interpositus nucleus. From there, excitatory drive via the red nucleus to premotor elements and then motor neurons in the accessory abducens and the facial nuclei produce the nictitating membrane and eyelid responses, respectively. The AIP also sends an inhibitory projection to the inferior olive.

& Yeo, 1992; Gruart & Yeo, 1995). That the performance of CRs and URs can be so clearly dissociated with cortical lesions is strong evidence against a simple, performance hypothesis for cerebellar function in conditioning.

A further problem with the "performance" hypothesis is the assumption that deficits in the UR must arise from excitability changes in the reflex output stages that are common to both the CR and the UR. This may not be true. Through projections via the red nucleus, the cerebellum can also modulate transmission within the trigeminal nuclei (Davis & Dostrovsky, 1986), which form part of the afferent limb of the blink reflex but may not be part of the CR pathway.

Overall, the arguments underlying the performance deficit hypothesis have been valuable in clarifying our analysis of the role of the cerebellum in conditioning. Permanent lesions, including those seen in clinical cases, have been vital for localizing circuitry that is necessary for conditioning, but cannot determine its mechanism. To do that, we need quite different tools.

Reversible Cerebellar Inactivations Define Circuitry That Mediates NMR Conditioning

Permanent lesion studies alone cannot resolve whether the cerebellum actively contributes to the acquisition and storage of motor memories or whether the cerebellum only regulates the performance of movements learned in other neural circuitry. Recent studies have addressed this problem using reversible lesion techniques. Conditioning trials can be given during a localized, functional inactivation of the cerebellum. If, after the inactivation is lifted, conditioned responses are present then they must have been formed during the inactivation and it may be concluded that the cerebellum is not essential for acquisition of the learning. If conditioned responses are not evident after the inactivation, then the cerebellum is implicated in the acquisition of conditioning; there may be essential plasticity within the cerebellum itself or in circuitry supplied by cerebellar output.

Most studies to date have employed localized inactivation of the AIP, which we presume disturbs function within all affected olivo-cortico-nuclear compartments and is thus a test of general cerebellar involvement (Welsh & Harvey, 1991; Clark et al., 1992; Nordholm et al., 1993; Krupa et al., 1993; Ramnani & Yeo, 1996; Hardiman et al., 1996; Yeo et al., 1997). The Welsh and Harvey (1991) study inactivated the AIP with lidocaine and indicated that learning to a tone CS was not impaired if the subjects had previously learned to a light CS, apparently indicating that the cerebellum is not essential for conditioning (Welsh & Harvey, 1991). However, we have shown how general transfer effects during incomplete inactivation of the AIP may have produced this apparent learning (Yeo et al., 1997). All subsequent studies have shown that de novo conditioning to a tone CS is prevented using cold block, lidocaine, or muscimol infusion techniques to inactivate the AIP (Clark et al., 1992; Nordholm et al., 1993; Krupa et al., 1993; Ramnani & Yeo, 1996; Hardiman et al., 1996).

However, there are two reasons why failure to produce CRs after conditioning during AIP inactivation might not indicate a failure to condition. First, there may be a performance problem induced by long-term drug effects carrying over to the postdrug testing phase of the experiments. We tested this possibility and found that, in addition to acquisition learning, extinction learning is also completely prevented by AIP inactivation (Ramnani & Yeo, 1996; Hardiman et al., 1996). Because a failure to extinguish during the drug condition was indicated by high levels of CRs in the postdrug phase, we could be sure that there were no simple performance deficits after the inactivation either during acquisition or extinction learning. Second, there could be general, state-dependent learning (SDL) effects induced by diffusion of low levels of the drug into other brain areas directly or via the circulation. Hence, during the postdrug testing phase, the sensory properties of the CS and US will be different and, even if the subjects had learned during the drug state, no CRs would be produced. SDL effects were tested using drug infusions in a variety of sites around the AIP that would have allowed similar diffusions out of the cerebellum. Such infusions did not prevent conditioning, so SDL effects can be ruled out.

The AIP inactivation studies unequivocally provide the most powerful evidence that the cerebellum is essential for eyeblink/NMR conditioning, but they do not, in themselves, indicate that the critical plasticity is within the cerebellum. This final piece of evidence derives from the recent work by Krupa and Thompson (1995), who showed that reversible inactivation with tetrodotoxin of cerebellar efferents in the brachium conjunctivum did *not* prevent NMR conditioning. If inactivation of the cerebellum prevents learning but inactivation of cerebellar efferents does not, we have the best evidence that the essential neural mechanisms of eyeblink/NMR conditioning are within the cerebellum. However, not all of the deep nuclear efferents run forward through the more rostral parts of the brachium conjunctivum. Some turn ventrally to brain stem targets. If any such efferents escaped inactivation, it will be important to determine whether there is additional plasticity in CS and US processing circuitry afferent to the cerebellum but regulated by ventrally directed AIP efferents.

The results of AIP inactivation should not be taken as conclusive evidence for essential plasticity in the cerebellar nuclei because, as we suggested earlier, they would probably disturb normal function within the entire olivo-cortico-nuclear circuitry. This idea is consistent with the finding that lidocaine inactivation of the inferior olive also prevents acquisition (Welsh & Harvey, 1998) and with the suggestion that there is essential cerebellar cortical plasticity. However, a necessary step toward confirming the CCC model would be to show that cortical inactivation also prevents acquisition. In recent studies, we have used the ionotropic glutamate receptor antagonist 6-cyano-7-nitroquinoxaline-2,3-dione (CNQX) to block glutamatergic transmission in the cerebellar cortex. This antagonist is relatively specific for the AMPA subtype of glutamate receptors, thought to be the main excitatory receptor at parallel fiber–Purkinje cell synapses and at mossy fiber–granule cell synapses. We have now shown that local inactivations of lobule HVI with CNQX abolish the performance of previously established CRs (Attwell et al., 1999a) and, with inactivation prior to conditioning, they prevent acquisition (Attwell et al., 1999b).

Testing Cerebellar Learning Models: Recording Studies

To evaluate the CCC model, it is important to know how neuronal elements in the cerebellum and in the afferent and efferent pathways behave during conditioning. Because the efferent pathway from the cerebellum to the blink motoneurons is excitatory, the model predicts that neurons in the anterior interpositus nucleus increase their firing just before and during the CR. As the interpositus neurons are under inhibitory control from Purkinje cells in the cortex, the latter would be expected to show a reverse firing pattern, that is, a depression of firing just before, and during, the CR.

Methodological Considerations

Single and multiunit recordings have been made from many CNS structures during conditioning. Neural activity elicited by the CS, which is correlated with the conditioned response and often preceding it, has been described for several brain locations such as the interpositus nucleus, the cerebellar cortex, the trigeminal nucleus, the red nucleus, the pontine nuclei, the thalamus, and the hippocampus. The finding of a neural CR correlate in some structure has often been proclaimed as evidence that that particular structure is involved in the learning. However, although the results of many recording studies tend to be consistent with the predictions of the CCC model, it must also be said that all these studies suffer from a number of severe methodological problems that make them difficult to interpret.

Neuronal recordings, taken in isolation, can at best reveal a pattern of correlation between behavior and neural activity, not a causal relation. For example, unit activity in the hippocampus is correlated with conditioned responses during standard delay conditioning (Berger & Thompson, 1978; Berger et al., 1980, 1983), but lesion studies have convincingly shown that such conditioning is not dependent upon the hippocampus. Thus, although we do not know why the hippocampal neurons correlate with the CR, they do not cause it.

If a certain behavior is correlated with some neuronal activity, the behavior may be the cause or the effect of that activity or both the behavior and the neuronal activity may be the effects of a common, third variable. There are several ways in which such spurious correlations between neural and overt behavior might arise. For instance, CR-correlated neural activity may reflect feedback information from the actual movement or feedback from a motor command. The latter possibility is important because such activity may precede the CR. It is often thought that if some neural activity precedes the CR, it cannot be the result of feedback information from the CR. This idea is correct, but the neural activity can be due to feedback from activity in motor or premotor elements that generate the CR. Ideally, activity in cerebellar neurons should be compared, not only to overt movement of the eyelid or NM, but also to activity in motor neurons and muscle cells that precedes the overt movement. Eyelid EMG recordings have revealed characteristic patterns in response topography that may be reflected in central command signals (Ivarsson & Svensson, 2000; Gruart et al., 2000). CR-correlated neural activity could also reflect commands to perform accompanying movements, such as a contraction of synergist or relaxation of antagonist muscles, and it could be caused by a simultaneous but independent learning of the same association by a different learning mechanism.

There is no simple method of circumventing these methodological problems, and several different techniques have been employed, but there are a couple of basic requirements that should be satisfied before any meaningful interpretation can be put on recording data. First, the neuronal elements should be identified with respect to cell type and functional role. In the case of cerebellar cortex recordings, this means that the recorded cells should be shown to be located in a microzone that controls eyeblink. Second, in the case of recordings purporting to demonstrate learning, there should be some evidence that the observed firing pattern actually results from the training. Third, there should be some information relevant to the question of the causal relation between the neuronal firing pattern and the CR. As we shall see, no studies satisfy all these requirements although some studies satisfy some of them.

Recordings from the Interpositus, Pontine, and Red Nuclei

Neuronal activity in the interpositus or dentate nucleus precedes and correlates strongly with conditioned NM responses (McCormick & Thompson, 1984b), but, for the reasons just indicated, this does not show that interpositus neuron activity caused the CRs. More persuasive evidence for causality was provided by the discrimination learning study of Berthier and Moore (1990). Two frequencies of tone were used. The CS+ was paired with the US, but the CS− was not. Some interpositus neurons responded with short-latency responses to both the CS+ and the CS−, suggesting that they were unlearned, tone-evoked responses. Other neurons responded with longer-latency activity that preceded, and correlated strongly with, conditioned NM movement. In some of these neurons, the trial-by-trial latencies correlated with the latencies of the overt NM responses. Of particular interest is the observation that some neurons, which occasionally responded on CS− trials, did so on the same trials on which there were overt, but erroneous CRs; on trials in which the animal erroneously failed to emit CRs, these same interpositus neurons also failed to respond. These findings strongly suggest that the responses were learning dependent and functionally related to the CR (Figure 4.10).

It does not follow that all the CR-correlated cells were actually driving the CR. In fact, there is reason to believe that some were not. First, some interpositus neurons increased their firing during the CR whereas others decreased their firing, so it is unlikely that both groups of neurons could be directly involved in generating the CR. Second, some neurons responded with quite strong short-latency responses as well as with long-latency, CR-related activity. If these neurons were controlling the blink, they should have generated a short-latency blink in addition to the long-latency CR, but no such responses were observed.

Neural activity strongly correlated with the CR has also been recorded in the red nucleus (Desmond & Moore, 1991). Such activity might reflect local plasticity or might be driven by cerebellar inputs. To disentangle the causal relations between neural activity at different sites, some studies have blocked activity in one location while recording activity in the other. Correlates of the CR recorded in the interpositus nucleus are not affected by lidocaine inactivation or cooling of the red nucleus (Chapman et al., 1990; Clark & Lavond, 1993). These inactivations prevent expression of the CR, so the modulation of interpositus activity by movement feedback from the periphery can be ruled out. Similarly, interpositus nucleus correlates of the CR are not affected by trigeminal inactivation, ruling out feed-forward effects from the US pathway (Clark & Lavond, 1996). Although

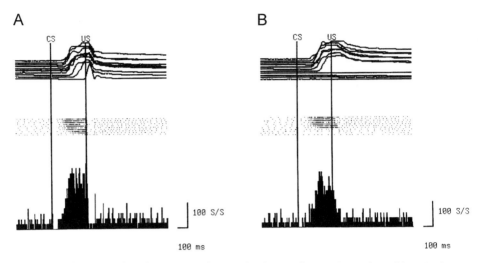

FIGURE 4.10. Activity of neurons in the anterior interpositus nucleus of conditioned rabbits. CRs, raster dot diagrams, and spike histograms from CS+ (**A**) and CS− (**B**) trials. Notice absence of spiking on the trials without CRs in (**B**). (From Berthier & Moore, 1990.)

these inactivations produce effects consistent with the CCC model they do not, of course, rule out that there may be some plasticity contributing to the CR or the red nucleus or cerebellar nuclear level.

The pontine nuclei are the major source of visual and auditory mossy fiber inputs to the cerebellum and so have been assigned a role of CS channel for these CS modalities in the CCC model. However, the pontine nuclei also receive somatosensory information via the cerebral cortex, superior colliculus, and brain stem relays. Thus, they potentially receive convergent input from the CS and the US and should be considered as a possible alternative to the cerebellum as an associative site. Recordings from the pontine nuclei in tone-conditioned subjects show short-latency, tone-evoked firing in many neurons and also long-latency, CR-correlated activity in others (McCormick et al., 1983). These findings are consistent with the CCC model but also with a model in which the site of learning is located in the pontine nuclei. Here, the pontine nuclei send a learned signal to the cerebellum, which then performs the CR. This hypothesis has been rejected by inactivation experiments. When the interpositus nucleus was inactivated by cooling while multiple-unit activity was simultaneously recorded in the pontine nuclei, Clark et al. (1997) found that short-latency CS-evoked activity remained but that the CR-related activity disappeared.

One interpretation of this finding is that the late CR-related activity is derived from the cerebellum via interposito-pontine projections (Brodal et al., 1972) and so supports the CCC model. However, the presence of ponto-interpositus connections, for long a matter of some dispute (Dietrichs et al., 1983) but now resolved in favor of such connections (Shinoda et al., 1992), means that these reciprocal excitatory connections can set up a reverberatory circuit. This loop can be interrupted by an external inhibitory input, for instance, from the Purkinje cells. Indeed, when the cerebellar cortex is removed and Purkinje cell inhibition of the interpositus nucleus is lost, electrical stimulation of part of this circuit can elicit a sustained firing (Tsukahara & Bando, 1970; Tsukahara et al., 1971). Given this circuitry, it is possible that plasticity for CR generation is located in the pontine nuclei and that interpositus inactivation abolishes CR-related activity in the pontine nuclei because it prevents setting up the necessary positive feedback loop.

In a model proposed by Houk, Buckingham and Barto (Houk et al., 1996), the learning site is located in the cerebellar cortex as in the CCC model, but they suggest that sustained firing in the interpositus during the CR depends on a nucleopontine reverberating circuit. This model was rejected by Hesslow (1996). A brief electrical stimulus applied to the area of the cerebellar cortex that controls the eyelid activates inhibitory Purkinje cells of the cerebellar cortex and briefly suppresses activity in the interpositus nucleus. When such a stimulus was applied during a conditioned eyeblink response, there was a brief pause in the eyelid EMG CR, but it quickly recovered so that the later part of the conditioned response was relatively unaffected. It therefore seems that the interpositus does not need feedback from its own output to generate or maintain the CR.

Recordings from Purkinje Cells

The CCC model predicts that Purkinje cells should decrease their firing during the later part of the CS–US interval. Several attempts have been made to test this prediction both with multiple-unit and with single-unit recordings from the cerebellar cortex. Multiple-unit recordings are insufficient for two reasons. First, the most important criterion by which Purkinje cells can be differentiated from other cell types is the presence of complex spikes. To identify complex spikes requires single-unit recordings of good quality; they cannot be seen in multiunit recordings or in 'noisy' single-unit recordings. Second, the Purkinje cells must be characterized with respect to their inputs. With a multiunit recording, it is impossible to know if reactions to various stimuli are generated by the same cell or by different cells.

Assuming that the recordings are of sufficient quality to identify units as Purkinje cells, there remains the task of demonstrating that these cells are in control of the conditioned response. As has been described, four cerebellar cortical eyeblink control areas have been identified in the cat. The two main regions in lobule HVI constitute only a small fraction of the HVI cortex. Clearly, the probability of finding the relevant Purkinje cells when randomly sampling lobule HVI is vanishingly small. Even those cells that respond with complex spikes to the US, as predicted by the model, are not necessarily the correct ones. Some Purkinje cells in the C2 zone respond with complex spikes to somatosensory input from the face, but they also respond to stimulation of virtually the whole body (and to auditory stimulation) and they need have no particular relationship with eyeblink.

The first NMR conditioning study with recordings that reliably identified simple and complex spikes is that of Berthier and Moore (1986). Many Purkinje cells responded with complex spikes to the US and some also to the CS. The recording sites were all within HVI but were not further specified, so it cannot be known what proportion of these cells were actually controlling eyeblink. Some Purkinje cells decreased their simple spike firing during the CS–US interval and in advance of the CR, consistent with the CCC model, but other cells increased their firing. This result need not contradict the model if those cells were not the ones controlling blink.

Two observations suggested that at least some of these neuronal responses were dependent upon learning the blink CR. First, the animals had been conditioned to discriminate between two different tone CSs, a CS+ that was reinforced by the US and a CS–, which was not. Many cells responded only to the CS+. Second, Berthier and Moore noted that sometimes an animal would erroneously emit a CR on a CS– trial. Some of their Purkinje cells tended to respond on these trials in accordance with the CR rather than with the CS. This result would be unlikely if the neuronal responses were unrelated to the CR.

More recently, activity in identified Purkinje cells has also been studied (Hesslow & Ivarsson, 1994; Hesslow, 1995). Using the decerebrate ferret preparation, a small part of the lobule HVI C3 zone was characterized as controlling conditioned eyeblink responses (Figure 4.11). Electrical stimulation of the cerebellar surface, which activates Purkinje cells, causes inhibition of the interpositus neurons that is followed by a rebound excitation. A narrow strip in the C3 zone of lobule HVI, which receives climbing fiber input from the periocular region, was stimulated. Delayed eyeblinks were elicited, indicating that Purkinje cells in this area control the orbicularis oculi muscle. Furthermore, a single electrical stimulus pulse to this area during an ongoing CR completely suppressed the response for as long as 100 ms.

In untrained animals, many Purkinje cells did not respond at all to the CS or they responded with a weak (<10%) increase or decrease in simple spike firing during the CS–US period. The most common pattern was an increased firing during the first 100 ms of the 300-ms interstimulus interval. After training, the majority of the Purkinje cells responded as in untrained animals, but up to 20% of them responded with a strong, sometimes complete, suppression of simple spike firing in advance of the CR.

It some cases, the subjects were shifted from trials with paired CS–US presentations to CS only trials and then back to paired trials again. In a few cases, the firing pattern of the Purkinje cells also changed, so that the simple spike suppression was weakened after a period of CS-alone trials and reappeared after paired CS–US presentations had been resumed. This result strongly suggests that the Purkinje responses were really learned. However, the fact that this was not always observed is puzzling.

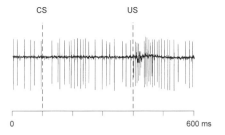
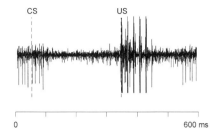
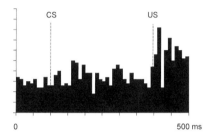
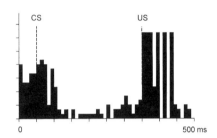

FIGURE 4.11. Activity of Purkinje cells in conditioned ferrets. Two examples are shown of neurons in the eyeblink area of the C3 zone of conditioned decerebrate ferrets. *Upper panels* are sample records; *lower panels* are spike histograms based on 20 trials. (**A**) was classified as nonresponsive whereas the neuron shown in (**B**) behaved in accordance with the CCC model. (From Hesslow & Ivarsson, 1994, Copyright 1994, Lippincott Williams & Wilkins.)

The data from these recording studies clearly demonstrate that there are learning-related changes in Purkinje cell simple spike activity that conform to predictions from the CCC model. They do not exclude, however, that such activity reflects learning elsewhere, especially at early levels of CS processing.

CS and US Pathways

Is the CS Transmitted by Cerebellar Mossy Fibers?

We have reviewed a considerable body of evidence that strongly indicates that the learning trace for eyeblink conditioning is in the cerebellum, but this does not reveal the mechanisms underlying the learning. A crucial feature of the CCC model is the assumption that CS information is transmitted to the cerebellum via mossy and parallel fibers and US information is transmitted via the inferior olive and its climbing fibers. A number of studies have attempted to investigate these hypotheses by replacing the peripheral CS or US with direct electrical stimulation of the mossy fibers and the climbing fibers, respectively.

Electrical stimulation of the pontine nuclei, or the middle cerebellar peduncle (MCP), can serve as a CS (Steinmetz et al., 1986a, 1989). The MCP contains mossy fibers projecting to the cerebellum from the pontine nuclei, which themselves receive convergent auditory, visual, and somatosensory inputs (Brodal & Bjaalie, 1992). Rabbits acquired CRs normally to a pontine stimulation CS paired with either a corneal air puff US or stimulation in the olivary region of the brain stem. Subsequent unpaired presentations of the pontine CS resulted in extinction, suggesting that the responses were authentic CRs. This evidence is consistent with the CCC model but it is far from conclusive. As we have mentioned previously, stimulation of almost any part of the brain can serve as a CS. A stronger indication that pontine inputs to the cerebellum might constitute the normal CS pathway is a transfer effect that has been observed in some experiments. After rabbits had been conditioned with stimulation of the pontine nuclei as the CS, they then learned to respond to a tone CS much more quickly than normal (Steinmetz et al., 1986b); there were major savings in conditioning through general transfer effects.

Studies using pontine stimulation must be evaluated cautiously. The pontine nuclei are situated in the base of the pons and a large number of descending and ascending fibers run through this region, including, in particular, the descending corticospinal and corticobulbar tracts. Thus, electrical stimulation in the region of the pontine nuclei unavoidably activates these fibers. It will also cause antidromic activation of all sources of input to the pontine nuclei, that is, the cerebral cortex, the spinal cord, and the brain stem. Where these fibers have branching collaterals, other, noncerebellar structures will be activated.

Because the middle cerebellar peduncle exclusively contains pontocerebellar mossy fibers, stimulation here is less problematic. Even so, stimulation can activate these fibers orthodromically and antidromically. Antidromic activation could, in principle, enable an associative mechanism within the pontine nuclei. Such effects would cast doubt on the idea that the critical association involves a mossy fiber action within the cerebellum.

Immediate transfer effects after substitution of a MCP stimulation CS for a peripherally applied CS would be a much stronger indication of CS transmission through the mossy fiber route. Such immediate transfer was recently demonstrated in decerebrate ferrets (Figure 4.12) (Hesslow et al., 1999). Eyelid conditioning was firmly established using electrical forelimb stimulation as the CS and periocular shock as the US. When middle cerebellar peduncle (MCP) mossy fiber stimulation was substituted for the peripheral CS, subjects responded immediately to the new CS without any further training, consistent with the suggestion that the MCP stimulus activated the same CS pathway as the previous forelimb stimulation CS. MCP stimulation CRs were authentic because unpaired MCP stimulation produced extinction; when paired presentations were resumed, conditioned responses gradually reappeared. Furthermore, if CRs to the forelimb CS were extinguished, using unpaired forelimb CS presentations, then CRs to the MCP CS were seen to have extinguished immediately upon testing. Immediate transfer of extinction to the forelimb CS was also seen after unpaired presentations of the MCP CS. The ability of the peripheral CS to substitute for the mossy

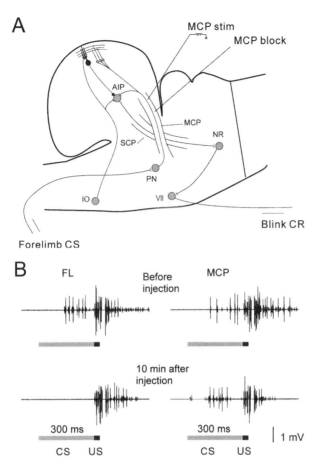

FIGURE 4.12. MCP stimulation experiment demonstrating that pontine mossy fibers constitute a CS pathway. (**A**) Experimental setup in the decerebrate ferret. The proposed CS pathway is indicated from pontine nuclei (*PN*) via the middle cerebellar peduncle (*MCP*) to the Purkinje cells and the anterior interpositus nucleus (*AIP*). The proposed US pathway from is via the inferior olive (*IO*). The efferent pathway is from the AIP via the superior cerebellar peduncle (*SCP*) to the red nucleus (*NR*) and facial nucleus (*VII*). Electrical stimulation was applied to the MCP while transmission was blocked by lignocaine ventral to the stimulation site. (**B**) EMG records from the eyelid before and after blocking transmission through the MCP. The forelimb CS (*FL*) and the MCP-CS elicit similar CRs. After the lignocaine injection, the FL-CS no longer elicits a CR but the MCP-CS does.

fiber CS was not, however, complete. Initial conditioning to the MCP stimulation CS did not transfer to the peripheral, forelimb CS. This last finding indicates that cerebellar association produced by peripheral application of a CS is not indiscriminate and is consistent with pattern-recognizing capacities maintained in the decerebrate, as predicted in the CCC model.

Before concluding that the stimulus substitution effects confirmed CS transmission via mossy fibers, Hesslow et al. (1999) ruled out possible antidromic activation effects. Transmission was reversibly blocked, using the local anesthetic lidocaine, in the MCP just ventral to the stimulation site. This blockade prevents normal transmission from the pontine nuclei reaching the cerebellum and blocks antidromic effects from MCP stimulation traveling back into the pontine nuclei. However, the blockade does allows normal transmission of the MCP stimulus through the mossy fibers into the cerebellum. During the lidocaine block, CRs elicited by the forelimb CS disappeared, but responses elicited by MCP stimulation remained. A further possibility, that MCP stimulation caused cerebellar output through the brachium conjunctivum to activate an extracerebellar CS pathway, was also rejected because a wide range of patterns of stimulation in the brachium conjunctivum failed to elicit CRs.

These stimulus substitution experiments provide direct and persuasive evidence that information relating to a peripherally applied CS is normally transmitted to the cerebellum via the mossy fiber system and supply crucial support for a cerebellar learning mechanism.

Information Processing in the CS Pathway

Direct stimulation of the mossy fibers has also been used to test whether CS information needs to be processed before entering the cerebellum. For instance, in trace conditioning there may be a temporal gap of several hundred milliseconds between the CS and US. Presumably, there must be some ongoing nervous activity, which bridges this gap. One possibility is that a brief CS may cause a sustained firing in the pontine neurons or in some other structure that projects back via the mossy fibers to the cerebellum.

However, this possibility could be rejected by an experiment by Svensson and Ivarsson (1999).They trained ferrets to respond to a 300-ms forelimb CS that was then replaced by mossy fiber stimulation CS as previously described. This CS did not need to consist of a sustained train of stimuli. In fact, a single 0.2-ms pulse was sufficient to elicit a normal CR, adaptively timed to the US. This experiment demonstrates that the cerebellum is itself able to provide a temporal bridge and suggests that only the initial part of the CS is necessary to generate a CR. It does not mean, however, that the cerebellum only utilizes the initial part of the CS, because, as described next, if the remainder of the CS train has a frequency higher than normal, the CR latency is shortened. (Traces longer than 300 ms do seem to require an extracerebellar mechanism. See following section on hippocampus.)

It may be argued that the brief mossy fiber stimulus could activate interpositopontine fibers, creating a temporal bridge by setting up a reverberating circuit between the pontine and interpositus neurons. Although this cannot be excluded, such a reverberating circuit is not normally needed. As noted, stimulating Purkinje cells and interrupting interpositus activity would break a reverberating circuit, but such stimulation only causes a brief pause in the CR (Hesslow, 1994a, 1996). Furthermore, direct stimulation of the mossy fibers in the middle cerebellar peduncle works even when the pathway is blocked by lidocaine ventral to the stimulus (Hesslow et al., 1999).

In experiments on rabbits, Moyer et al. (1990) showed that hippocampectomized subjects could not learn in a trace-conditioning paradigm when the trace was longer than 500 ms. Although it is clear that the hippocampus is not necessary for normal delay conditioning with shorter traces, this finding suggests that it may be instrumental in providing a prolonged CS signal to the cerebellum.

The mechanism for the adaptive timing of the CR also seems to reside in the cerebellum. An important aspect of classical conditioning is that the animal learns the temporal relationship between the CS and US and that the CR is delayed until just before the expected onset of the US. One possible mechanism for the adaptive timing of the CR could be that there are temporal delays in the CS pathway (Desmond & Moore, 1988), for instance, in the mossy fibers, so that different afferents to the site of plasticity would be activated at different times after the CS onset. If only inputs with a close temporal proximity to the US cause synaptic changes, only those afferents that are active toward the end of the CS–US interval would induce synaptic plasticity and then contribute to eliciting the CR. The CR would then automatically occur close to the US in time.

Although such a delay mechanism may exist within the cerebellum, we can be confident that the timing of the CR does not require precerebellar processing. In experiments in decerebrate ferrets, Svensson et al. (1997) found completely normal CR latencies when stimulating the mossy fibers directly. These authors also found that manipulating a peripheral forelimb CS had effects on the CR latency, which could be mimicked with similar manipulations of a mossy fiber CS. When they increased the frequency of the forelimb CS from 50 to 100 Hz, they observed a marked reduction in CR latency. This latency change also occurred when the frequency of the mossy fiber CS was increased. With continued training at the higher frequency, the CR latency gradually adapted until it was timed as before the frequency change. This adaptation was observed both for the peripheral and for the mossy fiber CS. The fact that the effects of manipulating the CS on the temporal properties of the CR are the same for a mossy fiber CS as for a peripheral CS shows that these properties are being generated after the CS signal has passed the mossy fiber system.

It would seem then, that most of the mechanisms for the basic features of conditioning are located within the cerebellum, although this does not mean that the cerebellum does not utilize information from other parts of the brain. There are prominent projections to the pontine nuclei from the sensory parts of the cerebral cortex, and it would seem likely that conditioning with more subtle CSs requires preprocessing in these areas.

The Role of Mossy Fiber Input to the Interpositus Nucleus

It has been known for a long time that many mossy fibers send collaterals to the deep nuclei of the cerebellum, but it has been difficult to determine if this is true of the mossy fibers from the pontine nuclei. There is now definitive evidence for such collateral projections, at least to the dentate nucleus (Shinoda et al., 1992; Brodal & Bjaalie, 1992). Because there is now clear evidence that mossy fibers transmit the CS, it is anatomically possible that the plasticity mediating NMR conditioning is located in the interpositus nucleus as well as in the cortex. There are, however, physiological reasons for doubting the importance of this mossy fiber input.

Because the Purkinje cells are inhibitory, something must drive the activity of the nuclear cells. There could be an internal spike-generating mechanism, but there could also be input from mossy fiber collaterals. In the experiments by Hesslow et al. (1999), the mossy fibers were stimulated directly and probably produced a stronger and more synchronous excitation of the nuclear neurons than natural CSs normally do. If the mossy fiber collaterals are able to drive the interpositus neurons in spite of tonic Purkinje cell inhibition, as assumed by the nuclear learning hypothesis, one would therefore expect the mossy fiber stimulus to excite these neurons and cause short-latency EMG activity in the eyelid. In contrast to the cerebellar cortex, there is little circuitry in the nuclei that could cause a delay in the input signal necessary for the delayed CR. However, in most cases in which mossy fibers were stimulated directly, there were no short-latency EMG responses. This fact suggests that collaterals of pontine mossy fibers are unlikely to exert a powerful drive on the interpositus neurons and that it is the projection to the cortex that is important.

Is US Information Transmitted by Climbing Fibers?

The CCC model of conditioning postulates that the US signal is transmitted to the cerebellum via climbing fibers from the inferior olive to provide a reinforcing input that induces changes in synaptic efficacy. This proposal is one of the most clearly defined components of the model, and testing it should provide the best evidence for, or against, the model. Both anatomic and physiological results confirm that there are face-related climbing fiber inputs to areas in the cerebellar cortex known to control the conditioned eyeblink, lobule HVI in the rabbit and the C1 and C3 zones in lobule HVI in the ferret. It should be noted, however, that the US can activate substantial mossy fiber inputs to these areas and there are also serotonergic and noradrenergic projections to the cerebellum, so it cannot be assumed on anatomic grounds alone that the US signal must be transmitted by climbing fibers. Two approaches have been used to test the role of the climbing fiber input in NMR conditioning: one has been to lesion or inactivate the inferior olive, and the other is to replace the US by direct electrical stimulation of the olive.

Olivary Lesion/Inactivation Studies

It might be thought that interrupting the putative US pathway by inactivating the olive would be equivalent to turning off the US peripherally and would therefore lead to extinction of the learned response during continued paired presentations of the CS and US. In line with this assumption, it has been reported that mechanical lesions of the olive cause a gradual extinction of a previously acquired NM response (McCormick et al., 1985). This result appears to confirm the CCC model, but it is in fact inconsistent with previous knowledge of cerebellar physiology. As described earlier, climbing fiber input has a nonspecific tonic influence on the excitability of Purkinje cells. Olivary inactivation causes a rise in tonic firing of Purkinje cells and a virtual shutdown of cerebellar output equivalent to a complete cerebellar inactivation (Colin et al., 1980; Montarolo et al., 1982). It is more likely, therefore, that olivary lesions should cause an immediate abolition of CRs, as was found by Yeo et al. (1986). The reasons for the difference in findings have not been resolved (Yeo, 1989), but a recent paper by Welsh & Harvey (1998) confirms the latter finding. Olivary inactivation by local lidocaine infusion immediately abolished previously acquired conditioned responses. These latter studies clearly implicate the olive in the performance of CRs, but what, if any, is its role in learning? Olivary inactivation during training completely prevents acquisition, as measured by postinactivation performance (Welsh & Harvey, 1998). Clearly, olivary function is essential for learning but, as olivary function is also essential for normal cerebellar cortical activity, this finding confirms the essential nature of the climbing fiber input without assigning it any specific role in information transmission. These findings neither confirm nor refute the hypothesis that climbing fibers signal US information and provide a reinforcing input.

Olivary/Cerebellar Stimulation Studies

Another approach is to use direct stimulation of the inferior olive as the US, and in two studies in which this was done, the subjects acquired CRs (Mauk et al., 1986; Steinmetz et al., 1989). As with the mossy fiber stimulation work just described, such studies raise concerns over stimulus spread and antidromic activation effects. Stimulation of the inferior olive is likely to antidromically activate the trigeminal nucleus, which receives afferent information from the cornea and periocular region. This effect probably occurred in these two studies because the stimulation elicited blinks and it is unlikely that pure climbing fiber activity will do this. No such olive-mediated reflex components have been described previously (Ito, 1984). So, if an olivary stimulus antidromically activates a crucial part of the unconditioned reflex pathway, it may not differ crucially from a peripherally applied US and so cannot reveal whether the inferior olive is a critical relay in the US pathway. Similarly, stimulation within the cerebellum can serve as a US (Brogden & Gannt, 1948; Swain et al., 1992), but it activates many different neuronal elements and so does not test the climbing fiber hypothesis directly.

Nucleo-Olivary Inhibition

With reference to the short-term excitatory action of the climbing fiber input on the Purkinje cells, the nucleo-olivary projection is unusual in that it constitutes a positive feedback loop. If Purkinje cells were excited by climbing fiber input, these would inhibit the interpositus neurons, which would then remove inhibition from the olive and increase the excitation of the Purkinje cells further. Positive feedback loops are inherently unstable, and are unusual in biological organisms. The nucleo-olivary connection makes better sense when one considers depressing effects of climbing fiber input on the Purkinje cells. The nucleo-olivary pathway could serve as a negative feedback system to regulate the spontaneous firing rate of the Purkinje cells and also regulate the magnitude of the synaptic changes underlying conditioning (Andersson et al., 1988). When output from the deep nuclei increases as learning proceeds, there would be an increased inhibition of the inferior olive consistent with observations that the reinforcing effect of the US gradually diminishes as training proceeds. Eventually, all further learning would be blocked. An inhibitory effect of the CR on the olive has later been confirmed by two studies. Recordings from groups of olivary cells suggest that, as the animals learn to perform a CR, the effect of the US on the olive is decreased (Sears & Steinmetz, 1991). This change could be caused by a suppression of other parts of the US pathway. However, it has also been shown that there is a direct inhibition of the olive by a preceding CR and that the magnitude of the US-elicited excitation of the olive is inversely proportional to the size of a preceding CR (Hesslow & Ivarsson, 1996).

Nucleo-olivary inhibition might also be the explanation for the blocking. If an animal is trained to two CSs separately, say a light and a tone, it will emit CRs in response to both CSs. However, if the animal is first trained to a light CS until it responds reliably and a tone CS is then added on a number of trials, the animal will not learn to respond to the tone. The previous learning is said to 'block' the second association. If the CR-generating activity in the interpositus nucleus inhibits the inferior olive when the animal has learned to respond to the first CS, the US would be blocked during later trials in which a second CS has been added (Andersson et al., 1988). This suggestion has recently been confirmed. In experiments in which a GABA-receptor blocker was injected into the olive, presumably preventing nucleo-olivary inhibition, the blocking effect was abolished (Kim et al., 1998). A problem with this suggestion is the report that hippocampal lesions prevent blocking (Solomon, 1977). If blocking was mediated by the nucleo-olivary pathway, it is difficult to understand why it would require an intact hippocampus unless it provides a superordinate mechanism of behavioral control.

Conditioning-Induced Plasticity in Other Motor Structures?

This chapter has focused on the cerebellum because of its critical importance in eyeblink/NMR conditioning. However there are cases of plasticity in noncerebellar structures induced by other skeletal muscle conditioning procedures. Most notable of these is the work of Woody and colleagues, discussed earlier, who demonstrated motor cortical plasticity relating to short-latency, eyelid responses in cats conditioned with a glabellar tap US (see Woody, 1982, for an extensive review). The longer-latency components of this response appear, however, to be cerebellar dependent.

In other studies, Tsukahara and his colleagues developed an experimental paradigm for electrophysiological analysis of leg flexion conditioning (Tsukahara et al., 1981; Tsukahara, 1981). The CS was a train of electrical stimuli to the cerebral peduncle. The latter had been sectioned caudal to the level of the red nucleus, so it was assumed that the stimulation would mainly excite corticorubral fibers. Stimulation at this site elicited a short-latency (12–32 ms) forelimb flexion response. The threshold for eliciting such responses was determined and the intensity of the CS was set slightly below this threshold. The US was an electrical stimulation of the forelimb that caused a clear flexion, and the interstimulus interval was usually 100 ms. After repeated paired presentations of CS and US, the cerebral peduncle CS could elicit a forelimb flexion at a lower intensity than before training. Thus, the effect of the training could be described as a reduction in the threshold for eliciting leg flexion with the CS or as an enhanced efficacy of synaptic transmission in the CS–CR pathway. As would be expected, if the same synapses were involved in producing the pre- and posttraining responses elicited by the CS, the CR latency was sometimes shorter (8–34 ms) than the latency of the responses elicited by a high intensity CS before training.

Because the cerebral peduncles had been lesioned caudal to the red nucleus, so that the CS would have mainly activated corticorubral fibers, it was suggested that the underlying plasticity was specifically at corticorubral synapses. This idea was supported by stimulation of the interpositus nucleus in the cerebellum, the second major source of input to the red nucleus. Interpositus nucleus stimulation could also elicit forelimb flexion, but the effectiveness of such stimulation did not change during training (Tsukahara et al., 1981). Both extra- and intracellular recordings from red nucleus cells before and after training revealed a conditioning-related development of a new, fast-rising component of the excitatory postsynaptic potential (EPSP) driven by the CS (Oda et al., 1988). Earlier work had shown that neuronal sprouting at corticorubral synapses could result in similar fast-rising EPSPs (Tsukahara et al., 1975), so it was suggested that this was the mechanism underlying the learning. Pananceau et al. (1996) applied a training paradigm that was similar to that employed by the Tsukahara group, except that the CS was a direct stimulation of the interpositus nucleus. In these experiments, they found evidence for synaptic plasticity in the interposito-rubral pathway.

In the work just described, randomly timed CS and US presentations caused extinction of the forelimb flexion CRs. This finding seems to rule out sensitization effects and suggests that the learning was truly associative. In other important respects, however, these results differ so much from those obtained with standard classical conditioning paradigms such as NMR/eyelid conditioning that it is very difficult to compare the two lines of research. The first, and most important, difference is that there was no demonstration in Tsukahara's work of any adaptive timing of the CR, a distinctive feature of eyeblink and most other forms of classical conditioning. A low-intensity CS to the cerebral peduncle elicited a short-latency response and the CR had a similar, sometimes even shorter, latency. In most instances, the maximum amplitude of the CR did not coincide with the US, nor was it shown whether CR latency could be changed by manipulating the CS–US interval, which was much shorter (usually 100 ms) than that normally used (250–500 ms) in other paradigms. A second difference concerns the rate of acquisition of the CR. In Tsukahara's paradigm, the animals were given 120 paired trials per day and typically reached asymptotic learning performance after 7 to 10 days of training. Extinction of the conditioned response, by presenting the CS alone or by backward pairing, also took about a week. In contrast, in typical eyeblink conditioning experiments, a CR can develop within 200 trials and extinction would usually be appreciable within fewer than 100 trials.

The use of direct stimulation of the brain as a CS in conditioning studies such as those described is a powerful technique to explore the potential for plasticity in the activated pathways, but it does not necessarily reveal mechanisms underlying conditioning using more natural stimulation at the periphery, as in eyeblink conditioning. For instance, axonal sprouting is a relatively slow process, and although it may account for the learning described in the Tsukahara paradigm, it is unlikely to mediate the faster development of eyeblink conditioning. Further work is needed to discover whether such plasticities are involved in natural forms of motor learning.

The Role of the Hippocampus in NMR Conditioning

The hippocampus became a major focus of much memory research after the first descriptions of the famous patient H.M., who became unable to form declarative memories after bilateral removal of the medial temporal lobe and hippocampus (Scoville & Milner, 1957). The hippocampus became a 'prime suspect' as a locus of classical conditioning. Further studies soon revealed, however, that rabbits with complete bilateral removal of the hippocampus (Schmaltz & Theios, 1972) and completely hemispherectomized cats (Norman et al., 1974) could acquire normal, delay-conditioned NM/eyeblink responses. Indeed, human amnesic patients can also acquire delay-conditioned eyeblink responses (Weiskrantz & Warrington, 1979).

Although the lesion studies pointed to no essential role of the hippocampus in delay eyeblink/NMR conditioning, electrophysiological recording studies indicated that the hippocampus does become engaged in the conditioning process. Hippocampal neurons develop responses to the CS after paired CS–US training, and an envelope of hippocampal multiunit activity models quite closely the movement of the NM during a CR (Berger et al., 1976, 1983; Berger & Thompson, 1978). Importantly, the neuronal activity precedes the CR in time and across conditioning trials, suggesting that it might indicate a causal relationship. But, as we discussed earlier, the NMR is only one of a constellation of responses that can become conditioned, and there is no reason to assume that it occurs earliest in time or across trials. Recording studies are purely correlative and constitute weak evidence for causality. So, hippocampal neurons may change their activity during conditioning because they are involved in the performance of an earlier response, or they may receive an efference copy of the command signal.

Behaviors other than the development of NM-conditioned responses may change during conditioning. The subject may learn to attend to the CS, to expect a certain duration of the CS, to be afraid of the CS, or may learn that the CS is not followed by other possible USs. In humans, a declarative memory of the training situation and the CS–US contingency will be formed, and presumably corresponding events occur in other animals as well. It is now generally accepted that the hippocampus is necessary for the formation of declarative memories, and conditioning-related neuronal activity could be a reflection of this.

Although it is clear that lesions to the hippocampus do not seriously interfere with standard delay conditioning, there are a large number of reports indicating that hippocampal lesions impair more complex forms of conditioning.

Trace Conditioning

Several studies show that lesions to the hippocampus can impair trace conditioning. Here, the CS is terminated before the onset of the US, leaving a CS-free 'trace' period between the CS and the US. Solomon et al. (1986) first reported impaired acquisition in trace conditioning after hippocampal lesions. Moyer et al. (1990) found that hippocampal animals acquired normally with a trace interval of 300 ms but showed a marked resistance to extinction. When the trace was increased to 500 ms, the animals did not learn at all. Other studies found altered CR topographies (Port et al., 1986; James et al., 1987) with 500-ms traces. Thus, there is clear evidence that normal trace conditioning is dependent upon an intact hippocampus.

Discrimination Reversal

Hippocampal lesions also interfere with discrimination reversal learning. Lesioned animals show normal discrimination learning, that is, they learn to respond to a reinforced CS+ and not to a nonreinforced CS−. When the contingencies are reversed so that the previous CS+ is now nonreinforced and the previous CS− becomes reinforced, however, hippocampal animals fail to produce the new discrimination. Specifically, responses to the new CS− do not drop away (Orr & Berger, 1985).

Latent Inhibition

When a normal animal receives repeated, nonreinforced exposures to the CS before a standard conditioning paradigm, subsequent acquisition is retarded. This effect, known as latent inhibition, was absent in rabbits with ablations of the dorsal hippocampus (Solomon & Moore, 1975).

Conditional Discrimination

An intact hippocampus also seems to be necessary for normal learning in a conditional discrimination paradigm. A CS is reinforced if preceded by a different stimulus, S, and unreinforced if presented alone. A normal subject will learn to respond to the CS only on S-CS trials. This ability was lost in rats and in human subjects with hippocampal lesions (Ross et al., 1984; Daum et al., 1991).

Sensory Preconditioning

In sensory preconditioning the animal is subjected to two stimuli, S1 and S2, neither of which elicits any overt response. S2 is then paired with a US until it elicits CRs. When S1 is later presented on its own, it too elicits the CR, indicating that an association has been formed between S1 and S2. This effect was not observed in animals with hippocampal lesions (Port et al., 1987).

Blocking

An animal can be trained to respond to two different CSs, such as a tone or a light, if these are paired with the US on separate trials. However, if the animal is first trained to respond reliably to a CS1 and a second stimulus CS2 is then added on the same trials, it will not learn to respond the CS2 alone. The CS1 is said to 'block' the CS2 (Kamin, 1969; Marchant & Moore, 1973). This effect is absent in rabbits with a bilateral dorsal hippocampal ablation (Solomon, 1977).

Although it is clear that lesions to the hippocampus can interfere with learning in some conditioning tasks, it is premature to conclude the exact nature of hippocampal involvement. One possibility is that there is important processing of the CS signal before it reaches the cerebellum. Complex or subtle CS information might not be transmitted directly through sensory relays to the pontine nuclei and the mossy fibers. A more complex CS probably requires information extraction by signal processing in forebrain structures before transmission to the pontine nuclei. Associative learning would then proceed in the cerebellum, as with other CSs. A similar argument may apply to trace conditioning. The cerebellum seems to be able to bridge traces as long as about 500 ms, but a prolonged mossy fiber input may be necessary for longer traces, and perhaps the maintenance of the CS signal is dependent on the hippocampus.

Some of the hippocampus-dependent tasks involve learning more than one contingency. In sensory preconditioning, the subject learns to associate two neutral stimuli and then to associate one of these with a response-eliciting stimulus, the US. The hippocampus could be critical for the first task and the cerebellum for the second. Conditional discrimination resembles the sensory preconditioning task and may also involve more than one kind of learning. Blocking, latent inhibition, and reversal learning are not obviously similar, but at present it cannot be excluded that these learning paradigms require either sensory preprocessing or a combination of two kinds of learning.

Currently, we view the cerebellum as the associative network capable of producing an accurately timed, adaptive response to a range of potential stimuli. When the contingencies or the stimuli are complex, forebrain mechanisms may be essential for CS preprocessing, but the cerebellum will still have its role in the formation of a final CS–US association for delivery of the conditioned response.

Conclusions

How well have the assumptions of the cerebellar cortical conditioning (CCC) model passed the empirical tests reviewed here? The hypothesis that associative learning could be a cerebellar function is still controversial (Bloedel & Bracha, 1995; De Schutter & Maex, 1996; Llinas et al., 1997) but, when taking all the evidence reviewed here into consideration, we suggest that the following conclusions are justified.

1. *The critical learning mechanisms are located within the cerebellum.* After the earliest lesion studies, it was possible to argue that the role of the cerebellum was mainly in facilitating or performing a response that had been learned elsewhere. However, the observations that inactivation of the cerebellum, but not of cerebellar output, blocks learning, that inactivation can block extinction, that a learned response can be elicited by direct stimulation of mossy fibers, and other results demonstrate the crucial role of the cerebellum. The experimental results that were thought to contradict this conclusion, such as recovery of CRs after cerebellar lesions, can now be given satisfactory alternative interpretations.

2. *There is critical plasticity for NMR conditioning in the cerebellar cortex.* There are strong theoretical arguments that there is essential plasticity in the cerebellar cortex, based on its information processing capacity and its convergent inputs. This idea is strongly supported by findings in lesion and inactivation experiments. It is also supported by experiments in which stimulation of mossy fibers elicit adaptively timed CRs but no short-latency responses. Although the hypothesis that the critical plasticity is located at the parallel fiber to Purkinje cell synapses is attractive, there is presently no direct evidence for this. The suggestion that there is critical cerebellar cortical plasticity is not exclusive. We

have reviewed evidence that there also may be plasticity at cerebellar nuclear sites.
3. *The CS pathway is through mossy fibers.* In addition to the theoretical and anatomic arguments for the CCC model, stimulation experiments demonstrate that mossy fibers from the pontine nuclei ascending through the middle cerebellar peduncle can transmit information related to a peripherally applied CS. It cannot be excluded at present that other mossy fiber pathways may also transmit CS signals.
4. *The US pathway may be through the climbing fibers.* This central assumption of the CCC model is less well supported than the others. It is based on theoretical considerations and also on the fact that climbing fiber activation can induce plasticity in coactivated parallel/fiber–Purkinje cell synapses. At present, there is little direct evidence that the US is transmitted by climbing fibers.

There are several different learning mechanisms in the CNS, and different behavioral tasks utilize various combinations of these mechanisms. The evidence reviewed in this chapter leads us to believe that the standard delay eyeblink conditioning only depends on the cerebellum, although there may be behavioral tasks that use more than one learning mechanism. Overall, the CCC model is consistent with almost all the known data, and some of its crucial assumptions have been confirmed.

References

Albus, J. (1971). A theory of cerebellar function. *Mathematical Biosciences, 10,* 25–61.
Andersson, G., Ekerot, C.F., Oscarsson, O., and Schouenborg, J. (1987). Convergence of afferent paths to olivo-cerebellar complexes. In M. Glickstein, C.H. Yeo, and J. Stein (Eds.), *Cerebellum and neuronal plasticity* (pp. 165–174). New York: Plenum.
Andersson, G., Garwicz, M., and Hesslow, G. (1988). Evidence for a GABA-mediated cerebellar inhibition of the inferior olive in the cat. *Experimental Brain Research, 72,* 450–456.
Andersson, G., and Hesslow, G. (1987a). Inferior olive excitability after high frequency climbing fiber activation in the cat. *Experimental Brain Research, 67,* 523–532.
Andersson, G., and Hesslow, G. (1987b). Activity of Purkinje cells and interpositus neurons during and after periods of high frequency climbing fiber activation in the cat. *Experimental Brain Research, 67,* 533–542.
Attwell, P.J., Rahman, S., Ivarsson, M., and Yeo, C.H. (1999a). Cerebellar cortical AMPA-kainate receptor blockade prevents performance of classically conditioned nictitating membrane responses. *Journal of Neuroscience, 19,* RC45.
Attwell, P.J., Rahman, S., Ivarsson, M., Gilbert, P.F., and Yeo, C.H. (1999b). Temporary, cerebellar cortical AMPA-receptor blockade prevents acquisition of nictitating membrane conditioning. *Society for Neuroscience, Abstracts, 25,* 84–84.
Bekhterev, V.M. (1932). *General principles of human reflexology.* New York: International Press.
Berger, T.W., and Thompson, R.F. (1978). Identification of pyramidal cells as the critical elements in hippocampal neuronal plasticity during learning. *Proceedings of the National Academy of Sciences of the United States of America, 75,* 1572–1576.

Berger, T.W., Alger, B., and Thompson, R.F. (1976). Neuronal substrate of classical conditioning in the hippocampus. *Science, 192,* 483–485.

Berger, T.W., Laham, R.I., and Thompson, R.F. (1980). Hippocampal unit-behavior correlations during classical conditioning. *Brain Research, 193,* 229–248.

Berger, T.W., Rinaldi, P.C., Weisz, D.J., and Thompson, R.F. (1983). Single-unit analysis of different hippocampal cell types during classical conditioning of rabbit nictitating membrane response. *Journal of Neurophysiology, 50,* 1197–1219.

Berkeley, K.J., and Hand, P.J. (1978). Projections to the inferior olive of the cat. II. Comparisons of input from the gracile, cuneate and spinal trigeminal nuclei. *Jorunal of Comparative Neurology, 180,* 252–264.

Berthier, N.E., and Moore, J.W. (1986). Cerebellar Purkinje cell activity related to the classically conditioned nictitating membrane response. *Experimental Brain Research, 63,* 341–350.

Berthier, N.E., and Moore, J.W. (1990). Activity of deep cerebellar nuclear cells during classical conditioning of nictitating membrane extension in rabbits. *Experimental Brain Research, 83,* 44–54.

Bloedel, J.R., and Bracha, V. (1995). On the cerebellum, cutaneomuscular reflexes, movement control and the elusive engrams of memory. *Behavioural Brain Research, 68,* 1–44.

Bracha, V., Webster, M.L., Winters, N.K., Irwin, K.B., and Bloedel, J. R. (1994). Effects of muscimol inactivation of the cerebellar interposed-dentate nuclear complex on the performance of the nictitating membrane response in the rabbit. *Experimental Brain Research, 100,* 453–468.

Brodal, A. (1981). The cerebellum. In *Neurological anatomy in relation to clinical medicine.* New York: Oxford University Press.

Brodal, P. (1992). The cerebellum. In *The central nervous system: structure and function* (pp. 262–282). New York: Oxford University Press.

Brodal, P., and Bjaalie, J.G. (1992). Organization of the pontine nuclei. *Neuroscience Research, 13,* 83–118.

Brodal, A., Destombes, J., Lacerda, A.M., and Angaut, P. (1972). A cerebellar projection onto the pontine nuclei. An experimental anatomical study in the cat. *Experimental Brain Research, 16,* 115–139.

Brogden, W.J. (1939). The effect of frequency of reinforcement upon the level of conditioning. *Journal of Experimental Psychology, 24,* 419–431.

Brogden, W.J., and Gannt, W.H. (1948). Intraneural conditioning: cerebellar conditioned reflexes. *Archives of Neurological Psychiatry, 48,* 437–455.

Bromiley, R.B. (1948). Conditioned responses in a dog after removal of neocortex. *Journal of Comparative Physiology and Psychology, 41,* 102–110.

Brons, J.F., and Woody, C.D. (1980). Long-term changes in excitability of cortical neurons after Pavlovian conditioning and extinction. *Journal of Neurophysioly, 44,* 605–615.

Burne, R.A., Azizi, G.A., Mihailoff, G., and Woodward, D.J. (1981). The tectopontine projection in the rat with comments on visual pathways to the basilar pons. *Journal of Comparative Neurology, 202,* 287–307.

Cegavske, C.F., Thompson, R.F., Patterson, M.M., and Gormezano, I. (1976). Mechanisms of efferent neuronal control of the reflex nictitating membrane response in rabbit (*Oryctolagus cuniculus*). *Journal of Comparative Physiology and Psychology, 90,* 411–423.

Chapman, P.F., Steinmetz, J.E., Sears, L.L., and Thompson, R.F. (1990). Effects of lidocaine injection in the interpositus nucleus and red nucleus on conditioned behavioral and neuronal responses. *Brain Research, 537,* 149–156.

Clark, R.E., and Lavond, D.G. (1993). Reversible lesions of the red nucleus during acquisition and retention of a classically conditioned behavior in rabbits. *Behavioral Neuroscience, 107,* 264–270.

Clark, R.E., and Lavond, D.G. (1996). Neural unit activity in the trigeminal complex with interpositus or red nucleus inactivation during classical eyeblink conditioning. *Behavioral Neuroscience, 110,* 13–21.

Clark, G.A., McCormick, D.A., Lavond, D.G., and Thompson, R.F. (1984). Effects of lesions of cerebellar nuclei on conditioned behavioral and hippocampal neuronal responses. *Brain Research, 291,* 125–136.

Clark, R.E., Zhang, A.A., and Lavond, D.G. (1992). Reversible lesions of the cerebellar interpositus nucleus during acquisition and retention of a classically conditioned behavior. *Behavioral Neuroscience, 106,* 879–888.

Clark, R.E., Gohl, E.B., and Lavond, D.G. (1997). The learning-related activity that develops in the pontine nuclei during classical eye-blink conditioning is dependent on the interpositus nucleus. *Learning and Memory, 3,* 532–544.

Colin, P., Manil, J., and Desclin, J.C. (1980). The olivocerebellar system. 1. Delayed and slow inhibitory effects: an overlooked salient feature of cerebellar climbing fibers. *Brain Research, 187,* 3–27.

Culler, F., and Mettler, F.A. (1934). Conditioned behavior in a decorticate dog. *Journal of Comparative Physiology and Psychology, 18,* 291–303.

Daum, I., Channon, S., Polkey, C.E., and Gray, J.A. (1991). Classical conditioning after temporal lobe lesions in man: impairment in conditional discrimination. *Behavioral Neuroscience, 105,* 396–408.

Davis, K.D., and Dostrovsky, J.O. (1986). Modulatory influences of red nucleus stimulation on the somatosensory responses of cat trigeminal subnucleus oralis neurons. *Experimental Neurology, 91,* 80–101.

De Schutter, E., and Maex, R. (1996). The cerebellum: cortical processing and theory. *Current Opinion in Neurobiology, 6,* 759–764.

Desmond, J.E., and Moore, J.W. (1982). A brain stem region essential for the classically conditioned but not unconditioned nictitating membrane response. *Physiology & Behavior, 28,* 1029–1033.

Desmond, J.E., and Moore, J.W. (1988). Adaptive timing in neural networks: the conditioned response. *Biology and Cybernetics, 58,* 405–415.

Desmond, J.E., and Moore, J.W. (1991). Single-unit activity in red nucleus during the classically conditioned rabbit nictitating membrane response. *Neuroscience Research, 10,* 260–279.

Dietrichs, E., Bjaalie, J.G., and Brodal, P. (1983). Do pontocerebellar fibers send collaterals to the cerebellar nuclei? *Brain Research, 259,* 127–131.

Disterhoft, J.F., Quinn, K.J., Weiss, C., and Shipley, M.T. (1985). Accessory abducens nucleus and conditioned eye retraction/nictitating membrane extension in rabbit. *Journal of Neurosciences, 5,* 941–950.

Donhoffer, H. (1966). The role of the cerebellum in the instrumental conditional reflex. *Acta Physiologica Academiae Scientarium Hungaricae, 29,* 247–251.

Doty, R.W. (1969). Electrical stimulation of the brain in behavioral context. *Annual Review of Psychology, 20,* 289–320.

Doty, R.W., Rutledge, L. T., and Larson, B. (1956). Conditioned reflexes established to electrical stimulation of cat cerebral cortex. *Journal of Neurophysiology, 19,* 401–405.

du Lac, S., Raymond, J.L., Sejnowski, T.J., and Lisberger, S.G. (1995). Learning and memory in the vestibulo-ocular reflex. *Annual Review of Neuroscience, 18,* 409–441.

Fanardjian, V.V. (1961). The influence of cerebellar ablation on conditioned motor reflexes in dogs. *Journal of Higher Nervous Activity, 11*, 920–926.

Flourens, P. (1824). *Recherches expérimentales sur les propriétés et les fonctions de systèeme nerveux, dans les animaux vertébrés.* Paris: Crevot.

Garcia, K.S., Steele, P.M., and Mauk, M.D. (1999). Cerebellar cortex lesions prevent acquisition of conditioned eyelid responses. *Journal of Neuroscience, 19*, 10940–10947.

Gellman, R., Houk, J.C., and Gibson, A.R. (1983). Somatosensory properties of the inferior olive of the cat. *Journal of Comparative Neurology, 215*, 228–243.

Gilbert, P. (1975). How the cerebellum could memorise movements. *Nature, 254*, 688–689.

Gilbert, P.F. (1974). A theory of memory that explains the function and structure of the cerebellum. *Brain Research, 70*, 1–18.

Gormezano, I. (1966). Classical conditioning. In A.H. Black and W.F. Prokasy (Eds.), *Classical conditioning II: Current research and theory* (pp. 385–420). New York: McGraw-Hill.

Gormezano, I., and Moore, J.W. (1969). Classical conditioning. In M.H. Marx (Ed.), *Learning: processes.* New York: Macmillan.

Gormezano, I., Schneiderman, N., Deaux, E.G., and Fuentes, I. (1962). Nictitating membrane: classical conditioning and extinction in the albino rabbit. *Science, 138*, 34.

Gormezano, I., Kehoe, E.J., and Marshall-Goodell, B. (1983). Twenty years of classical conditioning research with the rabbit. In J.M. Sprague and A.N. Epstein, (Eds.), *Progress in psychobiology and physiological psychology* (pp. 197–275). New York: Academic Press.

Granit, R., and Phillips, C.G. (1956). Excitatory and inhibitory processes acting on individual Purkinje cells of the cerebellum in cats. *Journal of Physiology (London), 133*, 520–547.

Gruart, A., and Yeo, C.H. (1995). Cerebellar cortex and eyeblink conditioning: bilateral regulation of conditioned responses. *Experimental Brain Research, 104*, 431–448.

Gruart, A., Schreurs, B.G., del Toro, E.D., and Delgado-Garcia, J.M. (2000). Kinetic and frequency-domain properties of reflex and conditioned eyelid responses in the rabbit. *Journal of Neurophysiology, 83*, 836–852.

Hardiman, M.J., and Yeo, C.H. (1992). The effect of kainic acid lesions of the cerebellar cortex on the conditioned nictitating membrane response in the rabbit. *European Jorunal of Neuroscience, 4*, 966–980.

Hardiman, M.J., Ramnani, N., and Yeo, C.H. (1996). Reversible inactivations of the cerebellum with muscimol prevent the acquisition and extinction of conditioned nictitating membrane responses in the rabbit. *Experimental Brain Research, 110*, 235–247.

Harvey, J.A., Land, T., and McMaster, S.E. (1984). Anatomical study of the rabbit's corneal-VIth nerve reflex: connections between cornea, trigeminal sensory complex, and the abducens and accessory abducens nuclei. *Brain Research, 301*, 307–321.

Harvey, J.A., Welsh, J.P., Yeo, C.H., and Romano, A.G. (1993). Recoverable and nonrecoverable deficits in conditioned responses after cerebellar cortical lesions. *Journal of Neuroscience, 13*, 1624–1635.

Hesslow, G. (1986). Inhibition of inferior olivary transmission by mesencephalic stimulation in the cat. *Neuroscience Letters, 63*, 76–80.

Hesslow, G. (1994a). Inhibition of classically conditioned eyeblink responses by stimulation of the cerebellar cortex in the decerebrate cat. *Journal of Physiology (London), 476*, 245–256.

Hesslow, G. (1994b). Correspondence between climbing fiber input and motor output in eyeblink-related areas in cat cerebellar cortex. *Journal of Physiology (London), 476,* 229–244.

Hesslow, G. (1995). Classical conditioning of eyeblink in decerebrate cats and ferrets. In W.R. Ferrell and U. Proske (Eds.), *Neural Control of Movement* (pp. 117–122). New York: Plenum Press.

Hesslow, G. (1996). Positive cerebellar feedback loops. *Behavioral and Brain Sciences, 19,* 455–456.

Hesslow, G., and Ivarsson, M. (1994). Suppression of cerebellar Purkinje cells during conditioned responses in ferrets. *Neuroreport, 5,* 649–652.

Hesslow, G., and Ivarsson, M. (1996). Inhibition of the inferior olive during conditioned responses in the decerebrate ferret. *Experimental Brain Research, 110,* 36–46.

Hesslow, G., Svensson, P., and Ivarsson, M. (1999). Learned movements elicited by direct stimulation of cerebellar mossy fiber afferents. *Neuron, 24,* 179–185.

Holstege, G., and Collewijn, H. (1982). The efferent connections of the nucleus of the optic tract and the superior colliculus in the rabbit. *Journal of Comparative Neurology, 209,* 139–175.

Holstege, G., and Tan, J. (1988). Projections from the red nucleus and surrounding areas to the brainstem and spinal cord in the cat. An HRP and autoradiographical tracing study. *Behavioural Brain Research, 28,* 33–57.

Holstege, G., Tan, J., van Ham, J.J., and Graveland, G.A. (1986a). Anatomical observations on the afferent projections to the retractor bulbi motoneuronal cell group and other pathways possibly related to the blink reflex in the cat. *Brain Research, 374,* 321–334.

Holstege, G., van Ham, J.J., and Tan, J. (1986b). Afferent projections to the orbicularis oculi motoneuronal cell group. An autoradiographical tracing study in the cat. *Brain Research, 374,* 306–320.

Houk, J.C., Buckingham, J.T., and Barto, A.G. (1996). Models of the cerebellum and motor learning. *Behavioral Brain Science, 19,* 368–383.

Ito, M. (1972). Cerebellar control of the vestibular neurons: physiology and pharmacology. *Progress in Brain Research, 37,* 377–390.

Ito, M. (1982). Cerebellar control of the vestibulo-ocular reflex—around the flocculus hypothesis. *Annual Review of Neuroscience, 5,* 275–296.

Ito, M. (1984). *The cerebellum and neuronal control.* New York: Raven Press.

Ito, M. (1998). Cerebellar learning in the vestibulo-ocular reflex. *Trends in Cognition Science, 219,* 321.

Ivarsson, M., and Hesslow, G. (1993). Bilateral control of the orbicularis oculi muscle by one cerebellar hemisphere in the ferret. *Neuroreport, 4,* 1127–1130.

Ivarsson, M., and Svensson, P. (2000). Conditioned eyeblink response consists of two distinct components. *Journal of Neurophysiology, 83,* 796–807.

Ivarsson, M., Svensson, P., and Hesslow, G. (1997). Bilateral disruption of conditioned responses after unilateral blockade of cerebellar output in the decerebrate ferret. *Journal of Physiology (London), 502,* 189–201.

James, G.O., Hardiman, M.J., and Yeo, C.H. (1987). Hippocampal lesions and trace conditioning in the rabbit. *Behavioural Brain Research, 23,* 109–116.

Kamin, L.J. (1969). Predictability, surprise attention and conditioning. In B. Campbell and R. Church, (Eds.), *Punishment and aversive behavior* (pp. 279–296). New York: Appleton-Century-Crofts.

Kawamura, K. (1975). The pontine projection from the inferior colliculus in the cat. An experimental anatomical study. *Brain Research, 95,* 309–322.

Kemble, E.D., Albin, J.M., and Leonard, D.W. (1972). The effects of amygdaloid lesions on a classically conditioned auditory discrimination in the rabbit (*Oryctalagus cuniculus*). *Psychonomic Science, 26,* 43–44.

Kim, J.J., Krupa, D.J., and Thompson, R.F. (1998). Inhibitory cerebello-olivary projections and blocking effect in classical conditioning. *Science, 279,* 570–573.

Krupa, D.J., and Thompson, R.F. (1995). Inactivation of the superior cerebellar peduncle blocks expression but not acquisition of the rabbit's classically conditioned eye-blink response. *Proceedings of the National Academy of Sciences of the United States of America, 92,* 5097–5101.

Krupa, D.J., Thompson, J.K., and Thompson, R.F. (1993). Localization of a memory trace in the mammalian brain. *Science, 260,* 989–991.

Larsell, O. (1970). *The comparative anatomy and histology of the cerebellum from monotremes through apes.* Minneapolis: University of Minnesota Press.

Larsell, O., and Jansen, J. (1972). *The comparative anatomy and histology of the cerebellum. The human cerebellum, cerebellar connections, and the cerebellar cortex.* Minneapolis: University of Minnesota Press.

Lavond, D.G., Hembree, T.L., and Thompson, R.F. (1985). Effect of kainic acid lesions of the cerebellar interpositus nucleus on eyelid conditioning in the rabbit. *Brain Research, 326,* 179–182.

Lavond, D.G., Steinmetz, J.E., Yokaitis, M.H., and Thompson, R.F. (1987). Reacquisition of classical conditioning after removal of cerebellar cortex. *Experimental Brain Research, 67,* 569–593.

Llinas, R., and Welsh, J.P. (1993). On the cerebellum and motor learning. *Current Opinion in Neurobiology, 3* 958–965.

Llinas, R., and Yarom, Y. (1981). Properties and distribution of ionic conductances generating electroresponsiveness of mammalian inferior olivary neurons in vitro. *Journal of Physiology (London), 315,* 569–584.

Llinas, R., Baker, R., and Sotelo, C. (1974). Electrotonic coupling between neurons in cat inferior olive. *Journal of Neurophysiology, 37,* 560–571.

Llinas, R., Lang, E.J., and Welsh, J.P. (1997). The cerebellum, LTD, and memory: alternative views. *Learning & Memory, 3,* 445–455.

Loucks, R.B. (1933). Preliminary report of a technique for stimulation or destruction of tissue beneath the integument and the establishing of conditioned responses with faradization of the cerebral cortex. *Journal of Comparative Physiology and Psychology, 16,* 439–444.

Loucks, R.B. (1935). The experimental delimitation of neural structures essential for learning. II. The conditioning of salivary and striped muscle responses to faradization of the sigmoid gyri. *Journal of Psychology, 1,* 5–44.

Luciani, L. (1915). The hind-brain. In G.M. Holmes, (Ed.) *Human physiology* (pp. 419–485). London: MacMillan.

Marchant, H.G., III, and Moore, J.W. (1973). Blocking of the rabbit's conditioned nictitation response in Kamin's two-stage paradigm. *Journal of Experimental Psychology, 101,* 155–158.

Marr, D. (1969). A theory of cerebellar cortex. *Journal of Physiology (London), 202,* 437–470.

Mauk, M.D. (1997). Roles of cerebellar cortex and nuclei in motor learning: contradictions or clues? *Neuron, 18,* 343–346.

Mauk, M.D., and Donegan, N.H. (1997). A model of Pavlovian eyelid conditioning based on the synaptic organization of the cerebellum. *Learning & Memory, 4,* 130–158.

Mauk, M.D., and Thompson, R.F. (1987). Retention of classically conditioned eyelid responses following acute decerebration. *Brain Research, 403,* 89–95.

Mauk, M.D., Steinmetz, J.E., and Thompson, R.F. (1986). Classical conditioning using stimulation of the inferior olive as the unconditioned stimulus. *Proceedings of the National Academy of Sciences of the United States of America, 83,* 5349–5353.

McCormick, D.A., and Thompson, R.F. (1984a). Cerebellum: essential involvement in the classically conditioned eyelid response. *Science, 223,* 296–299.

McCormick, D.A., and Thompson, R.F. (1984b). Neuronal responses of the rabbit cerebellum during acquisition and performance of a classically conditioned nictitating membrane-eyelid response. *Journal of Neuroscience, 4,* 2811–2822.

McCormick, D.A., Clark, G.A., Lavond, D.G., and Thompson, R.F. (1982a). Initial localization of the memory trace for a basic form of learning. *Proceedings of the National Academy of Sciences of the United States of America, 79,* 2731–2735.

McCormick, D.A., Guyer, P.E., and Thompson, R.F. (1982b). Superior cerebellar peduncle lesions selectively abolish the ipsilateral classically conditioned nictitating membrane/eyelid response of the rabbit. *Brain Research, 244,* 347–350.

McCormick, D.A., Lavond, D.G., and Thompson, R.F. (1983). Neuronal responses of the rabbit brainstem during performance of the classically conditioned nictitating membrane (NM)/eyelid response. *Brain Research, 271,* 73–88.

McCormick, D.A., Steinmetz, J.E., and Thompson, R.F. (1985). Lesions of the inferior olivary complex cause extinction of the classically conditioned eyeblink response. *Brain Research, 359,* 120–130.

Medina, J.F., and Mauk, M.D. (1999). Simulations of cerebellar motor learning: computational analysis of plasticity at the mossy fiber to deep nucleus synapse. *Journal of Neuroscience, 19,* 7140–7151.

Miles, F.A., and Lisberger, S.G. (1981). Plasticity in the vestibulo-ocular reflex: a new hypothesis. *Annual Review of Neuroscience, 4,* 273–299.

Miles, T.S., and Wiesendanger, M. (1975a). Organisation of climbing fiber projections to the cerebellar cortex from trigeminal cutaneous afferents and from the SI face area of the cat. *Journal of Physiology (London), 245,* 409–424.

Miles, T.S., and Wiesendanger, M. (1975b). Climbing fiber inputs to cerebellar Purkinje cells from trigeminal cutaneous afferents and the SI face area of the cerebral cortex in the cat. *Journal of Physiology (London), 245,* 425–445.

Montarolo, P.G., Palestini, M., and Strata, P. (1982). The inhibitory effect of the olivocerebellar input on the cerebellar Purkinje cells in the rat. *Journal of Physiology (London), 332,* 187–202.

Moore, J.W., Yeo, C.H., Oakley, D.A., and Steele-Russell, I. (1980). Conditioned inhibition of the nictitating membrane response in decorticate rabbit. *Behavioural Brain Research, 1,* 397–409.

Moyer, J.R., Deyo, R.A., and Disterhoft, J.F. (1990). Hippocampectomy disrupts trace eye-blink conditioning in rabbits. *Behavioral Neuroscience, 104,* 243–252.

Nelson, B., and Mugnaini, E. (1989). Origins of GABA-ergic inputs to the inferior olive. In P. Strata (Ed.), *The olivocerebellar system in motor control* (pp. 86–107). Berlin: Springer.

Nordholm, A.F., Thompson, J.K., Dersarkissian, C., and Thompson, R.F. (1993). Lidocaine infusion in a critical region of cerebellum completely prevents learning of the conditioned eyeblink response. *Behavioral Neuroscience, 107,* 882–886.

Norman, R.J., Villablanca, J.R., Brown, K.A., Schwafel, J.A., and Buchwald, J.A. (1974). Classical eyeblink conditioning in the bilaterally hemispherectomized cat. *Experimental Neurology, 44,* 363–380.

Oakley, D.A., and Russell, I.S. (1972). Neocortical lesions and Pavlovian conditioning. *Physiology & Behavior, 8,* 915–926.
Oakley, D.A., and Russell, I.S. (1975). Role of cortex in Pavlovian discrimination learning. *Physiology & Behavior, 15,* 315–321.
Oakley, D.A., and Russell, I.S. (1977). Subcortical storage of Pavlovian conditioning in the rabbit. *Physiology & Behavior, 18,* 931–937.
Oda, Y., Ito, M., Kishida, H., and Tsukahara, N. (1988). Formation of new cortico-rubral synapses as a possible mechanism for classical conditioning mediated by the red nucleus in cat. *Journal of Physiology (Paris), 83,* 207–216.
Orr, W.B., and Berger, T.W. (1985). Hippocampectomy disrupts the topography of conditioned nictitating membrane responses during reversal learning. *Behavioral Neuroscience, 99,* 35–45.
Pananceau, M., Rispal-Padel, L., and Meftah, E.M. (1996). Synaptic plasticity of the interpositorubral pathway functionally related to forelimb flexion movements. *Journal of Neurophysiology, 75,* 2542–2561.
Patterson, M.M., Cegavske, C.F., and Thompson, R.F. (1973). Effects of a classical conditioning paradigm on hind-limb flexor nerve response in immobilized spinal cats. *Journal of Comparative Physiology and Psychology, 84,* 88–97.
Pellegrini, J.J., Horn, A.K., and Evinger, C. (1995). The trigeminally evoked blink reflex. I. Neuronal circuits. *Experimental Brain Research, 107,* 166–180.
Perrett, S.P., and Mauk, M.D. (1995). Extinction of conditioned eyelid responses requires the anterior lobe of cerebellar cortex. *Journal of Neuroscience, 15,* 2074–2080.
Perrett, S.P., Ruiz, B.P., and Mauk, M.D. (1993). Cerebellar cortex lesions disrupt learning-dependent timing of conditioned eyelid responses. *Journal of Neuroscience, 13,* 1708–1718.
Popov, N.F. (1929). The role of the cerebellum in elaborating the motor conditioned reflexes. In D.S. Fursikov, M.O. Gurevich and A.N. Zalmanzon, (Eds.), *Higher Nervous Activity* (pp. 140–148). Moscow: Communist Academic Press.
Port, R.L., Romano, A.G., Steinmetz, J.E., Mikhail, A.A., and Patterson, M.M. (1986). Retention and acquisition of classical trace conditioned responses by rabbits with hippocampal lesions. *Behavioral Neuroscience, 100,* 745–752.
Port, R.L., Beggs, A.L., and Patterson, M.M. (1987). Hippocampal substrate of sensory associations. *Physiology & Behavior, 39,* 643–647.
Ramnani, N., and Yeo, C.H. (1996). Reversible inactivations of the cerebellum prevent the extinction of conditioned nictitating membrane responses in rabbits. *Journal of Physiology (London), 495,* 159–168.
Rawson, J.A., and Tilokskulchai, K. (1981). Suppression of simple spike discharges of cerebellar Purkinje cells by impulses in climbing fiber afferents. *Neuroscience Letters, 25,* 125–130.
Raymond, J.L., Lisberger, S.G., and Mauk, M.D. (1996). The cerebellum: a neuronal learning machine? *Science, 272,* 1126–1131.
Rosenfield, M.E., and Moore, J.W. (1983). Red nucleus lesions disrupt the classically conditioned nictitating membrane response in rabbits. *Behavioural Brain Research, 10,* 393–398.
Rosenfield, M.E., and Moore, J.W. (1995). Connections to cerebellar cortex (Larsell's HVI) in the rabbit: a WGA-HRP study with implications for classical eyeblink conditioning. *Behavioral Neuroscience, 109,* 1106–1118.
Rosenfield, M.E., Dovydaitis, A., and Moore, J.W. (1985). Brachium conjunctivum and rubrobulbar tract: brain stem projections of red nucleus essential for the conditioned nictitating membrane response. *Physiology & Behavior, 34,* 751–759.

Ross, R.T., Orr, W.B., Holland, P.C., and Berger, T.W. (1984). Hippocampectomy disrupts acquisition and retention of learned conditional responding. *Behavioral Neuroscience, 98,* 211–225.

Schmaltz, L.W., and Theios, J. (1972). Acquisition and extinction of a classically conditioned response in hippocampectomized rabbits (*Oryctolagus cuniculus*). *Journal of Comparative and Physiological Psychology, 79,* 328–333.

Schreurs, B.G. (1989). Classical conditioning of model systems: a behavioral review. *Psychobiology, 17,* 145–155.

Scoville, W.B., and Milner, B. (1957). Loss of recent memory after bilateral hippocampal lesions. *Journal of Neurology, 20,* 11–21.

Sears, L.L., and Steinmetz, J.E. (1991). Dorsal accessory inferior olive activity diminishes during acquisition of the rabbit classically conditioned eyelid response. *Brain Research, 545,* 114–122.

Shinoda, Y., Sugiuchi, Y., Futami, T., and Izawa, R. (1992). Axon collaterals of mossy fibers from the pontine nucleus in the cerebellar dentate nucleus. *Journal of Neurophysiology, 67,* 547–560.

Simpson, J.I., Wylie, D.R., and De Zeeuw, C.I. (1996). On climbing fiber signals and their consequences. *Behavioural Brain Science, 19,* 384–398.

Smith, A.M. (1970). The effects of rubral lesions and stimulation on conditioned forelimb flexion responses in the cat. *Physiology & Behavior, 5,* 1121–1126.

Smith, A M. (1996). Does the cerebellum learn strategies for the optimal time-varying control of joint stiffness? *Behavioural Brain Science, 19,* 399–410.

Smith, M.C., DiLollo, V., and Gormezano, I. (1966). Conditioned jaw movement in the rabbit. *Journal of Comparative Physiology and Psychology, 62,* 479–483.

Solomon, P.R. (1977). Role of the hippocampus in blocking and conditioned inhibition of the rabbit's nictitating membrane response. *Journal of Comparative and Physiological Psychology, 91,* 407–417.

Solomon, P.R., and Moore, J.W. (1975). Latent inhibition and stimulus generalization of the classically conditioned nictitating membrane response in rabbits (*Oryctolagus cuniculus*) following dorsal hippocampal ablation. *Journal of Comparative and Physiological Psychology, 89,* 1192–1203.

Solomon, P.R., Van der Schaaf, E.R., Thompson, R.F., and Weisz, D.J. (1986). Hippocampus and trace conditioning of the rabbit's classically conditioned nictitating membrane response. *Behavioral Neuroscience, 100,* 729–744.

Sotelo, C., Llinas, R., and Baker, R. (1974). Structural study of inferior olivary nucleus of the cat: morphological correlates of electrotonic coupling. *Journal of Neurophysiology, 37,* 541–559.

Steinmetz, J.E., Rosen, D.J., Chapman, P.F., Lavond, D.G., and Thompson, R.F. (1986a). Classical conditioning of the rabbit eyelid response with a mossy-fiber stimulation CS: I. Pontine nuclei and middle cerebellar peduncle stimulation. *Behavioral Neuroscience, 100,* 878–887.

Steinmetz, J.E., Rosen, D.J., Woodruff, P.D., Lavond, D.G., and Thompson, R.F. (1986b). Rapid transfer of training occurs when direct mossy fiber stimulation is used as a conditioned stimulus for classical eyelid conditioning. *Neuroscience Research, 3,* 606–616.

Steinmetz, J.E., Lavond, D.G., and Thompson, R.F. (1989). Classical conditioning in rabbits using pontine nucleus stimulation as a conditioned stimulus and inferior olive stimulation as an unconditioned stimulus. *Synapse, 3,* 225–233.

Steinmetz, J.E., Lavond, D.G., Ivkovich, D., Logan, C.G., and Thompson, R.F. (1992).

Disruption of classical eyelid conditioning after cerebellar lesions: damage to a memory trace system or a simple performance deficit? *Journal of Neuroscience, 12,* 4403–4426.

Svensson, P., and Ivarsson, M. (1999). Short-lasting conditioned stimulus applied to the middle cerebellar peduncle elicits delayed conditioned eye blink responses in the decerebrate ferret. *European Journal of Neuroscience, 11,* 4333–4340.

Svensson, P., Ivarsson, M., and Hesslow, G. (1997). Effect of varying the intensity and train frequency of forelimb and cerebellar mossy fiber conditioned stimuli on the latency of conditioned eye-blink responses in decerebrate ferrets. *Learning & Memory, 4,* 105–115.

Swain, R.A., Shinkman, P.G., Nordholm, A.F., and Thompson, R.F. (1992). Cerebellar stimulation as an unconditioned stimulus in classical conditioning. *Behavioral Neuroscience, 106,* 739–750.

Thach, W.T. (1996). On the specific role of the cerebellum in motor learning cognition: clues from PET activation and lesion studies in man. *Behavioral and Brain Sciences, 19,* 411–431.

Thompson, R.F. (1976). The search for the engram. *American Psychologist, 31,* 209–227.

Tsukahara, N. (1981). Synaptic plasticity in the mammalian central nervous system. *Annual Review of Neuroscience, 4,* 351–379.

Tsukahara, N., and Bando, T. (1970). Red nuclear and interposate nuclear excitation of pontine nuclear cells. *Brain Research, 19,* 295–298.

Tsukahara, N., Bando, T., Kitai, S.T., and Kiyohara, T. (1971). Cerebello-pontine reverbearating circuit. *Brain Research, 33,* 233–237.

Tsukahara, N., Hultborn, H., Murakami, F., and Fujito, Y. (1975). Electrophysiological study of formation of new synapses and collateral sprouting in red nucleus neurons after partial denervation. *Journal of Neurophysiology, 38,* 1359–1372.

Tsukahara, N., Oda, Y., and Notsu, T. (1981). Classical conditioning mediated by the red nucleus in the cat. *Journal of Neurosciences, 1,* 72–79.

Tyrell, T., and Willshaw, D. (1992). Cerebellar cortex: its simulation and the relevance of Marr's theory. *Philosophical Transactions of the Royal Society of London, 336,* 239–257.

van Ham, J.J., and Yeo, C.H. (1992). Somatosensory trigeminal projections to the inferior olive, cerebellum and other precerebellar nuclei in rabbits. *European Journal of Neuroscience, 4,* 317.

van Ham, J.J., and Yeo, C.H. (1996a). The central distribution of primary afferents from the external eyelids, conjunctiva, and cornea in the rabbit, studied using WGA-HRP and B-HRP as transganglionic tracers. *Experimental Neurology, 142,* 217–225.

van Ham, J.J., and Yeo, C.H. (1996b). Trigeminal inputs to eyeblink motoneurons in the rabbit. *Experimental Neurology, 142,* 244–257.

Voogd, J. (1992). The morphology of the cerebellum the last 25 years. *European Journal of Morphology, 30,* 81–96.

Voogd, J., and Glickstein, M. (1998). The anatomy of the cerebellum. *Trends in Neuroscience, 21,* 370–375.

Watson, J.B. (1916). The place of the conditioned reflex in psychology. *Psychological Review, 23,* 89–117.

Weiskrantz, L., and Warrington, E.K. (1979). Conditioning in amnesic patients. *Neuropsychologia, 17,* 187–194.

Wells, G.R., Hardiman, M.J., and Yeo, C.H. (1989). Visual projections to the pontine nuclei in the rabbit: orthograde and retrograde tracing studies with WGA-HRP. *Journal of Comparative Neurology, 279,* 629–652.

Welsh, J.P., and Harvey, J.A. (1989). Cerebellar lesions and the nictitating membrane reflex: performance deficits of the conditioned and unconditioned response. *Journal of Neuroscience, 9,* 299–311.

Welsh, J.P., and Harvey, J.A. (1991). Pavlovian conditioning in the rabbit during inactivation of the interpositus nucleus. *Journal of Physiology (London), 444,* 459–480.

Welsh, J.P., and Harvey, J.A. (1992). The role of the cerebellum in voluntary and reflexive movements: history and current status. In R. Llinas and C. Sotelo (Eds.), *The cerebellum revisited* (pp. 301–334). New York: Springer-Verlag.

Welsh, J.P., and Harvey, J.A. (1998). Acute inactivation of the inferior olive blocks associative learning. *European Journal of Neuroscience, 10,* 3321–3332.

Woodruff-Pak, D.S., Lavond, D.G., and Thompson, R.F. (1985). Trace conditioning: abolished by cerebellar nuclear lesions but not lateral cerebellar cortex aspirations. *Brain Research, 348,* 249–260.

Woody, C.D. (1970). Conditioned eye-blink: Gross potential activity at coronal-pericruciate cortex of the cat. *Journal of Neurophysiology, 33,* 838–850.

Woody, C.D. (1982). *Memory, learning and higher function.* New York: Springer-Verlag.

Yeo, C.H. (1989). The inferior olive and classical conditioning. In P. Strata (Ed.), *The olivocerebellar system in motor control* (pp. 363–373). Berlin: Springer.

Yeo, C.H., and Hardiman, M.J. (1992). Cerebellar cortex and eyeblink conditioning: a reexamination. *Experimental Brain Research, 88,* 623–638.

Yeo, C.H., Hardiman, M.J., and Glickstein, M. (1984). Discrete lesions of the cerebellar cortex abolish the classically conditioned nictitating membrane response of the rabbit. *Behavioural Brain Research, 13,* 261–266.

Yeo, C.H., Hardiman, M.J., and Glickstein, M. (1985a). Classical conditioning of the nictitating membrane response of the rabbit. I. Lesions of the cerebellar nuclei. *Experimental Brain Research, 60,* 87–98.

Yeo, C.H., Hardiman, M.J., and Glickstein, M. (1985b). Classical conditioning of the nictitating membrane response of the rabbit. II. Lesions of the cerebellar cortex. *Experimental Brain Research, 60,* 99–113.

Yeo, C.H., Hardiman, M.J., and Glickstein, M. (1985c). Classical conditioning of the nictitating membrane response of the rabbit. III. Connections of cerebellar lobule HVI. *Experimental Brain Research, 60,* 114–126.

Yeo, C.H., Hardiman, M.J., and Glickstein, M. (1985d). Classical conditioning of the nictitating membrane response of the rabbit. III. Connections of cerebellar lobule HVI. *Experimental Brain Research, 60,* 114–126.

Yeo, C.H., Hardiman, M.J., and Glickstein, M. (1986). Classical conditioning of the nictitating membrane response of the rabbit. IV. Lesions of the inferior olive. *Experimental Brain Research, 63,* 81–92.

Yeo, C.H., Lobo, D.H., and Baum, A. (1997). Acquisition of a new-latency conditioned nictitating membrane response—major, but not complete, dependence on the ipsilateral cerebellum. *Learning & Memory, 3,* 557–577.

5
Classical Conditioning: Applications and Extensions to Clinical Neuroscience

Paul R. Solomon

Most clinical research begins with animal models (and now increasingly in the dish) and progresses to clinical trials in human subjects. A variety of animal models are used, and the hope is that success in the animal model will predict success in the human condition. Applications of classical eyeblink (EB) conditioning to clinical medicine has followed a more circuitous path, from initial development in humans, where the basic behavioral parameters were established, to research in rabbits, where the neurobiological substrates were elucidated, and then a return to research in humans where clinical applications are beginning to be realized. Human EB conditioning was well established as a paradigm for studying learning as early as the 1930s (Hilgard & Marquis, 1936), but it was the pioneering work of Gormezano (Gormezano et al., 1962), demonstrating that EB conditioning could be studied in an animal preparation, that formed the basis for much of the progress in this paradigm.

More recently, considerable effort has been redirected toward humans. Classical EB conditioning can arguably be characterized as the paradigm in which more is known about both the behavioral and neurobiological aspects of learning and memory than any other. The exploitation of this paradigm has led to fundamental information about the basic laws of associative learning and about the neurobiological systems and mechanisms underlying this form of learning. More recently, this basic information has been applied to clinical situations. The purpose of this chapter is to review the clinical applications of classical EB conditioning including aging, Alzheimer's disease, neurological conditions, neurotoxicity, and developmental disorders. In doing so, my goal is to demonstrate how a thorough characterization of a basic form of learning and memory can be used to elucidate a variety of clinical conditions.

The Model Systems Approach

The model systems approach to the neurobiology of memory advocates studying a well-characterized learned response in a relatively simple and well-controlled preparation. The ultimate goal of this approach is to (1) characterize the behavioral response, (2) trace the neural circuitry controlling the response, (3) identify the site(s) of plasticity, and (4) elucidate the mechanism(s) of plasticity. The model systems approach has provided valuable information concerning the possible neural systems and mechanisms involved in learning and memory. The work in invertebrates has progressed to the point of beginning to identify possible mechanisms, and the work in mammalian preparations, and particularly EB conditioning, has made enormous strides in the past few years and is now at the point of identifying the circuitry for simple forms of learning. It was once thought that elucidation of the circuits and mechanisms of learning and memory would be the final goal, but this step may only represent the beginning. Understanding basic mechanisms of memory provides enormous opportunity to apply these principles to clinical conditions. For example, we have previously argued that the advantages of the model systems approach for studying memory would also apply to the study of age-related memory disorders (ARMD). Using a well-developed model system, it should be possible to both characterize the changes in learning and memory that accompany aging and investigate their neural substrate. With this information in hand, it should be possible to begin to develop interventions. Similarly, others have suggested that the model systems approach in general, and EB conditioning in particular, can be useful in elucidating the behavioral and neurobiological substrate of other clinical conditions including amnesia, neurotoxicity, and developmental disorders.

Age-Related Memory Disorders

Because so much is now known about both the behavioral and neurobiological aspects of EB conditioning, it has become possible to use this system to study age-related memory disorders (ARMD). There are a number of reasons why EB conditioning is especially well suited for this purpose (e.g., Solomon & Pendlebury, 1992);

1. The behavioral aspects of the response are well defined for both humans and rabbits. It is noteworthy that Gormezano (Gormezano et al., 1962) was the first to note the parallels between EB conditioning in humans and nictitating membrane (NM) conditioning in rabbits. Because the identical learned behaviors are used in the rabbit and human, animal-to-human extrapolation becomes more plausible. Indeed, Hilgard (Hilgard & Marquis, 1936) initially suggested that all mammals have similar neural circuitry for acquisition and retention of learned responses.
2. There are significant age-related differences in both humans and rabbits in the ability to learn and retain the conditioned response. These are reviewed in the next section.
3. The neural circuitry for the learned response is beginning to be well understood. It now appears that the cerebellum is the essential site of plasticity for the simple delay-conditioned response in both rabbits (Thompson & Kim, 1996) and humans (Topka et al., 1993; Solomon et al., 1989a; Woodruff-Pak et al., 1996a), whereas the hippocampus is involved in more complex types of learn-

ing such as trace conditioning (Moyer et al., 1990; Solomon et al., 1986) as well as modulation of the delay conditioning (Penick & Solomon, 1991). It is noteworthy that the hippocampus is a primary site of pathology in disorders of memory that accompany human aging (Ball, 1977; Hyman et al., 1984).
4. Although comparatively little is known about the pharmacology of this form of learning, what is known implicates the cholinergic system (Downs et al., 1972; Moore et al., 1976); this may be of some significance because of the hypothesized role of the cholinergic system in ARMD (Bartus et al., 1982; Collerton, 1986).
5. The rabbit provides an excellent preparation for studying the effects of experimentally induced neuropathology that may accompany ARMD. Specifically, aluminum exposure produces neurofilamentous degeneration in rabbits that is similar to the neurofibrillary tangles seen in Alzheimer's disease (AD) (Klatzko et al., 1965; Terry & Penna, 1965).
6. One major problem often associated with studying the memory disorders that accompany aging is that performance of the learned response can be undermined by factors that have nothing to do with learning or memory. Using a relatively simple form of learning such as the EB response, it is possible to rule out many of these factors, factors such as sensorimotor deficits, motivational differences, or fatigue (Durkin et al., 1993; Graves & Solomon, 1985).
7. Although classical conditioning is a relatively simple form of learning, it has been argued that both the behavioral and neurobiological mechanisms underlying Pavlovian conditioning are directly applicable to more complex forms of learning (Hawkins & Kandel, 1984).
8. Age-related deficits that are limited to certain conditioning tasks may help formulate hypotheses about the underlying neurobiological basis of these deficits. For example, problems with adult memory may be detectable earlier in a life span when using a trace conditioning protocol rather than a delay conditioning protocol. Such a finding could have implications for understanding the etiology of the deficit. As information regarding the neurobiological basis of conditioning in the rabbit becomes available, more precise statements about the anatomic substrates of ARMD becomes possible.
9. The EB preparation in rabbits and in humans has been used to begin to evaluate pharmacological agents that may alter learning and memory.

In summary, the advantages inherent to using a model systems approach to studying aging in young organisms may also apply to the study of age-related changes in learning and memory in aged organisms. Because the classically conditioned EB response is well characterized at the behavioral and neurobiological level, it may be the model system of choice for studying ARMD.

Conditioning in Animals Across the Life Span

It is now well established that a variety of species show age-related deficits in classical conditioning, including rabbits (Graves & Solomon, 1985; Powell et al.,

1981; Woodruff-Pak et al., 1987), cats (Harrison & Buchwald, 1983), and rats (Weiss & Thompson, 1991). Powell et al. (1981) reported that 3- to 5-year-old rabbits (mean age, 40 months) required significantly longer to condition to a criterion of 10 conditioned responses (CRs) in a block of 10 conditioned stimulus–unconditioned stimulus (CS–US) pairings than 6-month-old animals. They used a procedure in which a 500-ms tone CS terminated just as a 250-ms eye shock commenced. Graves and Solomon (1985) found no differences between 6-month-old and 36 to 60-month-old animals in delay conditioning in which the last 50 ms of the 500-ms tone CS overlapped with the eye shock US, but they reported a significant difference in a trace conditioning procedure in which there was a 500-ms interval between the CS offset and US onset. Similar results were reported by Woodruff-Pak et al., (1987), who found deficits in trace conditioning in animals beginning at 2.5 years old. Coffin and Woodruff-Pak (1993), using a delay procedure in which a 100-ms air puff US overlapped with the last 100 ms of a 500-ms tone CS, reported that 36-month-old animals acquired the CR significantly more slowly than 7-month old animals.

Although these studies suggest that EB conditioning in rabbits is impaired with aging, they leave several important questions unanswered. First, the exact ages of the animals are unknown, making comparisons across studies difficult. Second, because animals were often retired breeders who lived in the laboratory for only brief periods of time before conditioning, little was known about their health history or rearing. Third, there remains a question as to when conditioning deficits occur in various procedures.

The two most thoroughly studied procedures are delay conditioning, in which the CS and US overlap, and trace conditioning, in which there is a gap (trace interval) between the offset of the CS and the onset of the US. Because the hippocampus appears to play different roles in delay and trace conditioning (Moyer et al., 1990; Solomon et al., 1986), it is possible that these two types of conditioning degrade at different points in the life span. Graves and Solomon reported deficits in trace, but not delay, conditioning in rabbits between 36 and 60 months. Woodruff-Pak et al. (1987) have found deficits in trace conditioning as early as 30 months of age. A problem with these studies using the trace conditioning task is that they lack necessary controls. Specifically, in both the Graves and Solomon study, in which the CS and US are separated by 900 ms (the CS plus the trace period), and the Woodruff-Pak et al. study, in which the CS and US were separated by 700 ms, no "without trace" control was used. Thus, it is not clear whether a long interstimulus interval (ISI) or a long trace interval is crucial to observe the disrupting effects of hippocampal lesions.

To address these issues, Solomon and Groccia (1996) conducted a study in which four age groups of rabbits (0.5 years, 2+ years, 3+ years, and 4+ years) underwent acquisition of the classically conditioned nictitating membrane response (NMR) in either a delay (500-ms CS, 400-ms ISI), long-delay (1000-ms CS, 900-ms ISI), or trace (500-ms CS, 400-ms trace interval) procedure. Collapsing across age groups, there was a general tendency for animals to acquire trace conditioning more slowly than delay conditioning. Collapsing across conditioning procedures, there was a general tendency for aged animals to acquire the conditioned response more slowly than younger animals. Of greater significance, however, were the age differences in the different conditioning paradigms. In the delay and long-delay procedures, significant conditioning deficits first appeared in the group aged 4+ years. In the trace conditioning procedure, significant conditioning deficits became apparent in the animals that were 2+ years old. Nonassociative factors, including sensitivity to the air puff unconditioned stimulus (US) or the tone conditioned stimulus (CS) or general health, could not have accounted for these differences.

Differences in the time of onset of age-related deficits in the trace and delay paradigms may lead to hypotheses about the neurobiological bases for these two types of conditioning. As noted previously, trace conditioning appears to be hippocampally dependent in that hippocampal ablation disrupts conditioning in this procedure. Delay conditioning is not dependent upon an intact hippocampus, but the hippocampus does appear to play a modulatory role (Salvatierra & Berry, 1989; Solomon et al., 1983; Penick & Solomon, 1991), perhaps by affecting cerebellar function. Delay conditioning, however, appears to be mediated by the cerebellum (Thompson, 1986, 1990). Additionally, Woodruff-Pak et al., (1990a) have reported a positive correlation between the number of cerebellar Purkinje cells and rates of EB acquisition. They did not, however, correlate cell density in other brain areas with acquisition, raising the possibility that other cell densities in other brain areas would also be correlated. These studies raise the possibility that hippocampal neurons degenerate earlier in life than cerebellar neurons and that this could lead to the increased sensitivity of trace conditioning.

One interesting aspect of the data on aging and classical EB conditioning is that although aged animals on average showed deficits in conditioning, some aged animals acquired CRs as quickly as young animals (Solomon & Graves, 1985; Woodruff-Pak et al., 1987). To date, no one has investigated possible neurobiological differences between aged organisms that continue to learn with nearly the proficiency of their young counterparts and those that age less successfully. One such candidate mechanism is long-term potentiation (LTP). LTP is an increase in the excitability of neurons caused by high-frequency stimulation. LTP is considered to be a leading candidate mechanism for memory and has also been suggested to be important in ARMD. LTP has also been shown to be related to EB conditioning (Berger, 1984; Weisz et al., 1984; see also Schreurs & Alkon, this volume).

We (Yang et al., 1993) conducted a preliminary study examining LTP in the perforant path–dentate gyrus synapse in aged rabbits both before and after conditioning. These results indicated: (1) synaptic efficacy in the rabbit declines with aging; (2) the decay of LTP is more rapid in aged animals; (3) the rate of conditioning in aged animals is related to the degree of synaptic efficacy; specifically, the subset of old animals who condition at the same rate as young animals show LTP similar to that seen in young animals; and (4) following conditioning, there is an enhancement of synaptic efficacy that is significantly greater in young than

Retention of the Conditioned Response in Rabbits Across the Life Span

Although it is now clear that acquisition of the EB-conditioned response is disrupted in some aged subjects, much less is known about how long this response is retained. Several studies have suggested that retention is well preserved in aged rabbits. These studies have used reacquisition as a measure of retained information. Woodruff-Pak et al. (1987) presented data to suggest a positive correlation between initial CR acquisition and retention in five aged rabbits. Rabbits were initially trained in a trace conditioning procedure. They were then retested with a delay conditioning procedure in which the CS and US overlapped. Following a 2- to 5-month period, they underwent reacquisition as a retention test. Woodruff-Pak et al. (1987) reported a strong positive correlation between acquisition and reacquisition in the delay paradigms, and on the basis of this suggested that retention was good in aged animals.

In a subsequent study, Coffin and Woodruff-Pak (1993) also used reacquisition as a measure of retention. In this study, although they reported initial differences in the rate of acquisition in the delay paradigm as a function of age, they did not find any differences in reacquisition rates. Both young and aged rabbits reacquired at about the same rate. Based on these data, the authors suggested that aged rabbits actually showed superior retention of the conditioned response. Alternately, this rate of savings may be interpreted as poor retention; this is because the expectation for EB conditioning is that subjects will reacquire learning faster in subsequent testing sessions, presumably because they "retain" information from previous sessions. Unfortunately, we do not have sufficient data from young and, especially, aged animals to know what to expect in terms of reacquisition. Reacquisition can be influenced by a number of factors in addition to retaining the association between the CS and US. For example, simply habituating an animal to the apparatus with no CS or US presentations can lead to faster acquisition (Gormezano et al., 1983). There is also the issue of learning to learn, spontaneous recovery, and unextinguished responses to contextual stimuli following extinction sessions when multiple acquisition sessions are used (Napier et al., 1992). Each of these factors could contribute to changes in reacquisition rates without affecting retention of the CS–US association. Moreover, each could interact with the aging process.

In an attempt to circumvent some of the issues encountered in using reacquisition to evaluate the effects of aging on retention of the conditioned EB response, Solomon et al. (1995a) undertook a study using a retention paradigm in which animals initially acquired the CR, remained in their home cages during a retention interval, and then were tested with presentations of the CS alone to determine their retention of the CR (see Schreurs, 1993). Young (6–8 months) and aged (36–50 months) rabbits underwent classical conditioning of the NMR to a tone conditioned stimulus (CS) and a corneal air puff unconditioned stimulus (US) for 18 consecutive days (100 trials per day). Animals were then returned to their home cages for a 90-day period in which they received no further conditioning, but they were handled on a daily basis. On the 91st day, they underwent retention testing (extinction) during which the CS alone was presented 20 times. This step was immediately followed by reacquisition in which the CS and US were again paired for 100 trials. Reacquisition (100 trials) was repeated on the following day. As in previous studies, aged rabbits acquired the CR more slowly than young animals; however, by the end of acquisition, both groups reached similar levels of conditioned responding. Retention of the CR, considered either relative to performance on the last 6 days of acquisition or in terms of absolute number of CRs, was significantly lower for aged than young animals. Reacquisition was also retarded in aged versus young animals. Nonassociative factors including sensitivity to the stimuli or general health could not account for these differences.

The results of this study suggest that retention of the conditioned EB response is impaired in aged as compared to young rabbits. These data are generally consistent with the studies showing that acquisition is similarly retarded in aged versus young rabbits. Because these data suggest that retention deficits, a hallmark of cognitive disorders that accompany human aging, are also characteristic of EB conditioning, they further support the use of the classically conditioned EB response for studying age-related memory impairment.

Conditioning in Humans Across the Life Span

The first known demonstration that EB conditioning varies as a function of age was performed by Gakkel and Zinna in Russia in the 1950s (Jerome, 1959). This study reported a significant difference in CR acquisition favoring younger over older nursing home residents in a single 70- to 90-trial training session. Similar results have been reported in several subsequent studies of nominally healthy community-dwelling adults (Braun & Geiselhart, 1959; Kimble & Pennypacker, 1963). These studies, however, are subject to several alternative interpretations. For example, none of these studies attempted to determine if the subjects had age-appropriate cognitive abilities, nor was there any attempt to determine if nonassociative factors such as sensitivity to the CS or US or spontaneous blink rate could have contributed to the learning deficit. More recent studies have shown that EB conditioning declines across the age span, beginning in the fourth or fifth decade. Moreover, these deficits cannot be attributed to sensory or motor factors (Durkin et al., 1993), cognitive decline (Solomon et al., 1989b), or nonassociative factors (Solomon et al., 1989b; Woodruff-Pak & Thompson, 1988).

We (Solomon et al., 1989b) reported, in a cross-sectional study examining subjects aged 18 to 85 years, that significant differences began to appear as early as age 50 to 60. Moreover, there was a significant correlation (r = 0.59) between age and CR acquisition, with older subjects conditioning more slowly. Woodruff-Pak and Thompson (1988) reported similar results with significant declines in conditioning beginning at age 40.

Retention of the Conditioned Response in Humans Across the Life Span

As in the work with rabbits, studies of retention of the conditioned EB response in humans have typically used reacquisition as a measure of retention. Ferrante and Woodruff-Pak (1990) reported reacquisition at about the same rate as initial acquisition over a 12-month period in subjects aged 71 to 94. Deyo et al., (1990) had subjects aged 18 to 20 acquire the conditioned EB response and then tested retention 24 h, 7 days, or 28 days later. They reported no difference between initial acquisition and reacquisition for any of the retest intervals, but as in the case of the Ferrante and Woodruff-Pak study, there did not appear to be much evidence of faster reacquisition compared to initial acquisition.

There are few recent studies of long-term retention of conditioned responses in humans and animals (except see Schreurs, 1993). Skinner (1950) reported anecdotally that pigeons retained a key pecking response for 4 years. Several earlier studies using a variety of paradigms and preparations also suggested that long-term retention of the conditioned response is possible. For example, Marquis and Hilgard (1936) reported retention of the EB response in dogs for 16 months. Hilgard and Campbell (1936) reported good retention of the human EB response after 20 weeks, and Hilgard and Humphreys (1938) reported retention of this response after 19 months. In all cases, the subjects in these studies were young adults.

Solomon et al. (1998) recently had the unique opportunity to measure retention (not savings) in human subjects who 5 years earlier had participated in studies of acquisition of the conditioned EB response. The retention test consisted of 20 tone CS-alone presentations. Young subjects (23–31 years of age at the time of retention testing) showed good retention of the CR (45%), middle-aged subjects (45–52 years) showed reduced retention (28%), and aged subjects (69–78 years) showed little evidence of retention (<5%). Retention testing was followed by reacquisition of the CR in which the CS and US were once again paired. The ability to reacquire the CR also showed a decline with age. The data suggest that the CR can be retained over long intervals and that retention declines with age.

Alzheimer's Disease and Other Dementias

Solyom and Barik (1965) compared aged subjects (70–81 years), young subjects (20–43 years), and 17 hospitalized patients diagnosed as having either "senile dementia" or "cerebral arteriosclerosis." They found that young subjects conditioned most rapidly, followed by aged subjects, followed by "dementia-arteriosclerosis" patients.

More recent studies have found that patients with a diagnosis of probable Alzheimer's disease (AD) acquire CRs significantly more slowly than age-matched controls. We compared 15 patients with midstage AD to 15 age- and education-matched healthy controls in a single 70-trial delay EB conditioning session (Solomon et al., 1991). Although control patients readily acquired the CR, with evidence of CR acquisition beginning in the first 10 trials, AD patients showed no indication of CR acquisition. Similar results have been reported by Woodruff-Pak et al. (1989) in a single 90-trial EB conditioning session.

Although the data strongly suggest that conditioning is disrupted in AD patients, they do not address the question of whether such patients can acquire the CR given enough training. This information may be of some importance because it may begin to elucidate the neurobiological structures mediating the disrupted conditioning in AD patients. For example, if conditioning deficits are due to hippocampal damage, we would predict that AD patients would eventually acquire the CR because hippocampal damage is not sufficient to prevent acquisition of EB CRs. In contrast, if cerebellar damage were contributory, we would predict that AD patients should not acquire the CR. To address this question, we tested AD patients and age-matched controls over 4 days of EB conditioning (Solomon et al., 1995b). As in previous studies, AD patients performed significantly more poorly than controls on day 1, but by day 4 they were not significantly different than controls. Subsequent testing indicated that these effects were not caused by nonassociative factors such as changes in sensitivity to stimuli or disruption of the motor response. Additionally, we reported that neither AD patients nor controls showed any evidence of acquisition in an explicitly unpaired paradigm, suggesting that neither pseudoconditioning nor sensitization is contributory (Solomon et al., 1995b, p. 248).

Data regarding conditioning in patients with dementia from diseases other than AD are sparse. Woodruff-Pak et al. (1996b) found that EB conditioning in patients with vascular dementia was slower than age-matched controls but more rapid than in AD patients.

Animal Models of Disorders of Aging

There are currently no widely accepted animal models of AD. Nevertheless, it is possible to reproduce in animals some of the neurobiological deficits that are present in clinical conditions. One such approach is manipulation of neural transmitter systems.

Although there are multiple transmitter deficits in AD, the single most pronounced deficit is in acetylcholine (ACh). There is also considerable evidence that the cholinergic system is critical in EB conditioning. The septo-hippocampal cholinergic system is critically involved in EB conditioning. Early work by Berger and Thompson (1978) demonstrated increased firing of CA1 pyramidal cells that both preceded and modeled the behavioral conditioned response. We initially reported that blocking muscarinic cholinergic receptors with systemic scopolamine administration slowed acquisition of the CR in rabbits, but the animals did eventually acquire the CR and reached asymptotic levels (Moore et al., 1976). This finding has been independently replicated in several laboratories (Harvey et al., 1983; Woodruff-Pak & Hinchliffe, 1997). Similarly, we also initially demonstrated that acquistion of the EB response in humans is also disrupted following scopolamine administration (Solomon et al., 1993). In this study, 72 human volunteers received either saline, a low dose of oral scopolamine (0.6 mg), a high dose of oral scopolamine (1.2 mg), or a peripheral analogue (glycopyrolate). They then underwent classical EB conditioning in a delay conditioning paradigm. A dose-related decline in acquisition of the CR was demonstrated. Similar results have recently been reported by Bahro et al. (1995).

Our working hypothesis has been that the scopolamine-induced disruption of classical conditioning is mediated by the hippocampus. Several lines of evidence support this view. As noted earlier, hippocampal pyramidal cell firing predicts the emergence of the behavioral CR (Berger & Thompson, 1978). Additionally, spontaneous EEG activity predicts the rate of CR acquisition. Berry and Thompson (1978) reported the degree to which hippocampal EEG showed theta activity (4–8 Hz) before conditioning predicted the rate of CR acquisition. In a related study, Berry and Thompson (1979) reported that small lesions to the medial septum, the primary source of cholinergic projections to the hippocampus, disrupted both CR acquisition and hippocampal theta activity.

To directly test the hypothesis that scopolamine-induced disruption of classical conditioning is mediated by the hippocampus, Solomon et al. (1983) evaluated the effects of scopolamine injections in animals with hippocampal or neocortical ablations or unoperated controls. Although scopolamine administration disrupted animals with neocortical lesions and unoperated controls, the drug had no effect on rabbits with hippocampal ablations. These data suggest that the disruptive effects of scopolamine are mediated by the hippocampus. In support of this view, Salvatierra and Berry (1989) reported that both EB conditioning and conditioning-associated multiple-unit activity in hippocampus were disrupted in scopolamine-treated rabbits. Solomon and Gottfried (1981) reported that scopolamine administered directly into the medial septum disrupted acquisition of the EB response in a manner very similar to systemic administration.

Cholinergic afferents to the hippocampus originate in a single source in the medial septum (McKinney et al., 1983). Both muscarinic and nicotinic cholinergic receptors are present in the same target regions of these afferents, which raises the possibility that both systems are important in EB conditioning. In support of this, Woodruff-Pak et al. (1994) reported that the nicotinic receptor blocker mecamylamine disrupted acquisition of the EB CR.

Antidementia Compounds and Other Cognitive Enhancers

There are currently 4 million patients in the United States with Alzheimer's disease, and by the year 2040 it is estimated that there will be more than 14 million AD patients (Pendlebury & Solomon, 1996). Not surprisingly, there is considerable effort aimed at developing pharmacological treatments to ameliorate the symptoms of this disease. Because the primary cognitive symptom of AD is memory loss, many pharmacological agents are directed at enhancing memory. The general strategy for developing antidementia compounds is a "top-down" approach. Drugs for which the mechanisms of action may not be fully characterized are tested in human clinical populations. The U.S. Federal Drug Administration (FDA) will approve these compounds if they are deemed both safe and efficacious even if the mechanism is unknown. A second approach is "bottom up." This strategy advocates understanding basic mechanisms of learning and memory and then using these principles to develop pharmacological interventions.

The first drug approved to treat AD was tacrine (Cognex®), which was initially approved for medical use by the FDA in 1993. Development of this compound was based on the finding that AD patients have low levels of ACh. Tacrine is an acetylcholinesterase (AChe) inhibitor. The drug has beneficial effects on learning, memory, and other related cognitive processes (Knapp et al., 1994). Because of the work demonstrating disrupted acquisition of EB conditioning in both humans and rabbits with cholinergic antagonists, it might be reasonable to assume that cholinergic agonists such as tacrine would improve conditioning in AD patients. We have been collecting cases of AD patients who have been taking tacrine and comparing them to untreated controls matched for age, education, and stage of disease. The data indicate that tacrine does facilitate conditioning in a dose-dependent manner.

Metrifonate is a nonreversible Ache inhibitor that has been submitted to the FDA for approval for the treatment of AD. Like tacrine, it has been shown to improve cognitive abilities in AD patients (Morris et al., 1998). Disterhoft and colleagues have evaluated metrifonate in EB conditioning in rabbits. Kronforst-Collins et al. (1997a) reported facilitation of EB acquisition in aged rabbits pretreated for 1 week with metrifonate. In a subsequent study (Kronforst-Collins et al., 1997b), both acquisition and retention of the EB response were facilitated in aging rabbits pretreated for 3 weeks with metrifonate.

Stimulation of acetylcholine receptors can be accomplished in a variety of ways. In addition to competing with Ache, it is possible to directly stimulate receptors. Several drugs have been tested in AD patients that directly stimulate muscarinic receptors, but none to date have shown efficacy. The working hypothesis to explain this lack of efficacy is that muscarinic receptors downregulate following chronic stimulation. Currently, there are several drugs in various stages of clinical development that stimulate nicotinic receptors. Woodruff-Pak and colleagues have reported that the nicotinic agonist GTS-21 ameliorated the age-dependent conditioning deficits in acquisition of the rabbit's EB response (Woodruff-Pak et al., 1994). One interesting compound that has recently been approved for the treatment of AD is galantamine (Reminyl), which is both an AChe inhibitor and a nicotinic modulator. Nicotine has recently begun to receive considerable attention in AD. For example, the use of tobacco has now been associated with a reduced risk of AD and it is hypothesized that the neuroprotective effects may be mediated by nicotine (Lee, 1994). It is possible that the dual action of galantamine as both a muscarinic and nicotinic agonist may have potent effects on EB conditioning.

There are a variety of drugs in clinical development for AD that affect systems other than Ache. Several of these compounds have also been evaluated in EB conditioning. Disterhoft and colleagues have examined the role of nimodipine, a calcium channel blocker, in rabbit trace EB conditioning. They report that this compound significantly facilitates acquisition of the response in aged rabbits, elevating their performance to within the range of young controls (Deyo et al., 1989; Moyer & Disterhoft, 1994). Woodruff-Pak et al. (1997a) reported similar results in delay conditioning. Disterhoft and colleagues have suggested that calcium channel blockers affect conditioning at least in part by acting on the hippocampus. Biophysical studies in hippocampal slices of rabbits have demonstrated conditioning-specific reductions in the postburst afterhyperpolarization (AHP) in pyramidal cells of young rabbits (Disterhoft et al., 1986). Because this AHP is mediated by calcium-activated potassium currents, and because the AHP is increased in hippocampal neurons of aged animals, Disterhoft et al. (1989) have suggested that an increased calcium-mediated current is one potential mechanism of age-related learning deficits. Similarly, we (Solomon et al., 1995c) reported that retention of the CR was facilitated in aged rabbits in a dose-dependent manner in nimodipine-treated rabbits. In this study, aged rabbits initially underwent 18 days EB CR acquisition. They were then treated with a low or high dose of nimodipine or a vehicle control for 90 days. During this time, no further CS–US pairings were presented. Rabbits underwent testing for retention of the CR at 30 and 90 days. Retention testing consisted of 20 presentations of the CS alone. Rabbits in the control condition retained 46.4% of their predrug levels of conditioned responding and rabbits receiving the low dose of nimodipine retained 37.3% of their predrug levels after 30 days. After 90 days, retention in these animals declined to 8.1% and 14.1%, respectively. In contrast, rabbits receiving the high dose of nimodipine retained 85% of their predrug learning at 30 days with little decline at 90 days (77.1% retention).

Interestingly, despite these promising findings in animals, nimodipine was withdrawn from clinical development for AD because of lack of demonstrated efficacy. It is possible that the study designs used evaluate the drug were not appropriate to detect the types of changes seen in the animal studies. Specifically, the clinical research in human subjects was designed to detect symptomatic benefit over relatively short periods of time. The results in animals raise the possibility that the benefit of this class of agent may be on disease progression. Designs to detect changes in disease progression in human clinical trials tend to be longer in duration and use different outcome measures than studies to evaluate symptomatic change.

Nefiracetam is a nootropic agent in development for improved cognitive function in aging. It is hypothesized to stimulate Ach release via GABA-ergic mediation. Woodruff-Pak and Li (1994) reported that nefiracetam ameliorated impaired EB conditioning in aged rabbits. More recently, they demonstrated that this effect was mediated via the hippocampus (Woodruff-Pak et al., 1997b). Using the strategy initially proposed by Solomon et al. (1983), they reported that aged animals with bilateral hippocampal ablation did not benefit from nefiracetam treatment.

It is difficult to speculate about common neurobiological mechanisms of improved function with these compounds. It is possible, however, that cholinergic systems in hippocampus are important. This idea would certainly be consistent with the importance of the hippocampal cholinergic system in disorders of memory including AD and the central role of this system in classical EB conditioning. Because acquisition and retention of the EB response appear sensitive to the effects of antidementia compounds, EB conditioning could be helpful in evaluating these and other putative cognitive enhancers.

Neurological Damage

The basic work on acquisition of the classically conditioned EB response in animals with experimentally induced lesions has provided a basis for interpreting the effects of naturally occurring lesions on acquisition of the classically conditioned EB response in humans. The two neural systems that have been best characterized in animals studies of EB conditioning are the hippocampus and cerebellum. The work of Thompson and coworkers as well as that of numerous other groups (see Thompson & Krupa, 1994, for a review) has presented a compelling case that the cerebellum is necessary for learning, retention, and expression for classical conditioning of the EB and NMR responses in rabbits (Thompson et al., 1997b). The locus of the long-term memory trace is somewhat less clear, but emerging evidence suggests that a parallel, distributed system involving the interpositus nucleus and the cerebellar cortex is possible. Emerging evidence also suggests that long-term depression of Purkinje cells is a candidate mechanism of plasticity (Kim & Thompson, 1997; see also Schreurs & Alkon and Hesslow & Yeo, this volume). Emerging data on EB CR acquisition in patients with cerebellar damage also suggests that this structure is necessary for conditioning.

Two case studies provided the earliest evidence for cerebellar involvement in EB conditioning. Lye et al. (1988) trained a patient who had 6 years earlier experienced a right cerebellar hemisphere infarction. This patient readily acquired the EB CR in the eye contralateral to the infarct but did not acquire the CR in the ipsilateral eye. Solomon et al. (1989a) evaluated a patient with cerebellar dysfunction secondary to an atrial tumor. This patient also did not acquire an EB CR. More recent studies with larger groups of subjects confirm these results. Topka and colleagues (Topka et al., 1993) reported severe impairment in 12 patients with cerebellar atrophy. Daum et al. (1993) reported severely disrupted EB classical conditioning but intact autonomic conditioning in seven patients with cerebellar degeneration. Woodruff-Pak et al. (1996a) tested six patients with unilateral and seven patients with bilateral cerebellar lesions. They reported disrupted acquisition in the patients with bilateral lesions and in patients with unilateral lesions in the eye ipsilateral to the lesion.

The role of the hippocampus in NMR conditioning is more complex. Initial electrophysiological evidence demonstrated recruitment of hippocampal pyramidal cells during conditioning in a delay paradigm (Berry & Thompson, 1978);

bilateral ablation of this structure had no effect on CR acquisition (Schmaltz & Theios, 1972; Solomon & Moore, 1975). Nevertheless, although the hippocampus is not necessary for CR acquisition in the delay paradigm, it can play a modulatory role, perhaps by acting on the cerebellum. Additionally, the hippocampus appears necessary in trace conditioning. We initially reported that rabbits with bilateral hippocampal ablation did not acquire the EB CR with the trace procedure. Moreover, multiple-unit recording revealed hippocampal pyramidal cell activity during the trace interval. These results have been replicated and extended (James et al., 1987; Moyer et al., 1990). This activity disappears once the CR becomes robust, suggesting that the hippocampus is actively engaged during formation of the CS–US association but that it may not be essential for performance of a well-established and robust CR.

Research on patients with damage to the medial temporal lobes is somewhat more difficult to interpret. Weiskrantz and Warrington (1979) reported normal acquisition of the EB CR in a delay procedure in two amnesic patients, one with Korsakoff's disease and the second with postencephalitis syndrome. Daum et al. (1989) reported normal CR acquisition in a delay conditioning paradigm in three amnestic patients, one postencephalitic case and two patients with epilepsy. These patients were, however, impaired in discrimination learning and discrimination reversal. In a subsequent study, Daum et al. (1991) investigated delay EB conditioning as well as the electrodermal response in 17 normal subjects and 17 patients with resection of the right or left temporal lobe for treatment of intractable epilepsy. Daum et. al. (1991) reported that the unilateral hippocampal ablation did not affect EB CR acquisition, but did impair discrimination learning and discrimination reversal. It is noteworthy that acquisition of an initial discrimination is intact in rabbits with hippocampal ablation, but discrimination reversal is impaired in a differential EB conditioning procedure (Berger & Orr, 1983; Orr & Berger, 1985). Woodruff-Pak (1993) evaluated both delay and trace conditioning in patient H.M. and found that conditioning was impaired in both paradigms. It is unlikely, however, that H.M. represents a case of pure temporal lobe damage. Magnetic resonance imaging (MRI) results suggest that he also had extensive cerebellar damage, ostensibly due to prolonged treatment with the anticonvulsant Dilantin (Corkin et al., 1997). This damage could certainly explain the conditioning deficits. Gabrielli et al. (1995) found that 7 amnestic patients, 5 of whom had documented radiologic bilateral hippocampal damage, also acquired the CR normally in the delay conditioning paradigm. These studies would seem to suggest that in humans, as was the case for rabbits, the hippocampus is not necessary for acquisition of the CR in the delay conditioning procedure.

There are several studies investigating the effects of medial temporal lobe damage on trace conditioning. As noted earlier, Woodruff-Pak reported deficits in trace conditioning in patient H.M., but cerebellar damage could also be contributory in this paradigm. Gabrielli et al. (1995) evaluated trace conditioning in patients who had previously acquired the CR in the delay paradigm and found that the conditioning deficit increased as the trace interval increased, but previous acquisition in the delay paradigm makes these data somewhat difficult to interpret.

Numerous studies have shown that anoxia can produce damage limited to the hippocampus that results in global amnesia (Zola-Morgan & Squire, 1990).

In summary, parallels between conditioning in humans and rabbits with neurological damage are apparent in the research literature. Both rabbits with cerebellar lesions and humans with cerebellar damage cannot acquire an EB CR. In contrast, both rabbits and humans with hippocampal damage appear to readily acquire an EB CR in a delay conditioning paradigm (McGlinchey-Berroth et al., 1999). In trace conditioning, rabbits with hippocampal damage show impaired CR acquisition. Humans with hippocampal damage also appear to be impaired in the trace paradigm, but sufficient data to assert this point are lacking.

Neurotoxicity

Exposure to neurotoxins can play an important role in the etiology of memory impairment associated with neurodegenerative disorders. Because of this, significant effort has been devoted to developing animal models of these neurotoxin-induced disorders. One important aspect of any strategy for developing neurotoxin-induced models of neurodegenerative disorders is to reduce the uncertainty involved in generalizing from animal models to human conditions. One neurotoxin that has been evaluated using EB conditioning is aluminum. Aluminum is a particularly interesting substance because of its suggested relationship to a variety of degenerative neurological disorders, including amyotrophic lateral sclerosis and Guamanian Parkinson's disease (Perl & Pendlebury, 1986). Furthermore, aluminum may contribute to AD.

It has been known for some time that intrathecal (spinal) administration of aluminum salts in rabbit produces an acute encephalopathy characterized by the formation of neuronal cytoplasmic structures resembling neurofibrillary tangles at the light microscopic level (Klatzko et al., 1965). This result raised the possibility of using EB conditioning in the rabbit to characterize aluminum-induced neurofilamentous degeneration (NFD) at behavioral, pathological, neurochemical, and immunocytochemical levels. To this end, Pendlebury et al. (1988a) and Solomon and Pendlebury (1988) found the following:

1. Intraventricular injection of aluminum chloride produced NFD and disrupted both acquisition and retention of the CR. Animals that received aluminum before conditioning acquired the CR significantly more slowly than controls (Pendlebury et al., 1988b), and rabbits that first acquired the CR and then received aluminum showed disrupted retention of the CR relative to controls (Solomon et al., 1988).
2. Disrupted learning and memory did not appear to arise from either sensory or motor deficits. Motor responses were intact. Aluminum-treated rabbits were no different than controls in their ability to reflexively give the EB response when presented with an air puff US (i.e., the response amplitudes do not differ from

controls). Responses to the tone CS also appeared unaltered as threshold tests for tone were similar in control and aluminum rabbits. Furthermore, CR retention was disrupted in aluminum-treated rabbits when electrical brain stimulation to the medial geniculate body was used as a CS. This result indicated that memory deficits were not caused by pathology of the primary sensory pathways as they were bypassed using this technique (Solomon et al., 1987).
3. The deficits in learning and memory were related to the degree of pathology. We have evaluated the number of neurons containing neurofilamentous accumulation in five brain areas (frontal and parietal cortex, ventral hippocampus, cerebellum, and pons) and found that the overall degree of pathology was significantly correlated with the likelihood of CRs (Pendlebury et al., 1988; Solomon et al., 1988).
4. Quantitative neurochemical analysis of aluminum-exposed rabbits revealed similarities to neurochemical deficits seen in AD patients. The most striking parallel was the significant reduction in CAT activity in the entorhinal cortex and hippocampus in aluminum-exposed rabbits (Beal et al., 1989).

Developmental Disorders

Human EB conditioning does not require that the subject understand instructions, nor does it require that the subject emit a verbal response. Therefore, this task is well suited for very young subjects. Stanton and his colleagues have begun to study EB conditioning in the developing rat and infant humans. Stanton et al. (1992) evaluated acquisition of the EB CR in 17-, 18-, or 24-day-old rat pups. They found that both groups acquired the CR, but that the older pups learned more rapidly. In a parallel study in human infants, they reported that 4- and 5-month-old healthy infants demonstrated acquisition of the CS to a tone CS and air puff US (Ivkovich et al., 1999). The finding of robust CR acquisition in healthy infants suggests that this preparation may be useful for early identification of infants at risk for cognitive disorders (Solomon et al., 1994; Stanton & Freeman, 1994).

Sears et al. (1994) have begun to investigate EB conditioning in autistic patients because both cerebellar and limbic system pathology are present in autism. They evaluated acquisition of the EB CR in 11 autistic persons and 11 matched controls. Compared to controls, patients with autism acquired the CR faster. Their CR topographies and latencies, however, were different. Based on these data, Sears et al. (1994) hypothesized that autistic patients may be able to associate the CS and US but have difficulty with timing the response.

Summary and Conclusions

In this chapter, I have attempted to summarize the application of classical EB conditioning to clinical conditions including aging, Alzheimer's disease, amnestic disorders, neurotoxicity, and developmental disorders. A number of important points emerge from investigations of classical EB conditioning in both humans and animals. These points raise possibilities for clinical applications:

1. Acquisition of the conditioned response in delay conditioning begins to decline in rabbits as early as age 4 (years) and in the trace conditioning procedure as early as age 2.
2. Acquisition of the conditioned response in delay conditioning in humans begins to decline in the fourth decade. Age 40 in humans is analogous to age 4 in rabbits. The parallels between conditioning in humans and rabbits raise the possibility of using animals to model human clinical conditions using an identical behavioral paradigm.
3. Retention of the EB conditioned response is affected by age in both rabbits and humans.
4. Patients with a diagnosis of probable Alzheimer's disease acquire the conditioned response more slowly than age-matched controls.
5. Cholinergic blockade with scopolamine slows acquisition of the EB-conditioned response in both humans and rabbits. The cholinergic system is the most depleted transmitter system in AD and the target of current therapeutic interventions.
6. Treatment with antidementia compounds that boost synaptic acetylcholine levels facilitate acquisition of the EB response in rabbits. This finding raises the possibility of using this system to screen antidementia compounds.
7. Both rabbits and humans with cerebellar damage show disrupted acquisition of the conditioned EB response.
8. Both rabbits and humans with hippocampal damage readily acquire the conditioned EB response in the delay conditioning procedure but not in the trace conditioning procedure.
9. EB conditioning is highly sensitive to the toxic effects of aluminum at both the behavioral and neurobiological levels.
10. EB conditioning may be useful in studying developmental disorders in human infants.

It is difficult to imagine that Gormezano and the others who pioneered EB conditioning would have envisioned the application of this simple form of learning to a variety of clinical conditions. Perhaps, however, they would find these applications gratifying. Clinical research is often conducted in a top-down manner in which interventions (e.g., drugs) are adopted because they are efficacious and safe. Elucidation of the mechanisms by which these interventions operate is not crucial. The clinical application of the work on EB conditioning represents a bottom-up approach in which an understanding of basic behavioral and neurobiological mechanisms has led to clinical applications. Ultimately, the merging of both top-down and bottom-up approaches will lead to the most efficacious clinical outcomes. Because of the simplicity of the approach and the parallel findings in humans and animals, continuing work in classical EB conditioning should continue to make an important contribution to this process.

Acknowledgment. Preparation of this manuscript was supported by NIA grant AGO-5134–08S2, a grant from the Howard Hughes Medical Foundation, and a grant from the Essel Foundation.

References

Bahro, M., Schreurs, B.G., Sunderland, T., and Molchan, S.E. (1995). The effects of scopolamine, lorazepam, and glycopyrrolate on classical conditioning of the human eyeblink response. *Psychopharmacology, 122,* 395–400.

Ball, M.J. (1977). Neuronal loss, neurofibriallary tangels and granulovascular degeneration in the hippocampus with aging and dementia. *Acta Neuropathologica, 37,* 111–118.

Bartus, R.T., Dean, R.L., Beer, B., and Lipa, A.S. (1982). The cholinergic huypothesis of geriatric memory dysfunction. *Science, 217,* 408–416.

Beal, M.F., Mazurek, M.F., Ellison, D.W., Kowall, N.W. Solomon, P.R. and Pendlebury, W.W. (1989). Neurochemical characteristics of aluminum-induced neurofibrillary degeneration in rabbits. *Neuroscience, 29,* 339–346.

Berger, T. (1984). Long-term potentiation of hippocampal synaptic transmission affects rate of behavioral learning. *Science, 20,* 810–816.

Berger, T.W., and Orr, W.B. (1983). Hippocampetomy selectively disrupts discrimination reversal conditioning of the rabbit nictitating membrane response. *Behavioral Brain Research 8,* 49–68.

Berger, T.W., and Thompson, R.F. (1978). Neuronal plasticity in the limbic system during classical conditioning of the rabbit-nictitating membrane response. I. The hippocampus. *Brain Research, 145,* 323–346.

Berry, S.D., and Thompson, R.F. (1978). Prediction of learning rate from the hippocampal electroencephalogram. *Science, 200,* 1298–1300.

Berry, S.D., and Thompson, R.F. (1979). Medial septal lesions retard classical conditioning of the nicitating membrane response in rabbits. *Science, 205,* 209–211.

Braun, W.H., and Geislehart, R. (1959). Age differences in the acquisition and extinction of the conditioned eyelid response. *Journal of Experimental Psychology, 57,* 386–388.

Coffin, J.M., and Woodruff-Pak, D.S. (1993). Delay clasical conditioning in young and older rabbits: Acquisition and retention at 12 and 18 months. *Behavioral Neuroscience, 107,* 63–71.

Collerton, D. (1986). Cholinergic function and intellectual decline in Alzheimer's disease. *Neuroscience, 19,* 1–28.

Corkin, S., Amaral, D.G., Gonzalez, R.G., Johnson, K.A., and Hyman, B.T. (1997). H.M.'s medial temporal lobe lesion: findings from magnetic resonance imaging. *Journal of Neuroscience, 17,* 3964–3979.

Daum, I., Channon, S., and Canavan, A.G.M. (1989). Classical conditioning in patients with severe memory problems. *Journal of Neurology, Neurosurgery, and Psychiatry, 52,* 47–51.

Daum, I., Channon, S., Polkey, C.E., and Gray, J.A. (1991). Classical conditioning after temporal lobe lesions in man: impairment in conditional discrimination. *Behavioral Neuroscience, 105,* 396–408.

Daum, I., Schugens, M.M., Ackermann, H., Lutzenberger, W., Dichgans, J., and Birbaumer, N. (1993). Classical conditioning after cerebellar lesions in humans. *Behavioral Neuroscience, 107,* 748–756.

Deyo, R.A., Straube, K.T., and Disterhoft, J.F. (1989). Nimodipine facilitates associative learning in aging rabbits. *Science, 243,* 809–811.

Deyo, R.A., Gabrielli, I.D.E., and Disterhoft, J.F. (1990). Human associative learning: analysis of trace and delay eyeblink conditioning. *Neuroscience Abstracts, 16,* 841.

Disterhoft, J.F., Coulter, D.A., and Alkon, D.L. (1986). Conditioning-specific membrane changes of rabbit hippocampal neurons measured in vitro. *Proceedings of the National Academy of Sciences of the United States of America, 83,* 2733–2737.

Disterhoft, J., Deyo, R.A., Black, J., de Jonge, M.C., Straube, K.T., and Thompson, R.F. (1989). Associative learning and aging rabbits is facilitated by nimodipine. In J. Traber, W.H. Gispen (Eds.), *Nimodipine and central nervous system function: new vistas* (pp. 209–224). Stuttgart-New York: Schattauer.

Downs, D., Cardoza, C., Schneiderman, N., Yehle, A.L., Can Decar, D.H., and Zwilling, G. (1972). Central effects of atropine upon aversive classical conditioning in rabbits. *Psychopharmacology, 23,* 319–333.

Durkin, M., Prescott, L., Furchtgott, E., Cantor, J., and Powell, D.A. (1993). Concomitant eyeblink and heart rate classical conditioning in young, middle-aged, and elderly human subjects. *Psychology of Aging, 8,* 571–581.

Ferrante, L.S., and Woodruff-Pak, D.S. (1990). Longitudinal classical conditioning data in the old. *Gerontological Society Abstracts, 30,* 50.

Gabrielli, J.D.E., McGlinchey-Berroth, R., Carrillo, M.C., Gluck, M.A., Cermak, L.S., and Disterhoft, J.F. (1995). Intact delay-eyeblink classical conditioning in amnesia. *Behavioral Neuroscience, 109,* 819–827.

Gakkel, L.B., and Zinna, N.V. (1983). Changes of higher nerve function in people over 60 years of age. *Fiziolog Zhurnal, 39,* 533–539.

Gormezano, I., Schneiderman, N., Deaux, E., and Fuentes, I. (1962). Nictitating membrane: classical conditioning and extinction in the albino rabbit. *Science, 138,* 33–34.

Gormezano, I., Kehoe, E.J., and Marshall, B.J. (1983). Twenty years of classical conditioning with the rabbit. *Progress in Psychobiology and Physiological Psychology, 10,* 197-275.

Graves, C.A., and Solomon, P.R. (1985). Age-related disruption of trace but not delay classical conditioning of the rabbit's nictitating membrane response. *Behavioral Neuroscience, 99,* 88–96.

Harrison, J., and Buchwald J. (1983). Eyeblink conditioning deficits in the old cat. *Neurobiology of Aging, 4,* 45–51.

Harvey, J.A. Gormezano, I., and Cool-Hauser, V.A. (1983). Effects of scopolamine and methylscopolamine on classical conditioning of the rabbit nictitating membrane response. *Journal of Pharmacology and Experimental Therapy, 225,* 42–49.

Hawkins, R.D., and Kandel, E.R. (1984). Is there a cell biological alphabet for simple forms of learning? *Psychological Review, 91,* 375–391.

Hilgard, E.R., and Campbell, A.A. (1936). The course of acquisition and retention of conditioned eyelid responses in man. *Journal of Experimental Psychology, 19,* 227–247.

Hilgard, E.R., and Humphreys, L.G. (1938). The retention of conditioned discrimination in man. *Journal of General Psychology, 19,* 111–125.

Hilgard, E.R., and Marquis, D.G. (1936). Conditioned eyelid responses in monkeys, with a comparison of dog, monkey and man. *Psychological Monogram, 47,* 186–198.

Hyman, B.T., VanHoesen, G.W., Damasio, A.R., and Barnes, C.L. (1984). Alzheimer's disease: cell-specific pathology isolates the hippocampal formation. *Science, 225,*

1168–1170.

Ivkovich, D., Collins, K.L., Eckerman, C.O., Krasnegor, N.A., and Stanton, M.E. (1999). Classical delay eyeblink conditioning in 4- and 5-month-old human infants. *Psychological Science, 10,* 4–8.

James, G.O., Hardiman, J.J., and Yeo, C.H. (1987). Hippocampal lesions and trace conditioning in the rabbit. *Behavioral and Brain Research, 23,* 109–116.

Jerome, E.A. (1959). Age and learning—experimental studies. In J.E. Birren (Ed.), *Handbook of aging and the individual* (pp. 655–699). Chicago: University of Chicago Press.

Kim, J.J., and Thompson, R.F. (1997). Cerebellar circuits and synaptic mechanisms involved in classical eyeblink conditioning. *Trends in Neuroscience, 20,* 177–181.

Kim, J.J., Clark, R.E., and Thompson, R.F. (1995). Hippocampectomy impairs the memory of recently, but not remotely acquired trace eyeblink conditioned responses. *Behavioral Neuroscience, 109,* 195–203.

Kimble, G.A., and Pennypacker, H.S. (1963). Eyelid conditioning in young and aged subjects. *Journal of Genetic Psychology, 103,* 283–289.

Klatzo, I., Wisniewski, H., and Streicher, E. (1965). Experimental production of neurofibrillary degeneration: 1. Light microscopic observations. *Journal of Neuropathology and Experimental Neurology, 24,* 187–199.

Knapp, M.J., Knopman, D.S., Solomon, P.R., Pendlebury, W.W., Davis, C.S., and Gracon, S.I. for the Tracrine Study Group. (1994). A 30-week randomized controlled trial of high-dose tacrine in patients with Alzheimer's disease. *Journal of the American Medical Association, 271,* 985–991.

Kronforst-Collins, M.A., Moriearty, P.L., Ralph, M., Becker, R.E., Schmidt, B., Thompson, L.T., and Disterhoft, J.F. (1997a). Metrifonate treatment enhances acquisition of eyeblink conditioning in aging rabbits. *Pharmacology, Biochemistry and Behavior, 56,* 103–110.

Kronforst-Collins, M.A., Moriearty, P.L., Schmidt, B., and Disterhoft, J.F. (1997b). Metrifonate improves associative learning and retention in aging rabbits. *Behavioral Neuroscience, 111,* 1031–1040.

Lee, P.N. (1994). Smoking and Alzheimer's disease: a review of the epidemiological evidence. *Neuroepidemiology, 13,* 131–144.

Lye, R.H., O'Boyle, D.J., Ramsden, R.T., and Schady, W. (1988). Effects of a unilateral cerebellar lesion on the acquisition of eye-blink conditioning in man. *Journal of Physiology (London), 403,* 58P.

Marquis, D.G., and Hilgard, E.R. (1936). Conditioned lid responses to light in dogs after removal of the visual cortex. *Journal of Comparative Psychology, 22,* 157–178.

Mcdonald, M.P., and Overmier, J. B. (1998). Present imperfect: a critical review of animal models of the mnemonic impairments in Alzheimer's disease. *Neuroscience and Biobehavioral Reviews, 22,* 99–120.

McKinney, M., Coyle, J.T., and Hedreen, J.C. (1983). Topographic analysis of the innervation of the rat neocortex and hippocampus by the basal forebrain cholinergic system, *Journal of Comparative Neurology, 217,* 103–111.

Moore, J.W., Goodell, N.A., and Solomon, P.R. (1976). Central cholinergic blockade by scopolamine and habituation, classical conditioning, and latent inhibition of the rabbit's nictitating membrane response. *Physiological Psychology, 4,* 395–399.

Morris, J.M, Cyrus, P.M., Orazam, J., Mas, J., Bieber, F., Ruzicka, B.B., and Gulanski, B. (1998). Metrifonate benefits cognitive, behavioral, and global function in patients with Alzheimer's disease. *Neurology, 50,* 122–130.

Moyer, J.R., and Disterhoft, J.F. (1994). Increased calcium action potentials in aging rabbit

CA1 neurons are reduced by low concentrations of nimodipine. *Hippocampus, 4,* 11–18.
Moyer, J.R., Deyo, R.A. and Disterhoft, J.F. (1990). Hippocampectomy disrupts trace eyeblink conditioning in rabbits. *Behavioral Neuroscience, 104,* 241–250.
Napier, R.M., Macrae, M., and Kehoe, E.J. (1992). Rapid reacquisition of the rabbit's nictitating membrane response. *Journal of Experimental Psychology: Animal Behavior Processes, 18,* 182–192.
Orr, W.B., and Berger, T.W. (1985). Hippocampectomy disrupts the topography of conditioned nictitating membrane responses during reversal learning. *Behavioral Neuroscience, 99,* 35–45.
Pendlebury, W.W., and Solomon, P.R. (1996). Alzheimer's disease. *Clinical Symposia, 48(3).*
Pendlebury, W.W., Beal, M.F., Kowall, N.W., and Solomon, P.R. (1988a). Immunocytochemical, neurochemical, and behavioral studies in aluminum-induced neurofilamentous degeneration. *Journal of Neural Transmission.*
Pendlebury, W.W., Perl, D.P., Schwentker, A., Pingree, T.M., and Solomon, P.R. (1988b). Aluminum-induced neurofibrillary degeneration disrupts acquisition of the rabbit's classicly conditioned nictitating membrane response. *Behavioral Neuroscience, 102,* 615–620.
Penick, S., and Solomon, P.R. (1991). Hippocampus, context, and conditioning. *Behavioral Neuroscience, 105,* 611–617.
Perl, D.P., and Pendlebury, W.W. (1986). The neuropathology of dementia. *Neurological Clinics of North America, 4,* 355–368.
Powell, D.A., Buchanan, S.L., and Hernandez, L.L. (1981). Age-related changes in classical (Pavlovian) conditioning in the New Zealand allbino rabbit. *Experimental Aging Research, 7,* 453–465.
Powell, D.A., Buchanan. S.L., and Hernandez, L.L. (1991). Classical (Pavlovian) conditioning models of age-related changes in associative learning and their neurobiooloogical substrates. *Progress in Neurobiology, 36,* 201–228.
Salvatierra, A.T., and Berry, S.D. (1989). Scopolaimine disruption of septo-hippocampal activity and classical conditioning. *Behavioral Neuroscience, 103,* 715–721.
Schmaltz, L.W., and Theios, J. (1972). Acquisition and extinction of a classically conditioned response in hippocampectomized rabbits (*Oryctolagus cuniculus*) *Journal of Comparative Physiology and Psychology, 79,* 328–333.
Schneiderman, N., Fuentes, I., and Gormezano, I. (1962). Acquisition and extinction of the classically conditioned eyelid response in the albino rabbit. *Science, 136,* 650–652.
Schreurs, B.G. (1993). Long-term memory and extinction of the classically conditioned rabbit nictitating membrane response. *Learning and Motivation, 24,* 293–302.
Sears, L.L., Finn, P.R., and Steinmetz, J.E. (1994). Abnormal classical eye-blink conditioning in autism. *Journal of Autism and Developmental Disorders, 24,* 737–751.
Skinner, B.F. (1950). Are theories of learning necessary? *Psychological Review, 57,* 193–216.
Solomon, P.R., and Gottfied, K.E. (1981). Disruption of acquisition of the classically conditioned rabbit nictitating membrane response following injection of scopolamine into the medial septal nucleus. *Journal of Comparative and Physiological Psychology, 95,* 322–330.
Solomon, P.R., and Groccia, M.E. (1996). Classical conditioning in aged rabbits: delay, trace, and long delay conditioning. *Behavioral Neuroscience, 110,* 427–435.
Solomon, P.R., and Moore, J.W. (1975). Latent inhibition and stimulus generalization of

the classically conditioned nictitating membrane response in rabbits (*Oryctolagus cuniculus*) following dorsal hippocampal ablation. *Journal of Comparative and Physiological Psychology, 89,* 1192–1203.

Solomon, P.R., and Pendleberry, W.W. (1988). A model system approach to age-related memory disorders. *Neurotoxicology, 9,* 443–462.

Solomon, P.R., and Pendlebury, W.W. (1992). Aging and memory: a model systems approach. In L.R. Squire and N. Butters (Eds.), *Neuropsychology of memory* (pp. 262–276). New York: Guillford Press.

Solomon, P.R., Solomon, S.D. Vander Schaaf, E., and Perry, H.E. (1983). Altered activity in mammals. *Philosophical Transactions of the Royal Society of London, Biology, 329,* 161–170.

Solomon, P.R., Vander Schaaf, E.R., Weisz, D. J., and Thompson, R.F. (1986). Hippocampus and trace conditioning of the rabbit's classically conditioned nictitating membrane response. *Behavioral Neuroscience, 100,* 729–744.

Solomon, P.R., Koota, D., Kessler J.B., and Pendlebury, W.W. (1987). Disrupted retention of the classically conditioned eyeblink response in the aluminum intoxicated rabbit using brain stimulation as a conditioned stimulus. *Society for Neuroscience Abstracts, 13,* 642.

Solomon, P.R., Pingree, T.M., Baldwin, D., Koota, D., Perl, D., and Pendlebury, W. (1988). Disrupted retention of the classicially conditioned nictiating membrane response in rabbits with aluminum-induced neurofibrillary degeneration. *NeuroToxicology, 2,* 209–222.

Solomon, P.R., Stowe, G.T., and Pendleberry, W.W. (1989a). Disrupted eyelid conditioning in a patient with damage to cerebellar afferents. *Behavioral Neuroscience, 103,* 898–902.

Solomon, P.R., Pomerleau, D., Bennett, L., James, J., and Morse, D.L. (1989b). Acquisition of the classically conditioned eyeblink resposnse in humans over the lifespan. *Psychology and Aging, 4,* 34–41.

Solomon, P.R., Levine, E., Bein, T., and Pendlebury, W.W. (1991). Disruption of classical conditioning in patients with Alzheimer's disease. *Neurobiology of Aging, 12,* 283–387.

Solomon, P.R., Groccia-Ellison, M., Flynn, D., Mirak, J., Edwards, KL.R., Dunehew, A., and Stanton, M.E. (1993). Disruption of human eyeblink conditioning after central cholinergic blockade and scopolamine. *Behavioral Neuroscience, 107,* 271–279.

Solomon, P.R., Groccia-Ellison, M.E., Stanton, M.E., and Pendlebury, W.W. (1994). Strategies for developing animal models of neurotoxicant-induced neurodegenerative disorders. In M.I. Woodruff and A.J. Nonneman (Eds.), *Toxin-induced models of neurological disorders.* New York: Plenum Press.

Solomon, P.R., Barth, C.L., Wood, M.S., Velazquex, E., Groccia-Ellison, M.E., and Yang, B.-Y. (1995a). Age-related deficits in retention of the classically conditioned nictitating membrane response in rabbits. *Behavioral Neuroscience, 109,* 18–23.

Solomon, P.R., Brett, M., Groccia-Ellison, M.E., Oyler, C., Tomasi, M., and Pendleberry, W.W. (1995b). Classical conditioning in patients with Alzheimer's disease: a multiday study. *Psychology and Aging, 10,* 248–254.

Solomon, P.R., Wood, M.S., Groccia-Ellison, M.E., Yang, B.-Y., Fanelli, R.J., and Mervis, R.F. (1995c). Nimodipine facilitates retention of the classically conditioned nictitating membrane response in aged rabbits over long retention intervals. *Neurobiology of Aging, 16,* 791–796.

Solomon, P.R., Flynn, D., Mirak, J., Brett, M., Coslov, N., and Groccia, M.E. (1998). Five-year retention of the classically conditioned eyeblink response in young adult, middle-

aged, and older humans. *Psychology and Aging, 13,* 186–192.
Solyom, L., and Barik, H.C. (1965). Conditioning in senescence and senility. *Journal of Gerontology, 20,* 483–488.
Stanton, M.E., and Freeman, J.H., Jr. (1994). Eyeblink conditioning in the infant rat: an animal model of learning in developmental neurotoxicology. *Environmental Health Perspectives, 102,* 131–139.
Stanton, M.E., Freeman, J.H., Jr., and Skelton, R.W. (1992). Eyeblink conditioning in the developing rat. *Behavioral Neuroscience, 106,* 657–665.
Terry, R.D., and Penna, C. (1965). Experimental production of neurofibrially degeneration: 2. Electron microscopy, phosphatase histochemistry, and electron probe analysis. *Journal of Neuropathology and Experimental Neurology, 24,* 200–210.
Thompson, R.F., and Kim, J.J. (1996). Memory systems in the brain and localization of a memory. *Proceedings of the National Academy of Sciences of the United States of America, 93,* 13438–13444.
Thompson, R.F., and Krupa, D.J. (1994). Organization of memory traces in the mamalian brain. *Annual Review of Neuroscience, 17,* 519–549.
Thompson, R.F. (1986). The neurobiology of learning. *Science, 233,* 941–947.
Thompson, R.F. (1990). Neural mechanisms of classical conditioning in mammals. *Philosophical Transactions of the Royal Society of London, Biology, 329,* 161–170.
Thompson, R.F., and Kim, J.J. (1996). Memory systems in the brain and localization of a memory. *Proceedings of the National Academy of Sciences of the United States of America, 93,* 13438–13444.
Thompson, R.F., and Krupa, D.J. (1994). Organization of memory traces in the mamalian brain. *Annual Review of Neuroscience, 17,* 519–549.
Thompson, R.F., Bao, S., Chen, L., Cipriano, B.D., Grethe, J.S., Kim, J.J., Thompson, J.K., Tracy, J., Weninger, M.S., and Krupa, D.J. (1997a). Conditioned nictitating membrane response in aged rabbits over long retention intervals. *Neurobiology of Aging* 16:791–796.
Thompson, R.F., Bao, S., Chen, L., Cipriano, B.D., Grethe, J.S., Kim, J.J., Thompson, J.K., Tracy, J., Weninger, M.S., and Krupa, D.J. (1997b). Associative learning. *International Review of Neurobiology, 41,* 151–189.
Topka, H., Valls-Sole, J., Massaquoi, S.G., and Hallett, M. (1993). Deficit in classical conditioning in patients with cerebellar degeneration. *Brain, 116,* 961–969.
Weiskrantz, I., and Warrington, E.K. (1979). Conditioning in amnesic patients. *Neuropsychologia, 17,* 1187–194.
Weiss, C., and Thompson, R.F. (1991). The effects of age on eyeblink conditioning in the freely moving Fischer-344 rat. *Neurobiology of Aging, 12,* 249–254.
Weisz, D.J., Clark, G.A., and Thompson, R.F. (1984). Increased responsivity of dentate granule cells during nictitating membrane response conditioning in rabbit. *Behavioral Brain Research, 12,* 145–154.
Woodruff-Pak, D.S. (1993). Eyeblink classical conditioning in H.M.: delay and trace paradigms. *Behavioral Neuroscience, 107,* 911–925.
Woodruff-Pak, D.S. (1995). Evaluation of cognition-enhancing drugs: utility of the model system of eyeblink classical conditioning. *CNS Drug Reviews, 1,* 107–128.
Woodruff-Pak, D.S. (1997). Classical conditioning. *International Review of Neurobiology, 41,* 341–365.
Woodruff-Pak, D.S., and Hinchliffe, R.M. (1997). Scopolamine- or mecamylamine-induced leaning impairment: reversed by nefiracetam. *Psychopharmacology, 131,* 130–139.
Woodruff-Pak, D.S., and Li, T-Y., (1994). Nefiracetam (DM-9384): effect on classical

conditioning in older rabbits. *Psychopharmacology, 114,* 200–208.

Woodruff-Pak, D.S., and Papka, M. (1996). Huntington's disease and eyeblink classical conditioning: normal leaning but abnormal timing. *Journal of the International Neuropsychological Society, 2,* 323–334.

Woodruff-Pak, D.S., and Thompson, R.F. (1988). Classical conditioning of the eyeblink response in the delay paradigm in adults aged 18–83 years. *Psychology and Aging, 3,* 01–011.

Woodruff-Pak, D.S., Lavond, D.G., Logan, C.G., and Thompson, R.F. (1987). Classical conditioning in 3-, 30- and 45-month-old rabbits: behavioral learning and hippocampal unit activity. *Neurobiology of Aging, 8,* 101–108.

Woodruff-Pak, D.S., Finkbiner, R.G., and Katz, J.R. (1989). A model system demonstrating parallels in animal and human aging: extension to Alzheimer's disease. In A.M. Meyer, J.W. Simpkins, and J. Yamamoto (Eds.), *Novel approaches to the treatment of Alzheimer's disease* (pp. 351–375). New York: Plenum.

Woodruff-Pak, D.S., Cronholm, J.F., and Sheffeld, J.B. (1990a). Purkinje cell number related to rate of eyeblink classical conditioning. *NeuroReport, 1,* 165–168.

Woodruff-Pak, D.S., Finkbiner, R.G., and Sasse, D.K. (1990b). Eyeblink conditioning discriminates Alzheimer's patients from non-demented aged. *NeuroReport, 1,* 45–48.

Woodruff-Pak, D.S., Li, T-Y., Kazmi, A., and Ken, W.R. (1994). Nicotine cholinergic system involvement in eyeblink classical conditioning in rabbits. *Behavioral Neuroscience, 108,* 486–4493.

Woodruff-Pak, D.S., Papka, M., and Ivry, R. (1996a). Cerebellar involvement in eyeblink classical conditioning in humans. *Neuropsychology, 10,* 443–458.

Woodruff-Pak, D.S., Papka, M., Romano, S. and Li, Y.T. (1996b). Eyeblink classical conditioning in Alzheimer's disease and cerebrovascular dementia. *Neurobiology of Aging, 17,* 505–512.

Woodruff-Pak, D.S., Chi, J., Li, Y.-T., Pak, M.T., and Fanelli, R.J. (1997a). Nimodipine ameliorates impaired eyeblink classical conditioning in older rabbits in the long-delay paradigm. *Neurobiology of Aging, 18,* 641–649.

Woodruff-Pak, D.S., Li, Y.T., Hinchliffe, R.M., and Port, R.L. (1997b). Hippocampus in delay eyeblink classical conditioning: essential for nefiracetam amelioration of learning in older rabbits. *Brain Research, 747,* 207–218.

Woodruff-Pak, D.S., Green, J.T., Coleman-Valencia, C., and Pak, J.T. (in press). A nicotinic cholinergic drug (GTS-21) and eyeblink classical conditioning: acquisition, retention, and relearning in older rabbits.

Yang, B.Y., Kwak, M., Groccia-Ellison, M.E., and Solomon, P.R. (1993). Synergistic facilitation of hippocampal synaptic activation with high fequency stimulation and classical eyeblink conditioning in rabbits. *Society of Neurosciene Abstracts, 19,* 802.

Zola-Morgan, S., and Squire, L.R. (1990). Neuropsychological investigations of memory and amnesia: findings from humans and non-human primates. In A. Diamond (Ed.), In *The development and neural bases of higher cognitive functions* (pp. 434–456). New York: New York Academy of Sciences.

6
Fundamental Behavioral Methods and Findings in Classical Conditioning

E. James Kehoe and Michaela Macrae

This chapter is intended to provide a solid foundation in the methods and findings of classical conditioning at a behavioral level. We take readers from an introduction to the basics of classical conditioning through to its more complex and recent findings that appear related to higher-order processes that are commonly labeled as attention, perception, and even cognition. Many textbook descriptions of classical conditioning give the mistaken impression that it is an area of completed science. In fact, the methods of classical conditioning have continued to evolve, and the range of important phenomena has continued to expand. Although this chapter primarily describes the empirical methods and outcomes of classical conditioning, we introduce important theoretical principles along the way. These principles have proven crucial in organizing research findings and in linking classical conditioning to other learning phenomena. In addition, our introduction to behavioral theory should help the reader place the phenomena described in this chapter in the context of the theories described in more detail in the next chapter (Brandon, Vogel, & Wagner, this volume).

This chapter is organized in four sections. The first section describes the *basic procedures* for classical conditioning and the intellectual traditions that have shaped those procedures. The second section describes phenomena surrounding the *transfer, loss, and recovery* of classically conditioned behavior after it has been established. The third section describes the effects of *key variables* that affect classically conditioned behavior. The fourth section describes the results from experiments using *compound stimuli,* typically composed of multiple events in different sensory modalities. This last type of experiment has been crucial in driving theoretical developments during the past 30 years.

Basic Procedures and Findings

The prototype for all laboratory investigations of classical conditioning has been Pavlov's (1927) method of conditioned reflexes. The method entailed the presentation of two stimuli to a dog. First, the *conditioned stimulus* (CS) is usually a relatively innocuous event, for example, a bell. Second, the *unconditioned stimulus* (US) is an event of biological significance, for example, food in the mouth. Pavlov's method

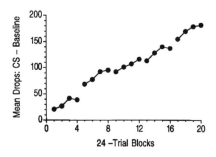

FIGURE 6.1. Acquisition of a salivary conditioned response (CR) in dogs (Fitzgerald, 1963).

takes advantage of the innate ability of the US to elicit reactions known as *unconditioned responses* (URs). Pavlov typically reported one of those responses, namely, salivation. Across successive presentations of a CS followed by a US, the CS itself comes to elicit a response that broadly resembles the UR but is rarely an exact copy. This learned response to the CS is labeled the *conditioned response* (CR).

The operations used to establish a CR often require bouts of training spaced out over several days. Each day contains one or more *training sessions*. In turn, each training session contains a series of observation periods called *trials*. The period between one trial and the next is called the *intertrial interval* (ITI). Each trial may contain one or more stimuli. For example, a trial containing a CS followed by a US is labeled as either a *CS–US trial* or a *CS–US pairing*. A trial containing only the CS is labeled as a *CS-alone trial*, and a trial containing only the US is labeled as a *US-alone trial*.

The growth of a CR that occurs across a series of CS–US trials is referred to as *CR acquisition*. This growth is often depicted graphically by plotting a measure of the response taken during the interval between the onset of the CS and the onset of the US as a function of the successive CS–US trials. In some cases, response measures are recorded only during CS-alone trials that are interjected as *test trials* to avoid any possible contamination of the CR measurement by the occurrence of the US and subsequent UR. Figure 6.1 shows an example of acquisition of a salivary CR in dogs (Fitzgerald, 1963). The ordinate axis shows the total number of drops of salivation during the interval between the onset of the CS and the onset of the US. Each point represents the mean number of drops excreted by six dogs during a block of 24 CS–US trials (abscissa). The points for successive blocks of trials are connected, and the unconnected points represent the gaps between successive days of training. Although the growth of the salivary CR in this experiment more or less followed a straight line, the growth of most CRs follows a curved, negatively accelerated function that reaches a final stable level, the *asymptote*. Hence, these depictions are usually labeled as *acquisition curves* or *learning curves*.

Classical Conditioning as an Exemplar of Learning

The acquisition of a CR is usually described as learning because it conforms to the common criteria for learning. In operational terms, an organism is said to have learned when it displays a relatively permanent change in behavior as a result of specific experience. On the basis of this definition, there are three criteria for asserting that learning has occurred. First, there must be an identified, specific experience. In classical conditioning, the presentations of the CS and US controlled by the researcher are the key elements in the experience. For historical reasons described here, joint presentations of the CS and US in close contiguity have been of particular interest for investigators using classical conditioning techniques. Second, there must be a change in behavior, which, in the prototypic classical conditioning experiment, is the acquisition of the CR to the CS. Third, the change in behavior must be long lasting. This criterion has never been well specified for any type of learning and may not be a crucial matter at a definitional level. At a theoretical level, a gross difference in the retention of a CR might indicate different underlying processes. As a rule of thumb, changes in behavior that last less than a few minutes would be of interest in understanding *short-term memory* or *working memory,* and changes in behavior that last a few hours or more would be of interest in understanding *long-term memory.* Traditionally, the emphasis in research using classical conditioning has been on the longer-lasting changes.

As may be apparent, the definition of learning is asymmetrical. On the one hand, if all the criteria are met, learning can be positively asserted by an investigator. On the other hand, if one or more of the criteria is not met, one cannot safely say that learning did not occur. It is only safe to say that there was a failure to observe learning. If, for example, CR acquisition does not occur despite repeated CS–US pairings, there always remains the possibility that the pairings had an as-yet-undetected impact on the animal. In fact, there are a number of phenomena described in this chapter in which the effects of exposure to the CS or US have latent effects that are not detected until much later. Furthermore, the expression of a CR can be hindered by a class of variables that are unrelated to learning, for example, effector fatigue. Such variables are typically referred to as nonassociative or performance factors.

Generality

It might seem easy to extend Pavlov's method of salivary conditioning in the dog to other species, stimuli, and responses, but there have been repeated disputes as to the defining features of classical conditioning. These disputes are more than semantic quibbling; they reflect a serious scientific concern. Specifically, a main aim of science is to discover principles that transcend particular situations. These principles range from low-level, single-variable regularities to complex theoretical axioms. A definition of a key phenomenon such as classical conditioning places boundaries on the range of experiments used to search for its principles. A narrow definition may yield detailed but specialized principles whereas a wider definition may yield broad but more superficial principles.

The generality of a definition comes into sharp focus when an empirical exception to a proposed principle appears. For classical conditioning, the question would become, "Is the principle false for the defined area, or is our definition too liberal, permitting us to stray into a different kind of learning?" Depending on how one answers this question, two things may happen. One may toss out the old principle, retain the original definition, and search for a new principle. Alternatively, one may retain the old principle but qualify it by narrowing the definition and hence the area of research.

Defining Features

Attempts to wrestle with the issue of generality in classical conditioning have been influenced by two traditions in the philosophy and physiology of the mind: the *reflex* tradition and the *associative* tradition (Gormezano & Kehoe, 1975).

Reflex Tradition

The reflex tradition is rooted in Descartes's attempt to provide a mechanistic account of actions by humans and animals. Descartes introduced the device of the stimulus. External forces provided by the stimulus flowed through the body and were reflected out from the brain in the form of muscular movements; thus, every action was a response to a stimulus. This concept of reflex is a general principle that all behavior is determined lawfully by physical antecedents. The reflex hypothesis gained scientific substance through laboratory preparations, starting with the scratch reflex in the frog (Whytt, 1751; cited in Boring, 1950, p. 35). To abolish the perceived influence of the will, these investigations were confined to preparations in which the spinal cord was severed from the brain. As a consequence, the concept of the reflex was narrowed to actions that were deemed "automatic, unconscious, and involuntary" (Hall, 1832; Whytt, 1751; both cited in Boring, 1950, pp. 35–38).

The Russian physiologist Sechenov (1863) attempted to recapture Descartes's original meaning of reflex as a general principle for all behavior. Although Sechenov acknowledged that humans have self-awareness, he rejected a fundamental distinction between conscious and unconscious behavior. Instead, Sechenov proposed two classes of reflex. The first class included inborn reflexes. The second class contained acquired reflexes, based on the organism's history of stimulation. However, the concept of the acquired reflex lay fallow until Pavlov's investigations.

In subsequent studies of classical conditioning, the reflex tradition has had a profound influence in two respects: response origin and response description. An acquired response may be either homogeneous or heterogeneous with respect to responses evoked by the US (Hilgard & Marquis, 1940, p. 70). In Pavlov's preparation, the salivary CR was selected from the response systems activated by the USs, which were food or acidic solutions. Thus, the CR and the UR were homogeneous, in that they originated in the same response system. This characteristic of homogeneity gives the experimenter the power to elicit the to-be-learned response whenever desired. In contrast, heterogeneous responses are unrelated to the URs. Heterogeneous responses are often used in studies of goal-directed *instrumental conditioning*, in which pressing a lever or traversing a maze are not themselves elicited by a food reward. In these situations, other means are needed to encourage the animal to emit the desired response.

Behaviors may be anatomically defined or outcome defined (Tolman, 1932). In traditional classical conditioning, anatomically defined responses include both glandular secretions, such as salivation, and muscular movements, such as the eyeblink. In contrast, responses used in instrumental conditioning are usually outcome defined. They are described in terms of how the movements of the animal operate on its surroundings, for example, pressing a lever. Whether the animal uses its paws, its nose, or another portion of its anatomy is tangential to the response definition.

Associative Tradition

Associationism stems from the commonplace observation that a current event often reminds us of something from our past. Hearing a certain song may take one back to a long-ago incident or personal encounter. Contiguity in time and space between the two events on previous occasions enables one event to revive the memory of the other on later occasions. Associative concepts first appeared in ancient methods for memorizing sets of objects. Later, the British Empiricist philosophers elaborated associationism to account for everything from the perception of causation (Hume, 1969) to the origin of chronic drunkenness (Brown, 1820/1977). The relation of Pavlov's work to associationism was quickly recognized (e.g., Lashley, 1916). As a result, the following features have become essential to operational definitions of classical conditioning.

Nonassociative Controls

CR acquisition occurs most reliably when the CS and US are presented in close succession. However, not all responses during the CS can be attributed to CS–US contiguity. Consequently, control methods have evolved for determining what other proportion of responding during the CS arises from CS–US contiguity. Such sources include any spontaneous occurrences of the target response, any innate tendency for the CS to evoke the target response, and pseudoconditioning, which is a sensitization-like effect arising from presentations of the US. Classical conditioning is said to occur only when the level of responding to the CS in the experimental group exceeds control levels. A detailed description of different nonassociative controls and their relative merits is included in a later section of this chapter.

Stimulus–Response Reinforcer Contingencies

Classical conditioning is sometimes defined by way of contrast to instrumental conditioning, in which the delivery of a reinforcer, either a reward or punishment, depends on a particular response by the subject (Hilgard & Marquis, 1940, p. 51). Thus, the defining feature of instrumental conditioning is said to be its response-reinforcer contingency. Conversely, classical conditioning is said to be defined by its stimulus-reinforcer contingency, because the stimulus (CS) and reinforcer (US) are delivered in a predetermined manner irrespective of the CR.

A reliance on the predetermined delivery of both stimuli as the sole criterion for defining classical conditioning is, however, mistaken. This control of relationships among stimuli is also seen in other methods for studying associative learning that are not ordinarily identified as classical conditioning, namely, the verbal methods of list learning and paired associate learning. Moreover, many instrumental procedures allow the experimenter to control at least the minimum time between the onset of a warning stimulus and the availability of a reinforcer.

As an aside, the reader should take note of the usage of the terms *reinforcement* and *reinforcer,* which have become identified with the use of rewards and punishments in instrumental contingencies. In fact, the term reinforcement originated in the English translation of Pavlov (1927, pp. 49–50, 112, 384). The term was used in an operational sense to denote the presentation of the US following a CS. A CS–US trial, hence, is called a *reinforced trial,* and, conversely, a CS-alone trial is called a *nonreinforced trial* or *unreinforced trial.*

Model Preparations

Neurobehavioral researchers have, to some extent, finessed the twin issues of generality and definition. They have relied increasingly on the in-depth study of a handful of relatively standardized *model preparations* (see the chapter by Schreurs and Alkon for further descriptions and discussion of model preparations.) This approach is similar to that taken in other natural sciences, for example, the use of fruit flies in the study of population genetics. In brief, each of these preparations is restricted to a particular species, response, and set of stimuli.

Model preparations are not models in any sense of being ideal. Far from it; each preparation has its own particular advantages and disadvantages for studying learned behavior and its neural substrates. The identification of the neural structures and mechanisms for each of these preparations has proceeded in parallel. Consequently, it has become a matter of empirical discovery, rather than definitional convention, whether behavior that occurs in each preparation relies on the same or different neural substrate.

Demonstrations of CR Acquisition

Figure 6.2 shows the learning curves obtained in demonstrations of classical conditioning in four preparations: (A) the rabbit eyeblink using a tone as the CS and an air puff to the cornea as the US (Deaux & Gormezano, 1963; Gormezano, 1966), (B) the marine mollusk *Aplysia* siphon retraction, using a tactile stroke to the siphon as the CS and tail shock as the US (Carew et al., 1981), (C) the potentiated startle response in the rat, using a tone-buzzer as the CS and foot shock as the US (Brown et al., 1951), and (D) the rabbit swallowing reflex, using tone as the CS and water injected into the mouth as the US, which elicits jaw movement responses (Sheafor, 1975).

To measure associative learning, it would be ideal if each response to a CS were based entirely on prior CS–US pairings. Not surprisingly, this ideal is not easily attained. Most response systems show a baseline of spontaneous activity

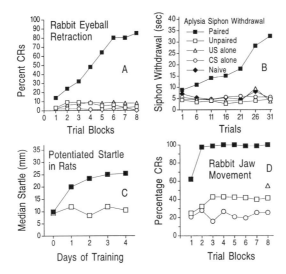

FIGURE 6.2. Acquisition in groups receiving either conditioned stimulus–unconditioned stimulus (CS–US) pairings or a nonassociative control condition for (**A**) eyeball retraction in rabbit (Deaux & Gormezano, 1963), (**B**) siphon retraction in *Aplysia* (Carew et al., 1981), (**C**) potentiated startle in rat (Brown et al., 1951), and (**D**) jaw movement in rabbit (Sheafor, 1975).

that produces spurious conjunctions of the CS and the target response. Moreover, responses to the CS can also result from two other nonassociative sources: alpha responding and pseudoconditioning.

Nonassociative Contributors

Alpha Responses

The alpha response is an unlearned, preexisting reaction to the CS in the same system as the target CR. For example, in conditioning *Aplysia,* the CS, which is a brief stroke of the siphon, innately produces a weak retraction of the siphon, which is classified as an alpha response and not a CR. Conditioning of an alpha response entails its strengthening by pairing the CS with an even stronger retraction of the siphon elicited by an electric pulse to the animal's tail (Carew et al., 1981). However, alpha responses are subject to sensitization and habituation, which can complicate the detection of learning specific to the CS–US pairings. (For discussions concerning the conditioning of alpha responses versus the acquisition of new responses to a CS, see Kandel et al., 1983; Schreurs, 1989; Skelton et al., 1988.)

Pseudoconditioning

Pseudoconditioning entails the acquisition of a response to the CS that depends only on prior presentations of the US and not on CS–US pairings. Pseudoconditioning can appear as sensitization of an alpha response. Alternatively, the pseudoconditioned response can be a new response that is indistinguishable from a CR (Gormezano, 1966; Sheafor, 1975).

Paired Conditions Versus Nonassociative Controls

Because total responding to a CS can reflect a mixture of associative and nonassociative contributions, a suite of control conditions has evolved to estimate the levels of nonassociative contributions to responding. To illustrate how these control methods are used, Figure 6.2 depicts the level of responding obtained from a variety of control groups. Consider first the rabbit eyeblink experiment, which contained one paired group and three control groups.

Paired Training

The experimental group received CS–US pairings; that is, each session contained 82 trials in which a 500-ms CS immediately preceded an air puff US, which elicited an eyeblink as a UR. The CS–US trials were separated by a minimum of 15 s. Eyeblinks that were initiated during the 500-ms CS–US interval were counted as potential CRs.

US-Alone

The US-alone control received only presentations of the US at times corresponding to CS–US trials in the paired group. Eyeblinks initiated during a 500-ms period before each US presentation provided an estimate of the baseline rate of eyeblinks plus any elevation in that rate produced by US presentations. In other experiments, the US-alone control is also used to measure pseudoconditioning and sensitization by presenting CS-alone tests at the end of training (Grant, 1943; Sheafor, 1975).

CS-Alone

The CS-alone control received only presentations of the CS. Eyeblinks during these presentations provided an estimate of the baseline rate of eyeblinks plus any alpha responses to the CS.

Unpaired

The unpaired control received separate presentations of the CS and US intermixed in a random order, with a minimum interval of 5 s between each event. The unpaired control matched the paired group in the number of exposures to both the CS and US, but with far less contiguity between the two stimuli. In principle, the unpaired control provided a joint estimate of baseline responding, alpha responding, sensitization, pseudoconditioning, and any unknown synergisms. However, there has been debate about the suitability of the unpaired control for estimating nonassociative contributions. This debate is summarized later in this section.

In the upper left panel of Figure 6.2, the response measure for the paired, unpaired, and CS-alone groups represents the mean proportion of trials on which an eye retraction of at least minimum amplitude occurred during the CS. This measure is commonly converted to a percentage and labeled as *percentage response* or *response likelihood*. (In the rabbit, retraction of the eyeball also causes closure of the inner eyelid, the nictitating membrane) (Gray et al., 1981). For the US-alone group, the response measure represents the mean proportion of trials in which an eye retraction occurred before the US during a period equivalent to the CS. In the paired group, the percentage of trials containing an eyeball retraction grew steadily over CS–US pairings. In the control groups, there were few eyeball retractions at any point during training. Accordingly, the bulk of responses in the paired groups were specific to the CS–US pairings and could be considered associative in origin.

The upper right panel of Figure 6.2 shows acquisition of a siphon retraction response in *Aplysia*. The CS was a weak tactile stimulus, namely, a stroke of the siphon, and the US was a shock to the animal's tail, which elicited retraction of the feeding siphon as the UR (Carew et al., 1981). In this preparation, CS–US pairings uniquely increased the duration of the siphon retraction, whereas the control conditions produced negligible changes in the response.

The lower left panel of Figure 6.2 shows the results for one group of rats with the CS and US paired, and for another group with the CS and US unpaired, in the *potentiated startle* procedure (Brown et al., 1951). In this procedure, the CS was a tone-buzzer and the US was a shock to the animal's feet. The effect of the CS–US pairings was measured by interspersing them with presentations of a brief, loud noise that elicited startle, a pronounced flinch of the whole body. On some of these startle trials, the noise was presented alone, and on other startle trials, the CS was presented just before the noise. The difference in the magnitude of the startle response on these two types of trials is construed to be an index of a hypothetical CR, namely, fear (Davis, 1986; Kiernan et al., 1995). As can be seen in Figure 6.2, the CS–US pairings produced an increase in the ability of the CS to potentiate the startle response whereas unpaired presentations of the CS and US did not.

The lower right panel shows the results from an experiment using conditioning of the rabbit's swallow reflex (Sheafor, 1975). The CS was a tone, and the US was water injected into the mouth, which elicited jaw movements. These results were included primarily to show an instance in which there was a substantial nonassociative contribution to response likelihood. Although the paired group showed the largest increase in response likelihood, the unpaired group showed an appreciable increase in the likelihood of a jaw movement during the CS. Similarly, in the US-alone group, tests with the CS at the end of training also revealed a high likelihood of responding to the CS. There was also a contribution from responses unrelated to the US, because the CS-alone group showed a response on approximately 20% of CS-alone presentations. In a subsequent experiment using a no-stimulus group, it was revealed that most of these responses during the CS-alone were spontaneous and not alpha responses to the CS.

The use of one or more control groups can increase the size and cost of an experiment. Separate control groups are highly recommended for estimating the single

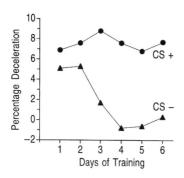

FIGURE 6.3. Differential conditioning of the heart rate in rabbit (Schneiderman, 1972).

and joint contributions from nonassociative sources. When these contributions have been shown to be small relative to the level of responding obtained from CS–US pairings, it is possible to dispense with the nonassociative control groups on most occasions. When, however, there are substantial nonassociative contributions, as in the rabbit jaw movement preparation, it remains necessary to retain at least one of the nonassociative controls.

Differential Conditioning as a Nonassociative Control

It is possible to use each animal as its own nonassociative control by using *differential conditioning* (Schneiderman, 1972; Schneiderman et al., 1987). Figure 6.3 shows changes in the heart rate of rabbits to two stimuli: a 500-Hz tone and 5000-Hz tone. One stimulus (CS+) was always paired with a shock US near the eye. The other stimulus (CS–) was always presented alone. Each daily session contained 15 trials of each type, which were randomly intermixed. For each animal, a baseline of heart rate was measured before the onset of the CS on each trial. Change in heart rate was expressed as a percentage relative to the animal's baseline. Each data point in Figure 6.3 represents the mean percentage deceleration over the first five beats following CS onset.

As seen in other examinations of heart rate conditioning (Schneiderman, 1972), the first day of training yielded a deceleration in heart rate to both CS+ and CS–. Across subsequent days, the deceleration to CS+ was maintained while the response to the CS– gradually declined. At the end of training, the substantial differentiation between CS+ and CS– indicated the specific effects of pairing the CS+ with the US.

Excitation, Inhibition, and the Truly Random Control

The development of nonassociative controls has been fueled by the associative tradition, specifically, the importance attached to distinguishing changes in responding based on contiguity between the CS and US from any other sources of responding. At times, in the Western psychological literature, the terms classical conditioning, association, and learning have been used almost synonymously (Gormezano & Kehoe, 1981; Guthrie, 1930). Researchers in this tradition certainly consider increases in CRs to the CS to be an index of increases in the strength of an underlying association. Conversely, declines in CRs tend to be attributed to nonassociative mechanisms, for example, fatigue and loss of motivation.

Researchers within the reflex tradition and reflex physiology have approached changes in responding to a CS from a different perspective. Specifically, Pavlov (1927) imported the concepts of *excitation* and *inhibition* into his interpretation of changes in CRs to CSs. Increases in responding as a result of CS–US pairings were attributed to a growth in neural excitation. Conversely, decreases in responding to a CS were attributed to a growth in neural inhibition. Pavlov identified both transient and more permanent forms of inhibition.

The more permanent form of inhibition, often denoted *conditioned inhibition,* has attracted considerable attention in Western research and theory during the past 30 years (Britton & Farley, 1999; Rescorla, 1969; Rescorla & Wagner, 1972). The incorporation of a learned form of inhibition as a negatively valued association into theories of conditioning is described elsewhere in this volume. As a consequence of this blending of inhibition with traditional associative concepts, there has been some concern that the traditional controls, particularly the unpaired condition, may underestimate nonassociative contributions. This concern arises from the possibility that the widely separated presentations of the CS and US in an unpaired condition may be conducive to the acquisition of an inhibitory association because the CS is, in fact, a signal that the US is not imminent (Rescorla, 1967). If so, responding in an unpaired condition would express the net effect of the positive contribution from nonassociative sources and the negative contribution from an inhibitory association.

This concern about conditioned inhibition arising from the explicit separation of the CS and US prompted an attempt to develop an associatively neutral condition known as the *truly random* control (Rescorla, 1967). This condition ideally contains the full range of temporal relationships between a CS and US that are possible given the total number of CSs and USs to be presented within a training session of a given duration. Presumably, the excitatory associative effects of shorter CS–US intervals would be canceled out by the inhibitory associative effects of longer CS–US intervals, thereby exposing any contributions to responding from nonassociative sources. In practice, however, it has proved difficult to arrive at a temporal distribution of CSs and USs that reliably balances these two effects. For example, in randomly generated presentations of CSs and USs, the exact number and placement of short CS–US intervals has produced CR acquisition that presumably reflects a net excitatory effect (Ayres et al., 1975).

The difficulties with the truly random control and the unpaired control might seem fatal to estimating the specific contribution from CS–US pairings to responses during a CS. Although no single control group is perfect, the full set of traditional controls will guard against a gross overestimation of the excitatory associative effects of CS–US pairings. When the level of responding is low in all the control conditions and high in the paired condition, it would seem safe to conclude that responding in the paired condition expresses an excitatory association. However, when there is a high level of responding in any of the controls, more caution is obviously required. [See Gormezano and Kehoe (1975) for an extended discussion of the findings and issues surrounding the truly random control.]

Transfer, Loss, and Recovery

The associative tradition has led researchers to focus on the determinants of CR acquisition. This section describes the wider compass of phenomena in classical conditioning, many of which were first identified by Pavlov (1927). This section is divided into two subsections. The first defines and describes several forms of *transfer of training* that occur when the CS is changed. The second subsection defines and describes *extinction*, in which CS-alone presentations follow CS–US pairings, plus other phenomena entailing the loss and recovery of CRs.

Transfer of Training

In the broad domain of research in learning and memory, transfer of training is said to occur whenever experience in one situation influences behavior in a new situation (Ellis, 1965; McGeoch, 1952). The study of transfer has been significant for two reasons. From a biological perspective, it reveals the ability of a creature to use the experience to adapt behaviorally to changes in the exigencies of its environment. From a psychological perspective, the study of transfer has long been seen as vital to understanding how knowledge and skills acquired in one setting are applied in other settings. For example, in education and training, the transfer of learning from the classroom to the workplace and the wider environment is a key criterion for judging the success of a teaching program (Hesketh, 1997). Similarly, in clinical psychology and psychiatry, the transfer of new skills and attitudes acquired in the therapeutic setting to a client's ordinary environment is essential to the practical success of the treatment (Belasco & Trice, 1969).

Specific Transfer: Generalization

Generalization is said to occur when CRs are evoked by stimuli that differ physically from the original CS. Figure 6.4 shows the results of a generalization exper-

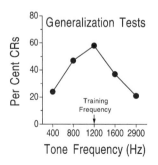

FIGURE 6.4. Generalization gradient for CR likelihood as a function of tonal frequency using the rabbit inner eyelid (nictitating membrane) (Moore, 1972).

iment using the rabbit's eyeblink response (Moore, 1972).[1] Initial training was conducted with a 1200-Hz tone CS. A series of tests were then conducted with tones of different frequencies. The figure shows the CR likelihood as a function of tonal frequency. Inspection of the figure reveals a *generalization gradient;* that is, the 1200-Hz tone CS yielded the highest level of responding. The stimuli physically most similar to the CS, namely, the 800- and 1600-Hz tones, produced somewhat less responding. Finally, the stimuli most different from the CS, the 400- and 2000-Hz tones, produced even less responding.

The generalization gradient can be used to determine how well a sensory dimension of the CS gains control over behavior. A sharp, relatively steep gradient demonstrates strong control by a dimension. However, a relatively flat gradient implies that the animal is insensitive to the dimension being tested. Generalization gradients can be sharpened through the use of differential conditioning, in which one value along the dimension (CS+) is always paired with the US and another value (CS−) is always presented alone. If the animal is sensitive to the difference between the two stimuli, then the animal should acquire CRs to CS+ but not CS−. For example, Figure 6.3 shows differential conditioning of heart rate using two tone frequencies, 500 and 5000 Hz. At first, the animals generalized widely, responding as strongly to CS− as to CS+. However, as training progressed, they gradually ceased responding to CS−.

Nonspecific Transfer: Learning to Learn

Generalization has two main features: specificity and immediacy. Generalization is specific in that it depends on the physical similarity between the original CS and the test stimuli. Generalization is immediate in the sense that responding can and does occur to test stimuli on their first presentation to the animal. However, there is another form of transfer that is neither specific nor immediate, known variously as *learning to learn* or *general transfer.* It is nonspecific, because it is seen using stimuli that can differ dramatically from each other; they can even be from different sensory modalities. Unlike generalization, learning to learn is not immediate, because responding to the new stimuli may not appear on their first presentation. Rather, learning to learn appears as an increase in the rate of acquisition of the response to the new stimuli (Ellis, 1965, p. 32; Harlow, 1949). In humans, for example, the number of repetitions needed for adults to memorize a list of unrelated word syllables decreases by about 50% over a series of a dozen different lists (Meyer & Miles, 1953).

---

[1]The remainder of this chapter uses examples drawn from studies using the rabbit eyeblink. We are not making any special claim about the rabbit eyeblink preparation. Rather, this is being done because one of the largest, coherent body of findings concerning classical conditioning has been generated using this preparation. Moreover, by focusing on a single preparation, it will be easier for the reader to see how the different procedures and findings relate to one another without having to mentally switch gears among species, stimuli, and response measures. No single preparation, however, can capture all the richness of classical conditioning, and other chapters in this book cover the findings from other preparations.

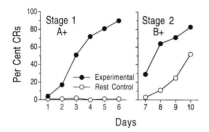

FIGURE 6.5. Learning to learn in cross-modal training using the rabbit inner eyelid (nictitating membrane). The experimental group received training with a CS from one sensory modality (A+) in *stage 1* (*left*) and training with a CS from a different sensory modality in *stage 2* (B+) (*right*). The rest control group received only B+ training.

Demonstrations of learning to learn use experimental designs that conform to the long-established procedure for detecting transfer in the laboratory. The basic demonstration involves two stages and two groups. In the first stage, the transfer group is given training in a particular task, say, pairings of a particular CS with a particular US. The other group, called the rest control group, is not given the CS–US pairings but may be exposed to the experimental apparatus. In the second stage, both groups receive training in a task that differs from the first task, for example, pairing a CS different from the one used in the first stage by the transfer group with the US. The behavior of the two groups is then compared during the common second-stage task. *Positive transfer* is said to occur when the transfer group responds at a higher level in the second stage than does the rest control group. *Negative transfer* is said to occur when the transfer group responds at a lower lever than the rest control group.

Learning to learn is a form of positive transfer that has been obtained in classical conditioning by switching training from one CS to another CS of a different modality, such as tone and light (Holt & Kehoe, 1985; Kehoe, 1992; Kehoe & Holt, 1984; Kehoe et al., 1995a; Kehoe & Macrae, 1997; Schreurs & Kehoe, 1987). Figure 6.5 shows an example of learning to learn as it has been observed in eyeblink conditioning. In stage 1, one group of rabbits (Experimental) received CS–US pairings. Other rabbits (Rest Control) were just restrained in the training apparatus. As seen in the left panel of Figure 6.5, experimentals showed acquisition to a high level in which CRs occurred on 90% of trials while rest controls showed only a few spontaneous responses.

At the end of stage 1, both groups were presented with a new stimulus that was from a different sensory modality than the original CS. If the original CS had been, say, a tone, the new CS was a light. These tests were conducted to determine whether there was any immediate transfer from the tone to the light. In fact, such tests have never revealed any immediate cross-modal transfer. Responding to the new stimulus has consistently been negligible.

In stage 2, both groups received CS–US pairings using the new CS (e.g., light). Learning to learn was evident almost as soon as pairings began. In the right panel of Figure 6.5, the experimentals showed rapid CR acquisition to the new CS. For example, they showed CRs on 29% of trials within the first block of pairings whereas rest controls showed CRs on only 3% of trials.

Loss and Recovery of CRs

The maintenance of CRs over time requires at least some sporadic CS–US pairings. In the absence of CS–US pairings, CRs can be lost in three ways. The most familiar is *extinction,* in which the CS is presented repeatedly without the US. The second is *counterconditioning,* in which an established CS is paired with a different US. The third way is *forgetting,* in which no training is conducted for an extended period of time. To the casual observer, a loss of CRs might suggest that the learning process has been reversed and the animal has been returned to its original state. However, as is described next, the impact of the previous CS–US pairings is largely preserved, and CRs can be quickly recovered.

Extinction

Extinction is a method for deliberately eliminating a CR. The CS is presented repeatedly without the US (Pavlov, 1927, p. 49). In the first session of extinction, responding to the CS progressively declines as the CS is presented without the US. Subsequent extinction sessions are punctuated by *spontaneous recovery* (Haberlandt et al., 1978). Spontaneous recovery is said to occur if the level of responding to the CS after a rest period is greater than at the end of the previous CS-alone trials. Spontaneous recovery is a reliable phenomenon but gradually fades over successive extinction sessions that are separated by rest periods. Nevertheless, even when CRs have completely disappeared during extinction training, they can show rapid *reacquisition* if CS–US pairings are resumed.

Figure 6.6 shows an example of acquisition, extinction, spontaneous recovery, and reacquisition of the eyeblink CR (Napier et al., 1992). One group of rabbits (Experimental) initially received 3 days of CS–US pairings. As seen in the top panel of the figure, they rapidly acquired CRs to levels at which CRs occurred on 90% of trials. A second group of rabbits was a rest control that only sat in the training apparatus.

Following CR acquisition, the experimental group received CS-alone trials. In the middle panel of Figure 6.6, the experimental group (stage 2) showed a decline in responding from CRs on 80% of trials at the start of the first day of extinction to 1% of the trials at the end of the first day. There was substantial spontaneous recovery in CR likelihood at the start of the second day of extinction. Across successive days, however, spontaneous recovery diminished, and by the fifth day CRs ceased to occur at any time during the session. The rest control group continued to sit in the conditioning apparatus.

The bottom panel of Figure 6.6 (stage 3) shows the CR acquisition curves when CS–US pairings were given for 2 days to both groups at the end of the experiment. For the experimental group, these pairings represented retraining, whereas for the rest control group the pairings represented initial training. Reacquisition was very rapid; the experimental group showed CRs on 80% of trials by the end of the first day of retraining, but the rest control group showed CRs on only 45% of its trials. Even when extinction training was extended for three times as many trials as

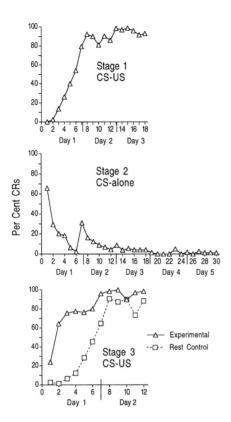

FIGURE 6.6. Acquisition (stage 1), extinction (stage 2), and reacquisition (stage 3) of CRs using the rabbit inner eyelid (nictitating membrane) (Napier et al., 1992).

shown in Figure 6.6, reacquisition was faster than in a rest control (Napier et al., 1992).

Rapid reacquisition is not the only evidence that extinction does not expunge the effects of previous CS–US pairings. In addition to rapid reacquisition, learning to learn appears after the original CR has been extinguished. In a three-stage experiment, rabbits were first given CS–US pairings using a light CS and, second, were given CS-alone presentations of the same light CS. Third, after extinction of the CR to the light CS, the animals were given pairings of a new CS, tone, with the US. Despite the extinction of the CR to light, CR acquisition to the new, tone CS was faster than in a rest control group (Kehoe et al., 1984). In fact, CR acquisition to the new CS was nearly as fast as in a group that had received the original CS–US pairings but had not received the extinction procedure (Kehoe et al., 1995a). Moreover, in a comparison of learning to learn and reacquisition (Macrae & Kehoe, 1999), CR acquisition to the new CS was nearly as rapid as reacquisition to the original CS.

Counterconditioning

In counterconditioning, animals trained with a particular CS–US pairing are subsequently presented with the same CS but paired with a different US. For example, if the first US is a shock near the eye, the second US could be delivery of water into the mouth. The two USs act on different receptors and elicit different URs. In the case of USs such as shock and water, the two USs are also thought to have a different emotional significance. Hence, at a theoretical level, there may be a mutual inhibition of their respective motivational systems when they are jointly activated by the tone CS during its pairings with the second US (Konorski, 1967).

This basic operation of switching the US produces the potential for at least three outcomes. First, with respect to the original CR, the extinction of the original CR could be accelerated as a result of interference from the new CR or motivational inhibition from the new US. Second, with respect to acquisition of the new CR, negative transfer could occur; that is, acquisition of the new CR could be retarded by interference from the original CR or inhibition from the motivational system of the original US activated by the CS. Third, positive transfer could also occur if acquisition of the new CR is sped up by the same process of learning to learn that operates when the CS rather than the US is switched. In fact, as is described next, all three outcomes have occurred in the rabbit, depending partly on motivational conditions.

Acceleration of Extinction of the Original CR

Scavio (1974, experiment 2) established eyeblink CRs in three groups of rabbits, all of which received initial pairings of a tone CS with a shock US. Subsequently, in a second stage of training, one group received counterconditioning in which the tone CS was paired with water delivered into the mouth. This water stimulus did not elicit eyeblinks. The second group received unpaired presentations of the CS and the water US. The third group was presented with CS-alone extinction trials.

During the second stage of Scavio's (1974) experiment, the group that received tone–water pairings showed the most rapid extinction of eyeblink CRs. However, this accelerated extinction arose largely from nonassociative sources, perhaps motivational inhibition, because extinction of the eyeblink CR was nearly as rapid in the group that received unpaired presentations of the tone and water. Nevertheless, both these groups showed a more rapid decline in responding than did the CS-alone extinction group. Across 150 trials, the counterconditioning group, the unpaired group, and the extinction group displayed CRs on 13%, 16%, and 25% of their respective trials.

Although extinction of the eyeblink CR has been accelerated during tone–water pairings, rapid reacquisition may occur when tone–shock pairings are reinstated (Bromage & Scavio, 1978; Scavio & Gormezano, 1980). For example, in an experiment by Bromage and Scavio (1978), rabbits required three sessions of initial tone–shock pairings to reach a level at which CRs occurred on 95% of trials. During tone–water counterconditioning, these rabbits showed partial extinction of the eyeblink CR. When tone–shock pairings were reinstated, the likelihood of the eyeblink CR returned to the 95% level within a single session. However, to confirm that rapid reacquisition occurs after counterconditioning, it would be necessary to conduct more thorough extinction of the eyeblink CR during tone–water counterconditioning.

Positive and Negative Transfer in Acquisition of the New CR

Although the counterconditioning procedure accelerates the extinction of the original CR, examination of the acquisition of the new CR has yielded instances of both positive transfer (Scavio & Gormezano, 1980) and negative transfer (Bromage & Scavio, 1978; Scavio & Gormezano, 1980; Scavio, 1974, experiment 1). In the case of transfer from tone–shock pairings to tone–water pairings, the acquisition of jaw-movement CRs based on the water US appears to depend on the animal's level of water deprivation. Under moderate water deprivation, positive transfer appeared; that is, acquisition of jaw-movement CRs during tone–water pairings after tone–shock pairings was faster than in a rest-control group, which received only the tone–water pairings. Under greater water deprivation, however, negative transfer appeared; acquisition of the jaw-movement CRs was retarded by prior tone–shock pairings (Bromage & Scavio, 1978).

Forgetting and Savings

Forgetting is said to occur when responding declines as a function of the length of the *retention interval.* In classical conditioning, the retention interval is usually defined as the period of time between one training session and the next session in which the CS is presented. Two recent investigations have revealed two important facts about forgetting in the rabbit eyeblink preparation. First, substantial forgetting does occur in rabbits, but it does not occur quickly. Schreurs (1993, 1998) found virtually perfect retention of a CR after a 1-month interval, but by the end of 6 months without training, the CR likelihood dropped to a level of 4%. Second, despite this evidence of forgetting, there were substantial *savings.* CS–US pairings yielded reacquisition of the CR that was significantly faster than the rate of original acquisition, even after the longest retention interval. These two observations parallel those seen in human memory; that is, initial recall after a retention interval can be very poor, but it can be quickly restored with some retraining (McGeoch, 1952, pp. 23–24).

Commentary

Generalization, extinction, and spontaneous recovery have received considerable theoretical attention since they were discovered by Pavlov (1927). However, the findings surrounding learning to learn and rapid reacquisition are less well understood. The data reveal that pairings of one CS with a US have a much wider and more permanent impact than might have been previously suspected. Until recently, these latter phenomena have resisted the development of rigorous theory at a behavioral level, but neural network models have begun to make testable predictions and perhaps provide conceptual insight into the structure of the neural pathways that mediate transfer effects (Kehoe, 1988; Klopf, 1988).

Key Variables in Classical Conditioning

Building on the ancient Law of Contiguity, the basic variables that control CR acquisition were first identified by the British Empiricist philosophers. The Scottish philosopher Thomas Brown (1820/1977) provided the Secondary Laws of Association, which would enable one to predict how associations would vary in their strength. Among other things, the strength of an association would depend on the intensity of the events, their duration, and the frequency of their pairing. This section focuses on the variables roughly corresponding to those identified by Brown.

CS–US Interval: The Construct of the Stimulus Trace

In line with the fundamental status of the Law of Contiguity in Associationism, the interval between the onset of the CS and the onset of the US has been the premier variable in classical conditioning. As a rule of thumb, most response measures show an inverse gradient. That is, as the CS–US interval increases, the rate and level of CR acquisition decline. However, the gradient can vary dramatically across response systems, even within the same species (Hall, 1976, pp. 111–115; Lennartz & Weinberger, 1992; Schneiderman, 1972). As is shown next, the rabbit eyeblink preparation shows a sharply defined gradient, which peaks and then falls over CS–US intervals of a few seconds. At the other extreme, conditioning over CS–US intervals of hours has been obtained in taste aversion learning (Garcia et al., 1966; Schafe et al., 1995), in which the CS is flavored water and the US is a nausea-inducing agent. Nevertheless, the strength of the aversion to the taste CS has been found to decline in a graded fashion as the CS–US interval is increased (Buresova & Bures, 1974).

The basic theoretical construct for explaining the graded effect of the CS–US interval on CR acquisition has evolved from two ideas formulated by Pavlov (1927). The first idea in the concept of the *stimulus trace,* which emerged from investigations using *trace conditioning.* Trace conditioning is a procedure in which the CS is usually brief and occupies only the early portion of the CS–US interval. Thus, there is a temporal gap between the end of the CS and the start of the US. Pavlov proposed that this gap, which he called the *trace interval,* is bridged by neural activity that is initiated by the CS but, more importantly, persists for some time after the physical CS itself has ceased (Pavlov, 1927, p. 39). The second idea emerged from the *delay conditioning* procedure, in which the CS occupies the entire CS–US interval and frequently overlaps the US. Across the duration of the CS, Pavlov proposed that there is a time-dependent pattern of neural activity that allows the animal to distinguish successive portions of the CS from each other. In theoretical terms, the CS can be treated as generating a sequence of stimulus elements, each acting as a CS in its own right. According to Pavlov, the elements occurring close to the US would acquire an excitatory value. Earlier elements in the sequence that were more remote to the US would acquire an inhibitory value (Pavlov, 1927, p. 103).

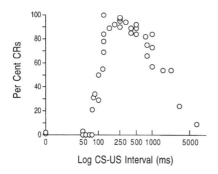

FIGURE 6.7. Scattergraph of CR likelihoods as a function of CS–US interval for the rabbit inner eyelid (nictitating membrane).

These two ideas of persistent, time-related activity generated by a CS have been blended into the single construct of the stimulus trace. This construct has been elaborated and refined in theories of conditioning (Gormezano, 1972; Gormezano & Kehoe, 1981; Hull, 1943) but particularly in *real-time models* of conditioning (Buonomano & Mauk, 1994; Desmond & Moore, 1988; Grossberg & Schmajuk, 1989; Moore et al., 1996; Sutton & Barto, 1981, 1990). The construct of the stimulus trace has been valuable in guiding research and theory concerning the effects of the CS–US interval manipulations, which are described later in this section.

As the basis for mapping the time course of the stimulus trace, it has been assumed that the strength of the CR reflects the intensity of the CS's trace at its point of contiguity with the US. Figure 6.7 summarizes the results from several studies using the rabbit eyeblink in which the CS–US interval was manipulated. Each point on the figure represents the mean CR likelihood achieved for a given CS–US interval. As can be seen, no CR acquisition occurs when CS onset and US onset occur simultaneously (CS–US interval = 0). As the CS–US interval approaches values between 200 and 400 ms, CR acquisition reaches progressively higher levels. CS–US intervals longer than 400 ms produce a downward gradient of progressively less responding, reaching a floor at an interval around 5000 ms. These findings have been construed as indicating that the trace recruits to its maximum intensity within 250 ms and then fades away over a few seconds. As is described next, subsequent research has confirmed the time course of the initial recruitment phase but has revealed that the decay phase persists for much longer than previously suspected.

The Minimum Effective CS–US Interval: Recruitment of the CS Trace

With regard to the recruitment phase of the CS trace, there are four lines of evidence that the CS trace requires 50 ms to become minimally effective, after which it quickly reaches its maximum within another 200 ms. First, CS–US intervals between 0 and 50 ms have yielded no evidence of CR acquisition, even when CS-alone test trials were inserted so that CRs with latencies longer than the CS–US interval could be detected (Kehoe et al., 1981a; Salafia et al., 1980; Smith et al., 1969). Second, there is no evidence of latent effects of CS–US pairings when animals are given initial training with CS–US intervals around 50 ms and then switched to a CS–US interval of 250 ms (Salafia et al., 1980). Third, the shortest, stable CR onset latency is 75 ms, pinpointing the minimal time needed for recruiting the CS trace, activating the neural pathways that subserve learning, and generating a CR (Ross et al., 1979; Salafia et al., 1980; Smith et al., 1969). Fourth, Patterson (1970) appeared to reduce the recruitment time for the trace by substituting a tone CS with stimulation of the inferior colliculus, a key auditory nucleus. Using a 50-ms CS–US interval, which otherwise yields no CR acquisition, Patterson (1970) obtained CR acquisition that reached a stable level at which the rabbits responded to the CS on 45% of the CS–US trials.

The Maximum Effective CS–US Interval: Decay of the CS Trace

Recent studies have revealed that the trace for auditory CSs persists in the eyeblink system for at least 20,000 ms, four times as long as found in earlier studies. There are three lines of evidence for a highly persistent trace.

Reduced Trials per Session

Conditioning at longer CS–US intervals rises when the number of CS–US trials per session is reduced (Kehoe et al., 1991; Levinthal, 1973; Levinthal & Papsdorf, 1970; Levinthal et al., 1985). For example, Kehoe et al. (1991) found that when 25 CS–US pairings were used in each session, a CS–US interval of 3200 ms yielded an asymptote at which CRs occurred during 15% of the CS–US pairings. However, when only 1 or 5 CS–US pairings occurred in each session, the 3200-ms interval yielded asymptotes at which CRs occurred on 50% or more of the CS–US pairings.

Serial CS Procedures

A later section of this chapter reviews the results of procedures in which a sequence of two or more CSs are paired with the US (Kehoe et al., 1979, 1987a; Schreurs et al., 1993). In brief, manipulations of the interval between a pair of CSs have provided evidence that the trace of the first CS can last up to 15,000 ms.

Heart Rate Conditioning

When heart rate is measured using the same CS and same US as in the rabbit eyeblink preparation, acquisition of a heart rate CR can be obtained at CS–US intervals up to 21,000 ms (Schneiderman, 1972).

Backward Trials: US–CS Pairings

Although it is conventional for the CS to precede the US, the US can be presented before the CS. From the perspective of trace theories, such "backward" US–CS pairings would be unlikely to yield an excitatory association unless a US trace overlaps the CS trace. On the other hand, a variety of theories predict that the

US–CS order should yield an inhibitory association, either because the CS is coincident with US offset (Rescorla, 1969; Sutton & Barto, 1981) or because the CS signals a long period without the US (Moscovitch & Lolordo, 1968).

Inhibition in the Rabbit Eyeblink Preparation

In the rabbit eyeblink preparation, there is no evidence of excitatory conditioning as a result of backward CS–US pairings; that is, test presentations of the CS after US–CS pairings do not elicit CRs. However, there is evidence of inhibitory conditioning. After backward pairings, the CS suppressed eyeblink CRs to a light CS that had received forward CS–US pairings (McNish et al., 1997). Moreover, when the order of the stimuli was switched from backward US–CS pairings to forward CS–US pairings, CR acquisition to the previously backward CS was slow relative to control groups that initially received either restraint-only, CS-alone presentations, US-alone presentations, or unpaired CS–US presentations (Plotkin & Oakley, 1975; Siegel & Domjan, 1971; Tait & Saladin, 1986).

Excitation in Fear Preparations

Although backward CS–US pairings appear inhibitory for the eyeblink CR in rabbits, two studies have demonstrated that the same backward pairings produce evidence of excitatory conditioning in other response systems (McNish et al., 1997; Tait & Saladin, 1986). For example, immediately after US–CS pairings, a tone CS potentiated a startle response to an air puff stimulus in rabbits. This potentiation suggests that there had been excitatory conditioning of a fear response.

Commentary

This behavioral dissociation between the eyeblink and fear response would be paradoxical if there were a single point of convergence between the neural traces of the CS and US that mediated all the effects of CS–US pairings. However, this dissociation can be reconciled in terms of neurobiological findings that there are multiple loci of convergence. Most notably, cerebellar damage prevents eyeblink conditioning but not fear conditioning (Thompson et al, 1987). Conversely, lesions of the amygdala hinder fear conditioning but do not prevent eyeblink conditioning (Weisz et al., 1992). With respect to trace theory, Wagner and Brandon (1989) have offered a theory in which the CS and US each have two traces with different time constants. The interactions among these traces allow for the concurrent acquisition of both excitatory and inhibitory associations, which, in turn, can be expressed in different response systems.

CR Timing

The CR is often depicted in textbooks as a replica of the UR. In fact, the time course of the CR has its own dynamical features. Although URs are typically short-latency responses with rapid recruitment and limited duration, CRs are long-latency responses that peak near the time of the US. Therefore, the timing of CRs depends on the CS–US interval. The upper panel of Figure 6.8 shows the

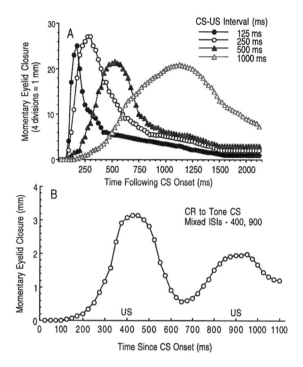

FIGURE 6.8. Average CR waveforms for fixed CS–US intervals (**A**) (Smith, 1968) and two intermixed CS–US intervals (**B**) (Kehoe et al., 1989) for the rabbit inner eyelid (nictitating membrane).

time course for eyelid CRs acquired by four groups of rabbits trained with different CS–US intervals, namely 125, 250, 500, and 1000 ms (Smith, 1968). Closure of the eyelid was initiated shortly following the onset of the CS, but peak eyelid closure occurred around the time of US presentation at the end of the CS–US interval.

CR Onset Latency

The most noticeable feature of the CR is its anticipatory nature. The acquisition of the anticipatory CR is seen as a progressive foreshortening of its onset latency (Gormezano et al., 1983). However, the onset latencies of even well-established CRs are dictated by the CS–US interval. For CS–US intervals of 400 ms or less, the distribution of CRs is roughly centered near the midpoint of the interval. For CS–US intervals longer than 400 ms, the distributions become more dispersed and asymmetrical, with a tail skewed toward CS onset and a mode about two-thirds of the way along the CS–US interval (Kehoe & Schreurs, 1986).

CR Peak Latency

Although the initiation of eyelid closure anticipates the US during training, the maximal closure of the eyelid, the CR peak, tends to remain near the US. This feature of the CR is not an artifact of US presentation. As shown in Figure 6.8, it appears on CS-alone trials in which neither the US nor the UR could intrude on the CR.

When the CS–US interval is shifted from one value to another, the CR peak disappears and then reappears at the new locus of the US (Coleman & Gormezano, 1971; Salafia et al., 1979). Moreover, two temporally distinct CRs can be acquired concurrently (Hoehler & Leonard, 1976; Kehoe et al., 1989, 1993; Mauk & Ruiz, 1992; Millenson et al., 1977).

The lower panel of Figure 6.8 shows the time course of CRs that appeared when a tone CS unpredictably signaled the US at one of two CS–US intervals, 400 ms and 900 ms. The CR developed two distinct peaks, one located at each locus of US delivery. Furthermore, rabbits can learn to generate two distinctive CRs when different CSs, such as a light and a tone, signal different CS–US intervals (Kehoe et al., 1989; Mauk & Ruiz, 1992).

Distribution of Training

In the general literature on learning and memory, it is widely supposed that practice spaced out over time produces faster learning than the same amount of practice massed into a shorter period of time (Baddeley, 1997, pp. 109–114). In classical conditioning, the distribution of training has been examined in two ways. One way has been to manipulate the intertrial interval (ITI) within training sessions. A second way has been to vary the number of trials within sessions. As discussed more fully next, the findings in classical conditioning resemble those found in other forms of learning: within-session ITI has weak effects on the rate of CR acquisition, but the number of trials per session has a large effect.

Trials per Session

For the rabbit eyeblink preparation, spreading a given number of CS–US pairings across several sessions considerably enhances the rate of CR acquisition on a trial-by-trial basis. Several studies have shown that when single CS–US pairings are given in sessions separated by 24 h, the first CR occurs after an average of 7 pairings (range, 2–18), that is, 7 days. When 5 pairings separated by 1 min are presented within each session, the first CR appears after an average of 19 pairings (range, 12–28). When 25 pairings are presented in each session, the first CR occurs after an average of 84 trials (range, 37–150) (see Kehoe & Macrae, 1994, for a summary of the available studies). In practice, it has been common to use many trials per session, often up to 100, which allows experiments to be completed in a few sessions and allows CS-alone trials to be inserted into the trial sequence with little impact on the overall level of responding.

Within-Session Trial Spacing

Manipulations of the interval between CS–US pairings within a training session have been repeatedly examined in the rabbit eyeblink preparation (Brelsford & Theios, 1965; Frey & Misfeldt, 1967; Mis, et al., 1970; Mitchell, 1974; Salafia et al., 1975). Intervals shorter than 10 s have yielded relatively low levels of CR acquisition (Nordholm et al., 1991; Salafia et al, 1973), and unpublished observations in our laboratory indicate that CR acquisition did not occur when the ITI was 4 s. Intervals of 60 s or so have tended to produce the highest levels of conditioned responding, but intervals of 300 s have yielded inconsistent results, with levels of responding sometimes higher and sometimes lower than intermediate ITIs. The use of variable ITIs tends to diminish differences in responding among groups receiving different mean ITIs (Mitchell, 1974).

Conditioned Stimulus Variables

CS Duration

Delay Versus Trace Conditioning

Delay conditioning usually produces faster and higher levels of CR acquisition than trace conditioning, particularly at longer CS–US intervals (Graves & Solomon, 1985; Kehoe & Napier, 1991; Kehoe & Schreurs, 1986; Kehoe, et al., 1995b; Manning et al., 1969; Schneiderman, 1966). For example, Kehoe and Schreurs (1986) conducted training in which the CS–US interval was fixed at 800 ms. The CS duration was either 50, 200, or 800 ms. The 800-ms delay CS produced the most rapid CR acquisition that reached an asymptote on which CRs occurred on 70% of CS–US trials after training for 300 trials. The 200-ms trace CS also produced CR acquisition to the same asymptote (70% CRs) but more slowly, requiring 420 trials. The 50-ms trace CS produced the slowest CR acquisition and only elicited CRs on 50% of CS–US trials within 600 trials. To explain the superiority of delay to trace conditioning, the continuing portions of the CS are assumed to maintain the intensity of the trace initiated at CS onset (Blazis & Moore, 1991; Gormezano, 1972; Hull, 1943).

Test CS Durations

When the duration of the CS is shortened after delay conditioning, there is a large decline in CR likelihood and a lesser impact on the amplitude and speed of the CR (Kehoe & Napier, 1991; Kehoe et al., 1995b, experiment 3). Specifically, Kehoe and Napier (1991) trained subjects with a 400-ms tone CS in a delay conditioning procedure. On test trials, responding to the CS was assessed at durations ranging from 400 ms down to 12 ms. Across these values, CR likelihoods declined from CRs on 94% of trials to only 44% of trials. In contrast, lengthening the duration of a trace CS has had a relatively slight impact on eyeblink CRs. In Kehoe and Napier's (1991) study, a group trained with a 25-ms CS showed only a slight, nonsignificant decline from CRs on 84% of trials to 72% of trials when the CS was lengthened from 25 ms to 400 ms. The asymmetrical results of shortening a delay CS versus lengthening a trace CS indicate that the ongoing, steady portions of a stimulus provide additional, conditioned elements. That is, shortening the delay CS reduced the total number of conditioned elements and hence reduced

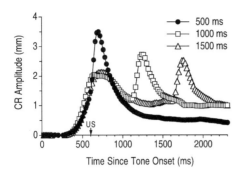

FIGURE 6.9. Demonstration that the onset and offset of a trace CS can elicit distinct CRs by CR waveforms seen in the rabbit inner eyelid (nictitating membrane) after trace conditioning when the CS duration was increased from 500 ms to 1000 ms and then 1500 ms.

the CR likelihood. However, lengthening a trace CS added elements that had no associative strength and therefore could not contribute to the CR. Such an assumption has been incorporated into real-time quantitative models (Desmond, 1990; Desmond & Moore, 1988; Kehoe et al., 1995b; Sutton & Barto, 1990).

Role of CS Offset

The offset as well as the onset of a continuing event can serve as an effective CS in eyeblink preparations (Gormezano, 1972, p. 157; Liu & Moore, 1969). Thus, the use of a trace conditioning procedure introduces two stimulus changes before the US, either of which is capable of being a CS. In fact, the offset of a trace CS does function as a CS separately from its onset. First, when a trace CS is long enough that its onset and its offset have substantially different CS–US intervals, the shorter CS–US interval for stimulus offset determines the rate of CR acquisition (Kehoe, 1979, 1983; Kehoe & Weidemann, 1999b). Second, when the duration of a trace CS has been manipulated in testing, the onset and offset of the CS each command a distinctive CR (Desmond & Moore, 1991).

A demonstration of the ability of a trace CS to evoke two distinct CRs is depicted in Figure 6.9. Rabbits were given training trials in which a 500-ms tone was followed by a 100-ms trace interval before the US. Thus, the CS–US interval for tone onset was 600 ms and the CS–US interval for tone offset was 100 ms. There were also periodic test presentations of the CS alone, in which its duration was either 500, 1000, or 1500 ms. The figure shows the time course of CRs on these test trials (Cox, 1990).

Examination of Figure 6.9 reveals there were CR peaks tied to both tone onset and tone offset. For the 500-ms tone duration, a single sharply defined peak appeared at a point that was 700 ms after tone onset and 200 ms after tone offset. As the tone duration was extended to 1000 and 1500 ms, the first peak remained located at 700 ms after tone onset but diminished in amplitude. Meanwhile, a second peak emerged at a point 200 ms after tone offset for both of the extended tones.

CS Intensity

The rate of eyeblink CR acquisition is a direct function of CS intensity (Kehoe 1982, 1983; Scavio & Gormezano, 1974; Solomon et al., 1974). This effect might appear intuitively obvious, but theoretically it is not so simple. It could arise from two sources. Philosophical associationism stresses the role of stimulus intensity in promoting associative linkages. On the other hand, reflex physiology stresses the role of stimulus intensity in energizing the performance of the CR. In addition, stimulus intensity could also constitute a sensory dimension for CS recognition and CR performance.

As yet, it has not been possible to separate the associative and energizing effects of CS intensity, but some progress has been made in confirming the existence of energizing and sensory processes. Following the establishment of the eyeblink CR, test trials with different CS intensities have been presented. With the associative strength of the CS fixed, the energizing and sensory effects on CR expression could vary. According to an energization hypothesis, reductions in CS intensity would lower responding, but increases would raise responding. According to a sensory hypothesis, however, changes in CS intensity in either direction would tend to lower responding as a consequence of incomplete generalization.

In agreement with both hypotheses, tests with a tone CS at intensities lower than the training value have yielded a sharp decrease in CR likelihood. However, for test intensities greater than the training value, the data suggest a trade-off between energizing effects and generalization decrements. Scavio and Gormezano (1974) observed a steady rise in responding over a 20-dB increase, which is consistent with energizing effects. In contrast, Cox (1990) found a small decline in responding for a 25-dB increase, indicative of a generalization decrement. In both cases, the changes in responding for increased test intensities were small compared to the large declines seen for lesser test intensities.

Unconditioned Stimulus Variables

The US is the foundation of classical conditioning: when combined with the CS, it evokes the neural activity necessary for the CS to gain access to the pathways that allow it to eventually evoke the CR. In addition, it is now clear that the CS influences the UR. Moreover, the US–UR reflex itself appears subject to associative modifications during CS–US pairings, some of which predate the appearance of CRs.

The Unconditioned Reflex

The two most widely used USs in the rabbit eyeblink preparation are an electrotactile shock delivered to the skin surrounding the eye and an air puff aimed at the cornea. However, not all stimuli that elicit an eyeblink can support conditioning. Bright lights can consistently elicit eyeblinks but do not produce CR acquisition when paired with a tone CS (Bruner, 1965; Frey et al., 1976). Unless otherwise mentioned, the studies described in this chapter have used electrotactile or airpuff USs.

Taken by itself, the US–UR relationship can be described according to reflex laws. That is to say, the vigor of the UR as measured by its likelihood, latency,

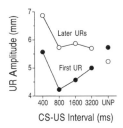

FIGURE 6.10. Reflex modification during early CS–US pairings as function of CS–US interval or CS/US unpairings using the rabbit inner eyelid (nictitating membrane) (Kehoe et al., 1991).

and amplitude varies as a function of US intensity, duration, repetition, and spacing. In brief, the vigor of the UR increases as a direct function of US intensity and duration (Marshall Goodell et al., 1992; Schreurs & Alkon, 1990; Schreurs et al., 1995). Repetition and spacing, however, interact strongly, demonstrating both habituation and sensitization. Within an experimental session, habituation is apparent; UR amplitude tends to decrease across successive presentations of the US (Saladin et al., 1989). Across successive sessions, however, sensitization occurs; the UR amplitude tends to increase across days (Mis & Moore, 1973; Saladin & Tait, 1986; Saladin et al., 1989).

Reflex Modification of the UR by the CS

Reflex modification is said to occur when presentation of an auditory CS alters the amplitude of the UR to the US (Ison & Leonard, 1971). Evidence suggests that associative changes in reflex modification can occur within the first half-dozen CS–US pairings and well before CRs begin to appear (Kehoe et al., 1991; Gormezano et al., 1983; Weisz & LoTurco, 1988; Weisz & McInerny, 1990). Figure 6.10 shows an example of reflex modification during early CS–US pairings. Rabbits were presented with pairings in which the CS–US interval was either 400, 800, 1600, or 3200 ms. Other rabbits received explicitly unpaired presentations of the CS and US (UNP). The lower function shows the UR amplitude on the very first trial, and the upper function shows UR amplitude during subsequent trials that preceded emergence of the first CR.

As may be apparent, facilitation of the UR was not evident on the first trial (the first US presentation). Compared to the UR in the unpaired group, the 400-ms CS–US interval did not appear to alter UR amplitude. The longer intervals appeared to diminish the UR. Thereafter, facilitation of the UR became more prominent in the paired conditions. As shown in the upper function, all CS–US intervals produced larger URs than did US-alone presentations in the unpaired condition. In addition to CS–US interval, the degree of UR facilitation is a direct function of CS intensity (Ison & Leonard, 1971; Weisz & Walts, 1990). UR facilitation is also greater with a longer CS than a shorter CS (Weisz & Walts, 1990).

Changes in reflex modification across early CS–US pairings could arise from two mechanisms (Weisz & McInerny, 1990). First, acquisition of an excitatory association may counteract the habituation of the ability of a CS to facilitate the UR. Second, the presence of the US after the CS may block the activity of the CS in the US–UR pathways, which would protect the CS from habituation (Pfautz et al., 1978). Whatever mechanisms underlie changes in reflex modification, they appear neurally separate from those that mediate CR acquisition. Weisz and LoTurco (1988) found that lesions of deep cerebellar nuclei abolished CR acquisition but failed to eliminate UR facilitation.

The UR After CR Emergence

Once a CR has become established, its trajectory often extends into the period of the UR. Nevertheless, except for CRs with large amplitudes, it is possible to measure the amplitude of the UR that appears superimposed on the CR. The amplitude of the UR on CS–US trials has been compared to the amplitude of the UR on US-alone trials by intermixing the two kinds of trial. Reflex facilitation has been said to occur when the response amplitude on CS–US trials is larger than the UR on US-alone trials. On this basis, there has been some evidence that reflex facilitation persists throughout training (Grevert & Moore, 1970; Leonard et al., 1972). In other cases, however, reflex facilitation has disappeared: the UR amplitude on CS–US trials fell below the UR amplitude on US-alone trials (Weisz & LoTurco, 1988; experiment 1). The intensity of the US may have been partly responsible for these divergent results. Donegan and Wagner (1987) reported that URs elicited by weak electrotactile USs (1 mA) were facilitated by a CS after CR acquisition, but URs to stronger USs (5 mA) were diminished by the same CS. In addition, increasing the CS–US interval on test trials can convert UR diminution into facilitation (Canli et al., 1992).

Any conclusions about the facilitation of the UR on CS–US trials are complicated by changes in the UR itself. Some instances of reflex facilitation can be largely attributed to declines in the UR on US-alone trials interspersed among CS–US trials (Hupka et al., 1970; Leonard et al., 1972). Furthermore, there may be associative changes in the UR that appear on US-alone trials. Specifically, when the UR on US-alone trials after CS–US pairings has been compared to the UR before training, the peak of the UR was delayed and the UR amplitude was increased (Schreurs et al., 1995). Similar, but far smaller, changes in the UR appeared in control groups that had received either unpaired CS–US presentations or restraint only.

CR Acquisition: US Intensity, Proximity, and Duration

Theories of conditioning generally predict that the rate of CR acquisition depends on the features of the US, inasmuch as they govern the vigor of the UR and any accompanying emotional reaction. For the electrotactile US, the rate of eyeblink CR acquisition increases with current intensity (Ashton et al., 1969; Ayres et al., 1984; Hupka et al., 1970; Smith, 1968) and the proximity of the stimulating electrodes to the eye (Salafia et al., 1974, 1979). In contrast, US duration has not had a consistent effect on CR acquisition. A modest positive effect of increased US duration has been observed when rabbits trained with a 50-ms US were compared to rabbits trained with a 350-ms US (Ashton et al., 1969). For longer USs, namely, 1500 ms and 6000 ms, a positive effect of duration has been seen only when a single CS–US pairing was presented in each session. When each session contained 90 CS–US pairings separated by an average ITI of 60 s, the effect was reversed. The 6000-ms duration reduced CR likelihood both within and between sessions (Tait et al., 1983).

Schedules of Reinforcement

Consistent CS–US pairings are the most reliable means for producing CR acquisition. Presentations of the CS-alone or the US-alone among CS–US pairings tend to reduce CR acquisition. Theoretically, this happens because CS-alone and US-alone presentations reduce the statistical contingency between the CS and US, thereby degrading the signal value of the CS (Prokasy, 1965; Rescorla, 1967). However, a variety of mechanisms appear to mediate the effects of CS-alone and US-alone presentations. This section describes the effects of such presentations before and during CS–US pairings.

CS Preexposure: Latent Inhibition

Latent inhibition is said to occur when the rate of CR acquisition during CS–US pairings is reduced because of prior exposure to CS-alone trials (Lubow & Moore, 1959). In qualitative terms, CS preexposure appears to reduce the associability of the CS (Schmajuk et al., 1996). Latent inhibition has been repeatedly demonstrated in the rabbit eyeblink preparation. Siegel (1969a) found that 1300 preexposures hindered CR acquisition but that 100 preexposures did not. In other studies, reductions in the rate of CR acquisition were obtained using the following numbers of preexposures: 100 (Clarke & Hupka, 1974), 108 (Robinson et al., 1993), 250 (Clarke & Hupka, 1974), 300 (Salafia & Allan, 1982), 450 (Solomon et al., 1974; Solomon & Moore, 1975), 1300 (Siegel, 1969b), and 1380 (Reiss & Wagner, 1972).

With respect to the features of the CS itself, latent inhibition has not been discernibly influenced by CS intensity (Solomon et al., 1974) or CS duration (Clarke & Hupka, 1974). However, latent inhibition has been shown to be relatively specific to both the preexposed CS and its context. When the acoustic frequency of a preexposed tone was changed at the start of CS–US pairings, the rate of CR acquisition to the tone became more similar to that of a nonpreexposed control condition. Specifically, a U-shaped generalization gradient appeared. The slowest CR acquisition occurred when the frequency of the tone during CS–US pairings was identical to that of the preexposed tone. As the tone frequency during CS–US pairings deviated further from the preexposed value, CR acquisition became progressively faster and approached control levels (Siegel, 1969b). Similarly, latent inhibition has been shown to be attenuated in relation to the number of changes in the olfactory, auditory, or visual background within the training apparatus (Kim, 1986).

A preexposed CS loses its capacity to inhibit responding when combined with another CS (Reiss & Wagner, 1972; Solomon et al., 1974, experiment 2; Solomon

& Moore, 1975). For example, Solomon et al. (1974) first trained rabbits to respond reliably to a light CS. Half the rabbits were then repeatedly exposed to a tone, and the other half were not given any tone exposures. Then, both groups were given tests in which the tone was compounded with the light CS. The addition of the tone produced a reduction in responding to the light CS. However, the size of the reduction was the same in both groups. This reduction in responding to the combined stimuli is an example of external inhibition. External inhibition has been attributed variously to a division of attention between the two stimuli (Solomon et al., 1974) or an incomplete generalization of CRs from the light to the compound of tone plus light (Kehoe & Gormezano, 1980). In either case, the failure of preexposure to add to the ability of a CS to reduce responding to another CS has been used to distinguish latent inhibition from a third type of inhibition, namely, conditioned inhibition. A CS is considered to be a conditioned inhibitor only if it can both hinder CR acquisition and reduce responding to another CS in excess of any external inhibition (Rescorla, 1969).

Interspersed CSs: Partial Reinforcement

CR Acquisition

The procedure of mixing CS-alone trials among CS–US pairings is commonly referred to as a *partial reinforcement* or, less commonly, *intermittent reinforcement.* In contrast, the consistent presentation of CS–US pairings with no CS-alone presentations is designated as *continuous reinforcement* or *100% reinforcement.* In the rabbit eyeblink preparation, intermixing CS-alone presentations with CS-US pairings on a 1:1 basis slows CR acquisition only slightly (Leonard, 1975; Leonard & Theios, 1967; Prokasy & Gormezano, 1979), sometimes hardly at all (Gormezano & Coleman, 1975; Thomas & Wagner, 1964) For example, in Gormezano and Coleman (1975), one group received 100 CS–US pairings per training session. A second group of rabbits received partial reinforcement, in which each session contained 100 CS-US pairings interspersed with an additional 100 CS-alone presentations. A third group also received partial reinforcement, but each session contained only 50 CS–US pairings interspersed with 50 CS-alone presentations. Thus, in the latter two groups, the CS was followed by the US in only half the trials. Nevertheless, all three groups required about 100 CS–US pairings to attain CRs on 90% of trials.

CR Maintenance

Well-established eyeblink CRs have been maintained when large numbers of CS-alone trials were substituted for CS–US pairings. Gibbs et al. (1978) initially established CRs using 100 CS–US pairings per session. By the end of three sessions, CRs occurred in virtually 100% of trials. Thereafter, one group of animals continued to receive 100 CS–US pairings per session (100% reinforcement), but other groups received various mixtures of CS-alone trials interspersed randomly among the remaining CS–US pairings (partial reinforcement). Specifically, CS-alone presentations were substituted for CS–US pairings on 50, 75, 85, and 95 of the 100 trials in each session, constituting CS–US pairings of 50%, 25%, 15%, and 5%, respectively. In the 50%, 25%, and 15% partial reinforcement groups, CR likelihood dropped only slightly; specifically, CRs occurred on 97%, 95%, and 80% of all trials, respectively. Large declines in responding appeared in the 5% partial reinforcement group. In that group, CR likelihood sank to a level of 45% by the sixth session.

CR Extinction

Subsequent to partial reinforcement in the Gibbs et al. study, all groups were given three sessions of extinction training in which all 100 trials were CS-alone presentations. Before extinction training, the continuous reinforcement group had 900 CS–US pairings, whereas the partial reinforcement groups had fewer CS–US pairings. In similar experiments in instrumental conditioning using food reward, partial reinforcement causes the response to persist in extinction training longer than after continuous reinforcement (Hall, 1976, pp. 286–297). This finding has been called the *partial reinforcement extinction effect.* In the experiment by Gibbs et al. (1978), a partial reinforcement extinction effect was observed. Across the three daily extinction sessions, the continuous reinforcement group showed CRs on only 25% of the CS-alone trials. In contrast, the 50% and 25% partial reinforcement groups showed CRs on 42% and 55% of their trials. In other studies using the rabbit eyeblink response, the partial reinforcement extinction effect was demonstrated following 50% partial reinforcement by Leonard (1975) (but see Gormezano & Coleman, 1975; Thomas & Wagner, 1964). In Gibbs et al. (1978), the partial reinforcement effect failed to appear in the group that had been reduced to 15% reinforcement. Although the 15% group sustained responding when there were still some CS–US pairings, its CR likelihood dropped during extinction training as quickly as in the continuous reinforcement group.

US Preexposure

Exposure to the US before CS–US pairings slows the rate of CR acquisition (Randich & LoLordo, 1979). In general, the rate of CR acquisition is progressively reduced as the number of US preexposures is increased, compared to control conditions in which no US-alone presentations occur before CS–US pairings (Hinson, 1982; Mis & Moore, 1973; Saladin & Tait, 1986; Saladin et al., 1989). For the rabbit eyeblink, 50 US preexposures slow CR acquisition only slightly, and 200 or more US preexposures are needed to reliably slow CR acquisition.

Although the effect of US preexposure on CR acquisition is reliable, it depends on at least two other variables. One variable is the distribution of US-alone presentations before training; the second variable is the context in which the US preexposures occur. CR retardation may only appear if the US-alone presentations are spread over several days. In Saladin et al. (1989), one group received 200 US-alone presentations in one session and another group received 200 US-alone presentations distributed over 10 days. Compared to control groups that had received no US preexposures, the rate of CR acquisition was significantly reduced only in the group that had received the distributed US preexposures.

Regarding context, reductions in the rate of CR acquisition may only appear if training occurs in the same context as the US-alone presentations. When US-alone presentations and CS–US pairings were conducted in different chambers, the rate of CR acquisition matched that of controls that were not preexposed to the US alone (Hinson, 1982; Saladin & Tait, 1986). When, however, the chambers remained the same, the effect of US preexposure was persistent; that is, CR acquisition was retarded during CS–US pairings that immediately followed the US-alone presentations. In addition, once CR acquisition was complete for the first CS, CR acquisition to a second CS paired with the US was also slow relative to control conditions (Saladin & Tait, 1986).

Interspersed USs: Partial Warning

In the rabbit eyeblink preparation, the interspersion of US-alone presentations among CS–US pairings has consistently slowed CR acquisition under a wide variety of conditions (Grevert & Moore, 1970; Hoehler & Leonard, 1981; Hupka et al., 1970; Leonard et al., 1972; Tait et al., 1983). Some theories imply that the interspersed US-alone presentations slow CR acquisition by degrading the statistical contingency between the CS and the US (Prokasy, 1965; Rescorla, 1967) or by disrupting the memory consolidation of the preceding CS–US pairing (Grevert & Moore, 1970). However, data from two studies indicate that the effect of interspersed USs on CR expression is temporary, perhaps through habituation of the response system. First, Grevert and Moore (1970) found that CR likelihood did not decline when the ITI between US-alone trials and CS–US trials was increased from 50 to 100 s, in comparison to control groups that received the same number and spacing of CS–US trials but without the interspersed US-alone trials (Leonard et al., 1972). Second, Hoehler and Leonard (1981) interpolated a single block of 40 US-alone presentations between two blocks of 50 CS–US trials. These rabbits showed only a temporary deficit in CR acquisition compared to a control group that did not receive the block of US-alone presentations.

Compound Conditioned Stimuli

Any organism must confront a multifaceted stream of stimulus events in its environment. Thus, there are a wealth of possible relationships among potential signals and biologically significant events. In classical conditioning, the search for the mechanisms by which animals select, integrate, and encode multiple signals has been conducted using the device of the compound CS. Typically, a compound CS is formed by using two distinctive components, often a tone and a light (Kehoe & Gormezano, 1980; Kehoe, 1998). By varying the conditions of training for the compound and its components in relatively small ways, experiments have demonstrated a startlingly diverse set of outcomes. These outcomes have fostered many of the major advances in conditioning theory over the past 30 years (Mackintosh, 1975; Rescorla, 1969; Rescorla & Wagner, 1972; see chapter in this volume by Brandon, Vogel, & Wagner).

Stimulus Compounding

In *stimulus compounding,* two CSs receive separate training and are periodically tested by presenting them simultaneously. When compounding a tone CS and a light CS for the eyeblink response, the usual outcome is *summation;* that is, the compound elicits a greater response than either component. In the rabbit eyeblink preparation, summation has been observed using measures of CR magnitude (peak extension), speed, and likelihood. For example, when a 500-ms tone CS was compounded with a 500-ms light CS, the mean magnitude of eyelid movements during the tone plus light compound was 1.82 mm as compared to 1.38 mm during the tone alone and 1.34 mm during the light alone. Similarly, using CR onset latency as a measure of CR speed, the mean latency of CRs during the tone + light compound was 237 ms versus means of 292 ms and 307 ms during the tone and light, respectively (Kehoe & Weidemann, 1999a). In the case of CR likelihood, summation follows a precise quantitative relationship. The CR likelihood for the test compound (P_C) is well predicted by the statistical sum of the CR likelihoods to the tone CS (P_T) and light CS (P_L): $P_C = P_T + P_L - (P_T P_L)$ (Kehoe 1982, 1998; Kehoe & Graham, 1988).

Although summation occurs reliably when a tone CS or noise CS is compounded with a light CS, summation is not a universal result for the rabbit eyeblink CR. Specifically, a pure auditory compound of tone plus noise has not yielded summation. The likelihood of a CR to an auditory compound has been equal to that of the individual components (Kehoe et al., 1994). Hence, summation may occur only when CSs from different sensory modalities are compounded.

Compound Conditioning

In stimulus compounding, the component CSs are paired with the US, and then they are combined during tests. In contrast, in *compound conditioning,* training is conducted by pairing the compound CS with the US. Testing then entails presentations of each CS component by itself. The compound conditioning procedure has yielded a variety of outcomes. In brief, these outcomes cannot be easily described in terms of the summation seen in stimulus compounding. Instead, the outcomes of compound conditioning have suggested that the components of a compound CS can interact with each other in complex ways that have been described conceptually in terms of selective attention, associative competition, and perceptual fusion. These conceptual mechanisms are outlined next.

Overshadowing

Overshadowing is said to occur when the rate and asymptote of CR acquisition to one component of a compound CS is reduced by a more salient component (Pavlov, 1927, p. 141). For example, the left-hand panel of Figure 6.11 shows the likelihood of CRs in three groups of rabbits that were given training in which a

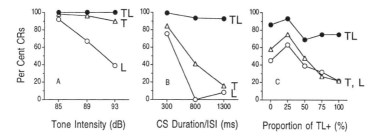

FIGURE 6.11. Compound conditioning using the rabbit inner eyelid (nictitating membrane): CR likelihood to a compound (TL), tone (T), and light (L) as a function of (**A**) tone intensity (Kehoe, 1982), (**B**) CS duration (Kehoe, 1986, experiment 1), and (**C**) the proportion of compound trials relative to component trials (Kehoe, 1986, experiment 2).

compound of light and tone was paired with the US. The light was constant in intensity, but the intensity of the tone differed across groups. Specifically, the tone intensities were 85, 89, and 93 dB. In addition to the compound CS–US pairings, the rabbits were also given occasional light-alone and tone-alone trials (Kehoe, 1982). When the tone was 85 dB, responding on light-alone trials (L) reached a 97% likelihood, which matched the CR likelihood on tone-alone (T) trials and compound trials (TL). Overshadowing appeared in the two groups trained with the 89-dB and 93-dB tones. That is, the level of CR acquisition to the light became progressively lower as tone intensity decreased. In follow-up experiments, it was found that for a less intense tone (73 dB) the direction of overshadowing could be reversed. Thus, a light can reduce CR acquisition to the tone.

Spontaneous Configuration

In some cases of compound conditioning, it has been possible to produce CR acquisition to a tone plus light compound, but when the components are occasionally presented alone, neither of them elicits a CR. The middle panel of Figure 6.11 shows an example (Kehoe, 1986). Three groups of rabbits were trained with compounds of tone and light in which both the CS durations and CS–US intervals were 300, 800, and 1300 ms, respectively. All three groups showed CRs to the compound (TL) on 90% or more of their trials. For the group trained with the 300-ms compound, CRs were elicited by both the tone and light on about 75% of their trials. However, for groups trained with the longer CSs, the level of responding on the tone and light dropped dramatically. For example, the group trained with the 1300-ms compound showed CRs on less than 20% of their tone-alone and light-alone trials.

The ability of a compound to elicit CRs in the absence of CRs to both its components has been described by some researchers as spontaneous configuration (Bellingham & Gillette, 1981; Razran, 1965). This description reflects a hypothesis that, during presentation of the compound, the components are perceptually fused, perhaps in polysensory neurons (Konorski, 1967, pp. 64–88). Consequently, the animal perceives the compound as an event distinct from its separate components (Bellingham et al., 1985; Heinemann & Chase, 1975; Pearce 1987, 1994). Consequently, the presentation of a single component is considered a test for stimulus generalization from the compound to that component. When animals fail to respond to components, there is said to be a *generalization decrement.*

Compound Conditioning and Stimulus Compounding

There might appear to be large discontinuity between spontaneous configuration obtained in compound conditioning, in which the CS components are jointly paired with the US (TL+), versus the summation outcome of stimulus compounding, in which the components are individually paired with the US (T+, L+). In fact, compound conditioning and stimulus compounding can be viewed as the endpoints of a continuum in which TL+ trials can be intermixed with component trials (T+, L+) in varying proportions. In fact, when intermixtures have been used, the results have revealed that there is a continuity in their outcomes.

The right-hand panel of Figure 6.11 shows responding in five groups of rabbits, in which the proportion of TL+ trials relative to T+ and L+ trials was varied from 0% (stimulus compounding) to 100% (compound conditioning) (Kehoe, 1986). The CSs were 800 ms in duration, and on paired trials, the CS–US interval was 800 ms. Regardless of the proportions, overall responding on TL test trials was at a constant high level across groups. In contrast, responding on T and L trials declined smoothly as the proportion of TL+ trials increased.

These results have been interpreted as evidence that a compound and its components activate a neural network that contains three sets of neuron-like units (Kehoe, 1988; Rescorla, 1973; Rudy & Sutherland, 1995; Schmajuk & DiCarlo, 1992; Sutherland & Rudy, 1989). One set contains polysensory units, also called configural units, that are only activated by a joint occurrence of the tone and light. Another set contains unisensory units activated only by the tone, and the third set also contains unisensory units activated only by the light. On T+ and L+ trials, only their respective sets of units would be activated, and only those activated sets would gain the capacity to elicit the CR. On TL+ trials, however, all three sets of units would be activated and may interact in a competitive fashion in gaining the capability to elicit the CR.

According to these network models, total responding to the compound CS would reflect the sum of the CR-generating capabilities acquired previously by the three sets of units. However, the proportion of total responding elicited by the different sets of units could vary dramatically. In the case of spontaneous configuration, in which CRs occurred with a high likelihood on TL trials but hardly at all on T and L trials, it would be inferred that the polysensory units had gained the lion's share of the capacity to elicit CRs. Conversely, in cases where CRs occurred with a high likelihood on T and L trials, it would be inferred that the unisensory units gained the majority of CR-eliciting capability.

Overprediction

There have been several studies of conditioning in which training was initially conducted by pairing the individual component CSs with the US (component training; T+, L+) (Kremer, 1978; Levitan, 1975; Rescorla, 1970, 1997). Training was then switched to compound trials (TL+). During the component training in stage 1, responding on T+ and L+ trials achieved high levels. During compound training, sporadic tests with the components revealed that responding to each of them rapidly declined to a level much lower than they had been before compound training.

In our laboratory, we have confirmed that the same phenomena occurs in rabbit eyeblink conditioning (White, 1998). Separate pairings of a tone CS and a light CS with the US yielded CRs on 90% of their respective CS–US trials. When a compound of the tone and light was then repeatedly paired with the US, CRs to the compound occurred with virtually 100% likelihood, but the likelihood of a CR on tone-alone and light-alone test trials rapidly dropped to 60%.

This decline in responding to the individual CSs might seem counterintuitive because they are always paired with the US. Even if the compound CS were perceived as a distinctive event, it is not immediately obvious why well-established responding to the individual CSs would be lost. In fact, this outcome has provided important support for one of the key theories of classical conditioning, specifically, the model of Rescorla and Wagner (1972). Their model supposes that, on any one CS–US pairing, the US can support only a fixed amount of a learning, which is usually denoted theoretically as associative strength. When two new CSs are presented in a compound and paired with the US, they effectively compete for the fixed amount of associative strength that the US is capable of supporting.[2] When the two CSs are equally salient, they would each ultimately gain half the associative strength supported by the US. According to the Rescorla–Wagner model, this same rule applies when CSs with established associative strengths are combined. Each CS must lose about half its associative strength so that their combined strength matches the amount supportable by the US. This decline in the associative strength of each CS is expressed behaviorally in the reduction in CR likelihood on tests of the individual CSs interspersed among TL+ trials.

Blocking

Perhaps the phenomenon that has played the most pivotal role in modern research and theory in classical conditioning is *blocking*. In the blocking procedure, one

---

[2]Rescorla and Wagner's (1972) incorporation of a competitive learning rule in their model of classical conditioning has been recognized as having wide generality to learning systems, especially neural networks (Sutton & Barto, 1981), under such names as the delta rule (Rumelhart et al., 1986) and least mean square rule (Gluck & Bower, 1988).

CS component is initially paired with the US (A+). Once CRs have been acquired to the A component, the animals are then trained to a compound CS consisting of the A component and another component (X). The AX compound is paired with the US in a second stage of training (AX+). Test trials of X alone are presented either sporadically during AX+ training or at the end of AX+ training (Kamin, 1968).

In an experiment by Marchant and Moore (1973), for example, group A-AX received A+ trials until CR likelihood exceeded 80%. At the same time, a rest control group (Sit-AX) did not receive exposure to either the CS or US. Then, both groups received AX+ training. Finally, extensive testing of X alone revealed a negligible CR likelihood to the X component in group A-AX (<1%) but many more CRs to the X component in group Sit-AX (32%). In descriptive terms, CR acquisition to the X component is said to be blocked by the prior pairings of the A component with the US. Blocking has been frequently reported in rabbit eyeblink conditioning (Kehoe et al., 1981b; Kinkaide, 1974; Maleske & Frey, 1979; Marchant & Moore, 1973; Schreurs & Gormezano, 1982; Solomon, 1977).

Blocking has been explained most commonly by appealing to a competitive hypothesis (but see Miller & Matzel, 1989). Most notably, Rescorla and Wagner's (1972) model contends that the initial pairings of the A stimulus with the US permits the A stimulus to capture the bulk of associative strength that can be supported by the US. In other words, the A stimulus gets a massive head start in the competition with X for associative strength. Consequently, when the X stimulus is added to form the compound, there is little remaining associative strength available for the X stimulus. A similar explanation has been proposed by the selective attention hypotheses. In brief, this view contends that the initial A+ trials bias the animals to attend to the A stimulus and to filter out added stimuli that provide no new predictive value for the US (Mackintosh, 1975; Moore & Stickney, 1980; Pearce & Hall, 1980; Schmajuk & Moore, 1988; Sutherland & Mackintosh, 1971, pp. 144–146).

Explicit Differentiation Between a Compound and Its Components

The procedures described in the previous section all entail pairings of either a compound CS or its separate components with the US. This section describes studies involving differential reinforcement of a compound and its components. The procedures used in these studies are extensions of differential conditioning described previously. Specifically, the procedures entailed a mixture of compound and component trials. Some types of trials are reinforced by pairing the CS with the US; other types of trials are not. These procedures originated with Pavlov (1927), who used them to demonstrate conditioned inhibition (pp. 68–87) and the "synthesis" of stimuli (p. 144). The latter anticipates the concept of configural learning.

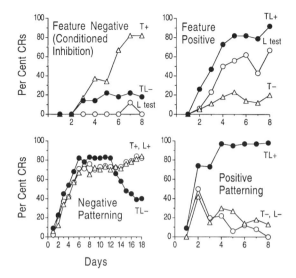

FIGURE 6.12. Differential conditioning of a compound (*TL*) versus its components, – tone (T) and light (L) –, using the rabbit inner eyelid (nictitating membrane) (Kehoe, 1988).

Conditioned Inhibition (Feature Negative Discrimination)

The upper left-hand panel of Figure 6.12 shows the acquisition curves for the rabbit eyeblink obtained from a *feature negative* procedure (Kehoe, 1988). Each session contained pairings of a tone with the US (T+) intermixed with trials on which a tone plus light compound was presented alone (TL–). The T+ and TL– trials were presented in equal numbers, and the light component was tested sporadically (L test). As can be seen, this schedule produces high levels of responding on T+ trials, low levels of responding on TL– trials, and virtually no responding on L tests (Kehoe, 1988; cf. Marchant et al., 1972; Solomon, 1977).

Since Pavlov (1927, p. 75), the low level of responding to a compound stimulus has been explained by assuming that the excitatory strength of the component paired with the US is counteracted by a inhibitory strength acquired by the other component. The inhibitory strength of such a stimulus has been demonstrated in the rabbit eyeblink preparation by Marchant et al. (1972). They conducted feature negative training in which a light CS was paired with the US and a compound of the tone plus light was presented alone (L+ versus LT–). In addition, another CS, white noise, was paired with the US (N+). When this training was completed, two standard tests of the conditioned inhibitory strength of the tone CS were conducted (Rescorla, 1969). First, a *summation test* was run in which the tone CS was compounded with the noise CS (NT–). This addition of the tone to the noise reduced the likelihood of a CR to the noise from 43% to 31%. (In a control group, the compounding of an excitatory tone CS with the noise CS did not noticeably change the likelihood of a CR to the noise.) Second, a *retardation test* was run, in which the tone CS was paired with the US. CR acquisition to the tone CS was slower than in control conditions.

Negative Patterning

The lower left-hand panel of Figure 6.12 shows *negative patterning* in rabbit eyeblink conditioning (Kehoe & Graham, 1988). The early training sessions were essentially a stimulus compounding procedure. That is, the sessions contained a mixture of trials in which the individual components of tone (T) and light (L) were each repeatedly paired with the US (T+, L+). Presentations of the compound without the US (TL−) were rare. Once the CR was established on T+ and L+ trials, more TL− trials were added to the trial mix. At the end of the experiment, each session contained 40 TL− trials, 10 T+ trials, and 10 L+ trials. As seen in Figure 6.12, responding on TL− trials gradually declined, while responding on T+ and L+ trials was maintained.

Negative patterning has been of considerable conceptual interest in recent years, because it is an example of an organism's solving the *exclusive-OR* (XOR) problem, in that the organism learns to respond to each of the separate inputs but not their joint occurrence (Barto, 1985; Rumelhart et al., 1986). The XOR problem is considered to be difficult, because it is nonlinear. That is, it is impossible to generate the appropriate reaction, namely, no response, to the joint inputs by a summation of the responses attached to the separate inputs (Minsky & Papert, 1969; Rumelhart et al., 1986). To convert a nonlinear problem into a solvable, linear form, it is necessary for the organism to possess some means for processing the compounded inputs as a functional unit that is distinct from the component inputs (Barto, 1985; Gluck & Bower, 1990; Kehoe, 1988, 1998). In the specific case of negative patterning, the drop in responding on TL− trials is thought to reflect the acquisition of a substantial inhibitory capacity by a dedicated configural unit activated by the TL compound. This inhibitory capacity overcomes the tendency toward summation of excitation when the T and L components are combined on TL− trials (Kehoe, 1998; Kehoe & Gormezano, 1980; Rescorla, 1973; Schmajuk & DiCarlo, 1992).

Feature Positive Discrimination and Positive Patterning

The right-hand side of Figure 6.12 shows the results of a *feature positive* procedure and a *positive patterning* procedure, which are the operational complements to the feature negative and negative patterning procedures, respectively (Kehoe, 1988). In the feature positive procedure, pairings of the compound with the US (TL+) were intermixed with presentations of one component (T−). CRs were acquired on TL+ trials, while a few CRs appeared on T− trials. Sporadic tests with the light CS yielded a level of responding nearly as high as on TL+ trials. Because the light CS had been consistently paired with the US on TL+ trials, the majority of CRs on TL+ trials could be attributed to a contribution from the light CS with the remainder being contributed by the tone CS. In the case of positive

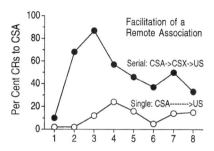

FIGURE 6.13. Serial compound conditioning (CSA→CSX→US) using the rabbit inner eyelid (nictitating membrane) (Kehoe et al., 1979): CR acquisition to CSA when CSA is temporally remote from CSX and the US.

patterning, TL+ trials were intermixed with frequent presentations of both CSs (T−, L−). Consequently, a high level of responding was observed only on TL+ trials and not on T− or L− trials. Like negative patterning, this discrimination between a compound and both its components has been construed as evidence for configural learning.

Serial Conditioned Stimuli

A sequence of CSs can be used to signal a US. Such a sequence mimics natural situations in which biologically significant events are signaled by an extended series of cues. In the rabbit eyeblink preparation, studies using sequential CSs have revealed that they engage mechanisms in addition to those seen in compounds of simultaneous CSs. When even a sequence of just two CSs, such as a tone and a light, is paired with the US (CSA→CSX→US), there have been three major findings, which are described in this section.

Facilitation of CR Acquisition to CSA

In some studies using a CSA→CSX→US sequence, CSA has been presented several seconds in advance of both CSX and the US (Kehoe et al., 1979, 2000; Schreurs et al., 1993). Figure 6.13 shows an example of CR acquisition to CSA when it is relatively remote to the US in two groups of rabbits (Kehoe et al., 1979; experiment 2). One group (Serial) was given training with CSA→CSX→US trials. The other group (Single) had only CSA as a signal for the US (CSA→US). The CSA preceded the US by 2800 ms in both groups, which ordinarily is a CS–US interval that yields only a low level of CR acquisition. In Group Serial, the CSX→US interval was 400 ms. As can be seen, responding to CSA in Group Serial reached a much higher level than in Group Single, which lacked CSX. Although the highest levels of responding to CSA in Group Serial were not sustained, the observed performance was hardly transient. In Group Serial, responding to CSA increased over the course of 6 days, which contained a total of 360 CSA→CSX→US trials, before any decline appeared. It has been possible to demonstrate that such facilitation of CR acquisition to CSA in a sequential CS occurs when the CSA→CSX interval has been extended to nearly 20 s (Kehoe et al., 1979, 1987a, 2000).

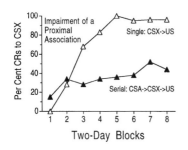

FIGURE 6.14. Serial compound conditioning (CSA→CSX→US) using the rabbit inner eyelid (nictitating membrane) (Kehoe et al., 1979): responding to CSX when CSA is temporally close to CSX and the US.

Second-Order Conditioning

The facilitation of responding to CSA in a serial compound may rely on second-order conditioning. In the rabbit eyeblink preparation, second-order conditioning has been conducted by giving the animals a mixture of *second-order* CSA→CSX trials and *first-order* CSX→US trials. As CRs are acquired on the CSX→US trials, CRs are also acquired to CSA on the CSA→CSX trials (Gibbs et al., 1991; Kehoe et al., 1981a). The same process can occur in a serial compound (CSA→CSX→US), because it contains both the CSA→CSX and CSX→US relationships.

Sensory Preconditioning

In addition to second-order conditioning, studies of sensory preconditioning have provided further evidence that CSX in a serial compound can contribute to responding to CSA. The sensory preconditioning procedure contains three stages (Brogden, 1939). First, the animals are given CSA→CSX pairings. In the absence of the US, no CRs are observed in this stage. Second, the animals are given CSX→US pairings, in which CRs are acquired to CSX. Third, CSA is tested to determine whether it gained any capacity to elicit CRs by virtue of its indirect association with the US via CSX. In rabbit eyeblink preparations, the overall levels of responding to CSA have been low; nevertheless, the likelihood of a CR to CSA has exceeded levels in control conditions lacking either the CSA→CSX or CSX→US pairings (Port et al., 1987; Tait et al., 1986).

Primacy Effect in Serial Compound Conditioning

When CSA is presented in temporal proximity to CSX and the US, CR acquisition to CSA proceeds as if CSX were not present. However, a primacy effect appears (Kehoe, 1979, 1983; Kehoe et al., 1979; Schreurs et al., 1993). Specifically, the level of CR acquisition to CSX is reduced, even though it is contiguous to the US. An example of this primacy effect is shown in Figure 6.14 (Kehoe et al., 1979; experiment 2).

Figure 6.14 shows responding on X-alone test trials in a group trained with a serial compound (CSA→CSX→US) versus a control group given only CSX→US pairings. In both groups, CSX preceded the US by 400 ms, and CSA preceded the US by 800 ms in the serial group. Thus, CSA immediately preceded CSX. As shown in the figure, the level of responding to CSX in the serial group initially grew as fast as in the control group. However, after the first few days, responding to CSX in the serial group stabilized at a level well below that of the control group.

Serial Conditional Discriminations

Sequential CSs have been used in a type of differential conditioning, known as a serial conditional discrimination or as *occasion-setting* (Holland, 1983, 1992; Moore et al., 1969; Schmajuk & Holland, 1998). Specifically, animals can learn to respond when a CS is paired with the US (X+) in the presence of one cue (A→X→US) but to withhold responding to the same stimulus without the US when it is preceded by another cue (B→X–). The two initial cues (A, B) are known variously as *conditional cues, feature cues,* or *occasion-setters.* The second stimulus (X) is known as the *target stimulus.*

For the rabbit eyeblink preparation, differential responding to the two serial compounds has been found when the onset of the A and B cues precedes the onset of the X stimulus by a relatively long interval, say, 1 s or more. The X stimulus is usually relatively brief, but the A and B cues can vary in their duration. In some cases, the A and B cues are short (e.g., 1 s or less) as in the serial compounds described previously (Kehoe et al., 1987a; Weidemann et al., 1999). In other cases, the A and B cues are longer (e.g., 60 s) and continue throughout the presentation of the X stimulus (Brandon & Wagner, 1991; Macrae & Kehoe, 1995; Rogers & Steinmetz, 1998; Weidemann & Kehoe, 1997). In still other cases, the A and B cues can arise from distinctive features of a conditioning chamber, referred to as either the *contextual cues* or *background cues,* present for an entire training session (Hinson, 1982).

The control of the A and B cues over responding during the common X stimulus has sometimes been attributed to a special type of learning. Specifically, they are thought to act indirectly on the ability of the X stimulus to elicit the CR by (a) exercising superordinate control over memory retrieval (Holland, 1992) or (b) evoking an appropriate motivational state for responding (Brandon & Wagner, 1991; Konorski, 1967). These hypotheses have stemmed in part from the observation that the onsets of CSA and CSB are often temporally remote from the US. This temporal relationship seems well beyond the range of CS→US intervals that ordinarily produce CR acquisition or the range of CSA→CSX intervals that yield CR acquisition to CSA in serial compounds. In fact, the A and B cues in a conditional discrimination often seem behaviorally silent. That is, where the CSA→CSX interval is longer than, say, 20 s, the CRs elicited by the X stimulus do not occur to the A cue (Kehoe et al., 2000; Macrae & Kehoe, 1995; Weidemann & Kehoe, 1997).

Based on these theoretical and empirical considerations, it has been thought that the initial cues (A, B) and the target stimulus (X) in a conditional discrimination have different functions (Holland, 1983; Macrae & Kehoe, 1995). The A and B cues signal whether or not the US will occur whereas the X stimulus signals

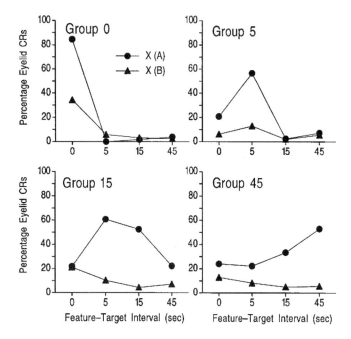

FIGURE 6.15. Proportion of test trials with CRs to CSX as a function of the duration of the test interval (Kehoe et al., 2000).

when the US occurs. However, it has been recently demonstrated that the A and B cues do more than signal the US; they also have a temporal function in signaling when the X stimulus will occur (Holland et al., 1997; Weidemann et al., 1999).

An example of findings that have revealed the temporal function of the A and B cues is shown in Figure 6.15. Four groups of rabbits were given A→X→US trials intermixed with B→X– trials. The A and B cues were 65 s (± 5 s) in duration; one was a 1000-Hz tone, and the other was white noise. The assignment of these stimuli as A and B cues was balanced across animals. The X stimulus was a 20-Hz flashing light of 400-ms duration. On A→X→US trials, the interval between the onset of the X stimulus and the onset of the US was 400 ms. The groups differed only in the interval between the onset of the A and B cues and the onset of the X stimulus. Specifically, in group 0, the initial cue (A or B) and the X stimulus had a simultaneous onset. In groups 5, 15, and 45, the onset of the initial cue (A or B) preceded the X stimulus by 5, 15, and 45 s, respectively. When the animals in all groups were differentially responding to X on A→X→US trials versus B→X– trials, all groups were tested with all four intervals on both the A→X and B→X trial types (Kehoe et al., 2000).

Figure 6.15 shows the results of the interval testing for each group. Each panel shows the proportion of trials on which a CR occurred during the X stimulus as a function of the duration of the test interval. Inspection of the figure reveals two key findings. First, when tested with the interval used in training, each group showed a conditional discrimination. That is, the level of responding during the X stimulus in A→X trials was higher than in B→X trials by more than 40 percentage points. Second, when the tested interval differed from the value used in training, the conditional discrimination in each group showed *temporal specificity;* that is, responding during the X stimulus, particularly on A→X trials, tended to decline as the interval departed from the training value. These temporal gradients have suggested that the concept of the CS trace, particularly as implemented in real-time models, may be usefully extended to conditional discriminations (see Buhusi & Schmajuk, 1999; Schmajuk et al., 1998; Zackheim et al., 1998, for models along these lines.)

Commentary

It might seem paradoxical that CR acquisition to CSA in a serial compound is facilitated by the presence of CSX, but, at the same time the CSA can reduce or modulate CR acquisition to CSX. When these phenomena were first discovered, they each seemed to require an appeal to a different process. At various times, for example, the facilitation of CR acquisition to CSA has been attributed vaguely to bridging of a hypothetical CSA→US association by CSX (Kehoe & Morrow, 1984; Kehoe et al., 1987a; Rescorla, 1982), the reduction in responding to CSX has been attributed to the capture of either attention or information value by CSA (Egger & Miller, 1962; Kehoe, 1979), and conditional discriminations have been attributed to a hierarchical memory system under the heading of occasion-setting (Bonardi, 1996; Holland 1983, 1992). Increasingly, however, these phenomena are being described in a unified fashion by the use of real-time, connectionist models (Buhusi & Schmajuk, 1999; Kehoe et al., 1987b; Schmajuk et al., 1998; Wagner & Brandon, 1989).

Conclusions

Take-Away Messages

Introductory textbooks in psychology and neuroscience have emphasized two empirical principles of classical conditioning, namely, (1) acquisition of CRs through CS–US pairings and (2) extinction of CRs through CS-alone presentations. Although these two principles do represent pervasive, fundamental findings, it should now be clear that they hardly capture the richness of phenomena that have been obtained through relatively simple manipulations of two or three stimuli and their interrelationships. At the same time, this richness can be overwhelming and may tempt the reader to wish for plainer fare, namely, a few, easily remembered, rules of thumb to help organize the mass of findings. Accordingly, in this final section, we propose some rules of thumb that the neuroscientist may apply in designing a study in classical conditioning:

Estimation of Associative Versus Nonassociative Contributions

When exploring a new response system or new species, the specific effects of CS–US pairings on changes in responding to the CS can be best distinguished from either positive or negative contributions from other sources by using the full array of nonassociative controls, namely, CS-alone, US-alone, and explicitly unpaired presentations of the two stimuli. The relative contributions from associative and nonassociative sources can also be estimated for each animal by using differential conditioning, in which one CS is paired with the US and the other CS is not.

CS–US Contiguity

The most rapid CR acquisition to the highest asymptote can be achieved by using the optimal CS–US interval for each preparation. This value varies dramatically from preparation to preparation. The range of CS–US intervals effective for conditioning can be broadened, at least in the rabbit eyeblink preparation, by using a CS composed of a sequence of distinctive components before the US and by reducing the number of CS–US pairings per session.

Basic Determinants of CR Acquisition

The rate and asymptote of CR acquisition are a direct function of (a) CS duration during the CS–US interval, (b) CS intensity, (c) US intensity, (d) the consistency of CS–US pairings, and (e) the spacing between CS–US pairings, particularly as manipulated by the number of pairings per session. Conversely, CS-alone and US-alone presentations before or during CS–US pairings can reduce responding (e.g., latent inhibition, partial reinforcement, US-preexposure effect, partial warning). Bear in mind, however, that CR acquisition is a robust phenomenon. It is only at the very low end of these dimensions that CR acquisition fails to occur.

Retention, Extinction, and Recovery of CRs

CRs reappear immediately after retention periods of several months, and, even when the CRs do not appear immediately, they do reappear typically with a few CS–US pairings (savings). Similarly, CRs can be recovered equally quickly by a few CS–US pairings after both extensive extinction or counterconditioning.

Transfer of Conditioning

After CRs have been established to one CS, they will transfer in a graded fashion to physically similar CSs (stimulus generalization). Although there is little or no stimulus generalization across CSs from different sensory modalities, there is more rapid CR acquisition to a CS from another sensory modality when it is paired with the same US (learning to learn). Conversely, preexposures to the CS-alone (latent inhibition) or the US-alone have negative transfer effects; CR acquisition during subsequent CS–US pairings occurs more slowly than would otherwise be the case.

CR Timing

The CS does not become a substitute US, and the CR does not mimic the UR in its time course. Rather, the time course of the CR is highly attuned to the CS–US interval. That is, a well-established CR usually begins to arise well before the US. Then, the CR is recruited in a graded fashion such that the maximal amplitude of the CR occurs around the time at which the US has occurred in the past following CS onset. If the CS–US interval changes, the time course of the CR will change accordingly.

UR Modification

The UR itself not an invariant reaction. In addition to showing habituation, the UR can also be modified as a function of the CS–US interval and the number of previous CS–US pairings.

Simultaneous Compound Stimuli

When two CSs from different modalities—typically, tone and light—are paired separately with the US and then tested in combination (stimulus compounding), summation of the CRs often occurs as an outcome. However, if a compound of two CSs is paired with the US, then a number of interactions appear. Typically, responding to both the individual CSs declines to lower levels (overprediction) and can disappear altogether (spontaneous configuration). Responding to one component of a compound can be elevated to the detriment of the other component by increasing the intensity of one component (overshadowing) or pairing one component individually with the US (blocking). By differentially pairing a compound and its components with the US, it is possible to produce an explicit discrimination in responding to a compound and its components (conditioned inhibition, negative patterning, feature positive discrimination, positive patterning).

Serial Compound Stimuli

Presenting a sequence of two CSs before the US (CSA→CSX→US) produces additional interactions among the components of the compound that depend on the interval between the onsets of the two stimuli. When the onsets are close, CRs are acquired to CSA to the detriment of CRs to CSX, even though CSX is positioned at the more favorable CS–US interval. When the onsets are temporally remote to each other, CR acquisition to CSA is facilitated by the presence of CSX in the sequence. This bridging effect can also occur, albeit more weakly, if CSA→CSX and CSX→US pairings occur in separate phases of training (second-order conditioning, sensory preconditioning). In addition, if one serial compound is paired with the US (CSA→CSX→US) and another compound with a different initial CS is not paired with the US (CSB→CSX−), the animals will show acquisition of a conditional discrimination, in which they respond to CSX following the onset of CSA but not following the onset of CSB.

Implications

This chapter was intended to guide readers beyond the basic features of classical conditioning to its more complex features, including some of the most recent findings. From a historical perspective, research in classical conditioning is fulfilling Pavlov's original intention to use classical conditioning as a laboratory method for illuminating the "higher nervous" functions of the brain that mediate learned behavior that, in turn, adapts the individual organism to the exigencies of its environment. In broad terms, research in classical conditioning is helping to illuminate not only basic learning mechanisms but also the workings of attention, short-term memory, perception, and even cognition. Learning-to-learn effects and the interplay among sequential stimuli certainly border on cognition. However, this is not to imply that these higher-order processes are merely a concatenation of CRs. In fact, to account for such findings as rapid reacquisition, latent inhibition, CR timing, and learning to learn, the seemingly simple matter of CR acquisition and extinction has increasingly been shown to require a sophisticated theoretical approach, namely, connectionist networks, often with multiple layers. Conversely, the complexity that is evident in classical conditioning with animals indicates that the higher-order processes are biologically fundamental and start to appear at a low level of neural organization.

Acknowledgment. Preparation of this manuscript was supported by Australian Research Council Grant A79600502. Correspondence should be sent to E. James Kehoe, School of Psychology, University of New South Wales, Sydney, NSW 2052, Australia; email: j.kehoe@unsw.edu.au.

References

Ashton, A.B., Bitgood, S.C., and Moore, J.W. (1969). Auditory differential conditioning of the rabbit nictitating membrane response: III. Effects of US shock intensity and duration. *Psychonomic Science, 15,* 127–128.

Ayres, J.B., Benedict, J.O., and Witcher, E. (1975). Systematic manipulation of individual events in a truly random control in rats. *Journal of Comparative and Physiological Psychology, 88,* 97–103.

Ayres, J.J., Moore, J.W., and Vigorito, M. (1984). Hall and Pearce negative transfer: assessments in conditioned suppression and nictitating membrane conditioning experiments. *Animal Learning & Behavior, 12,* 428–438.

Baddeley, A.D. (1997). *Human memory: theory and practice (revised edition).* Hove, England: Psychology Press.

Barto, A.G. (1985). Learning by statistical cooperation of self-interested neuron-like computing elements. *Human Neurobiology, 4,* 229–256.

Belasco, J.A., and Trice, H.M. (1969). *The assessment of change in training and therapy.* New York: McGraw-Hill.

Bellingham, W.P., and Gillette, K. (1981). Spontaneous configuring to a tone-light compound using appetitive training. *Learning and Motivation, 12,* 420–434.

Bellingham, W.P., Gillette-Bellingham, K., and Kehoe, E.J. (1985). Summation and configuration in patterning schedules with the rat and rabbit. *Animal Learning & Behavior, 13,* 152–164.

Blazis, D.E.J., and Moore, J.W. (1991). Conditioned stimulus duration in classical trace conditioning: test of a real-time neural network model. *Behavioral Brain Research, 43,* 73–78.

Bonardi, C. (1996). Transfer of occasion setting: the role of generalization decrement. *Animal Learning & Behavior, 24,* 277–289.

Boring, E.G. (1950). *A history of experimental psychology.* New York: Appleton-Century-Crofts.

Brandon, S.E., and Wagner, A.R. (1991). Modulation of a discrete Pavlovian conditioned reflex by a putative emotive Pavlovian conditioned stimulus. *Journal of Experimental Psychology: Animal Behavior Processes, 17,* 299–311.

Brelsford, J., Jr., and Theios, J. (1965). Single session conditioning of the nictitating membrane in the rabbit: effect of intertrial interval. *Psychonomic Science, 2,* 81–82.

Britton, G., and Farley, J. (1999). Behavioral and neural bases of noncoincidence learning in *Hermissenda. Journal of Neuroscience, 19,* 9126–9132.

Brogden, W.J. (1939). Sensory preconditioning. *Journal of Experimental Psychology, 25,* 323–332.

Bromage, B.K., and Scavio, M.J., Jr. (1978). Effects of an aversive CS+ and CS− under deprivation upon successive classical appetitive and aversive conditioning. *Animal Learning & Behavior, 6,* 57–65.

Brown, J.S., Kalish, H.I., and Farber, I.E. (1951). Conditioned fear as revealed by magnitude of startle response to an auditory stimulus. *Journal of Experimental Psychology, 41,* 317–328.

Brown, T. (1820/1977). Sketch of a system of the philosophy of the human mind, Edinburgh. In D. N. Robinson (Ed.), *Significant contributions to the history of psychology.* Washington, DC: University Press of America.

Bruner, A. (1965). UCS properties in classical conditioning of the albino rabbit's nictitating membrane response. *Journal of Experimental Psychology, 69,* 186–192.

Buhusi, C.V., and Schmajuk, N.A. (1999). Timing in simple conditioning and occasion setting: a neural network approach. *Behavioral Processes, 45,* 33–57.

Buonomano, D.V., and Mauk, M.D. (1994). Neural network model of the cerebellum: temporal discrimination and the timing of motor responses. *Neural Computation, 6,* 38–55.

Buresova, O., and Bures, J. (1974). Functional decortication in the CS-US interval decreases the efficiency of taste aversive learning. *Behavioral Biology, 12,* 357–364.

Canli, T., Detmer, W.M., and Donegan, N.H. (1992). Potentiation or diminution of discrete motor unconditioned responses (rabbit eyeblink) to an aversive Pavlovian unconditioned stimulus by two associative processes: Conditioned fear and a conditioned diminution of unconditioned stimulus processing. *Behavioral Neuroscience, 106,* 498–508.

Carew, T.J., Walters, E.T., and Kandel, E.R. (1981). Classical conditioning in a simple withdrawal reflex in *Aplysia. Journal of Neuroscience, 1,* 1426–1437.

Clarke, M.E., and Hupka, R.B. (1974). The effects of stimulus duration and frequency of daily preconditioning stimulus exposures on latent inhibition in Pavlovian conditioning of the rabbit nictitating membrane response. *Bulletin of the Psychonomic Society, 4,* 225–228.

Coleman, S.R., and Gormezano, I. (1971). Classical conditioning of the rabbit's (*Oryctolagus cuniculus*) nictitating membrane response under symmetrical CS-US interval shifts. *Journal of Comparative and Physiological Psychology, 77,* 447–455.

Cox, J. (1990). *Conditioned response topography in the rabbit nictitating membrane response as a function of conditioned stimulus intensity and duration: implications for models of adaptive timing.* Honours thesis, Sydney: University of New South Wales.

Davis, M. (1986). Pharmacological and anatomical analysis of fear conditioning using the potentiated startle paradigm. *Behavioral Neuroscience, 100,* 808–818.

Deaux, E., and Gormezano, I. (1963). Eyeball retraction: classical conditioning and extinction in the albino rabbit. *Science, 141,* 630–631.

Desmond, J.E. (1990). Temporal adaptive responses in neural models: the stimulus trace. In M. Gabriel and J. W. Moore (Eds.), *Learning and computational neuroscience* (pp. 421–456). Cambridge, MA: MIT Press.

Desmond, J.E., and Moore, J.W. (1988). Adaptive timing in neural networks: the conditioned response. *Biological Cybernetics, 58,* 405–415.

Desmond, J.E., and Moore, J.W. (1991). Altering the synchrony of stimulus trace processes: tests of a neural-network model. *Biological Cybernetics, 65,* 161–169.

Donegan, N.H., and Wagner, A.R. (1987). Conditioned diminution and facilitation of the UR: a sometimes opponent-process interpretation. In I. Gormezano, W.F. Prokasy, and R.F. Thompson (Eds.), *Classical conditioning.* Hillsdale, NJ: Erlbaum.

Egger, D.M., and Miller, N.E. (1962). Secondary reinforcement in rats as a function of information value and reliability of the stimulus. *Journal of Experimental Psychology, 64,* 174–184.

Ellis, H. (1965). *The transfer of learning.* New York: Macmillan.

Fitzgerald, R.F. (1963). Effects of partial reinforcement with acid on the classically conditioned salivary response in dogs. *Journal of Comparative and Physiological Psychology, 56,* 1056–1060.

Frey, P., Maisiak, R., and Dugue, G. (1976). Unconditional stimulus characteristics in rabbit eyelid conditioning. *Journal of Experimental Psychology: Animal Behavior Processes, 2,* 175–190.

Frey, P.W., and Misfeldt, T.J. (1967). Rabbit eyelid conditioning as a function of the intertrial interval. *Psychonomic Science, 9,* 137–138.

Garcia, J., Ervin, F.R., and Koelling, R.A. (1966). Learning with prolonged delay of reinforcement. *Psychonomic Science, 5,* 121–122.

Gibbs, C.M., Latham, S.B., and Gormezano, I. (1978). Classical conditioning of the rabbit's nictitating membrane response: effects of reinforcement and resistance to extinction. *Animal Learning & Behavior, 6,* 209–215.

Gibbs, C.M., Cool, V., Land, T., Kehoe, E.J., and Gormezano, I. (1991). Second-order conditioning of the rabbit's nictitating membrane response: interstimulus interval and frequency of CS-CS pairings. *Integrative Physiological and Behavioral Science, 26,* 282–295.

Gluck, M.A., and Bower, G.H. (1988). Evaluating an adaptive network model of human learning. *Journal of Memory and Language, 27,* 166–195.

Gluck, M.A., and Bower, G.H. (1990). Component and pattern information in adaptive networks. *Journal of Experimental Psychology: General, 119,* 105–109.

Gormezano, I. (1966). Classical conditioning. In J.B. Sidowski (Ed.), *Experimental methods and instrumentation in psychology* (pp. 385–420). New York: McGraw-Hill.

Gormezano, I. (1972). Investigations of defense and reward conditioning in the rabbit. In A.H. Black and W.F. Prokasy (Eds.), *Classical conditioning. II: Current research and theory* (pp. 151–181). New York: Appleton-Century-Crofts.

Gormezano, I., and Coleman, S.R. (1975). Effects of partial reinforcement on conditioning, conditional probabilities, asymptotic performance, and extinction of the rabbit's nictitating membrane response. *Pavlovian Journal of Biological Sciences, 10,* 13–22.

Gormezano, I., and Kehoe, E.J. (1975). Classical conditioning: some methodological-conceptual issues. In W.K. Estes (Ed.), *Handbook of learning and cognitive processes.* Hillsdale, NJ: Erlbaum.

Gormezano, I., and Kehoe, E.J. (1981). Classical conditioning and the law of contiguity. In P.M. Harzem and M.D. Zeiler (Eds.), *Advances in analysis of behavior, Vol. 2. Predictability, correlation, and contiguity* (pp. 1–45). New York: Wiley.

Gormezano, I., Kehoe, E.J., and Marshall, B.S. (1983). Twenty years of classical conditioning research with the rabbit. In J.M. Sprague and A.N. Epstein (Eds.), *Progress in psychobiology and physiological psychology* (pp. 197–275). New York: Academic Press.

Grant, D.A. (1943). Sensitization and association in eyelid conditioning. *Journal of Experimental Psychology, 32,* 201–212.

Graves, C.A., and Solomon, P.R. (1985). Age-related disruption of trace but not delay classical conditioning of the rabbit's nictitating membrane response. *Behavioral Neuroscience, 99,* 88–96.

Gray, T.S., McMaster, S.E., Harvey, J.A., and Gormezano, I. (1981). Localization of retractor bulbi motoneurons in the rabbit. *Brain Research, 226,* 93–106.

Grevert, P., and Moore, W. (1970). The effects of unpaired US presentations on conditioning of the rabbit's nictitating membrane response: consolidation or contingency. *Psychonomic Science, 20,* 177–179.

Grossberg, S., and Schmajuk, N.A. (1989). Neural dynamics of adaptive timing and temporal discrimination during associative learning. *Neural Networks, 2,* 79–102.

Guthrie, E.R. (1930). Conditioning as a principle of learning. *Psychological Review, 37,* 412–428.

Haberlandt, K., Hamsher, K., and Kennedy, A.W. (1978). Spontaneous recovery in rabbit eyelid conditioning. *Journal of General Psychology, 98,* 241–244.

Hall, J.F. (1976). *Classical conditioning and instrumental learning: a contemporary approach.* Philadelphia: Lippincott.

Harlow, H.F. (1949). The formation of learning sets. *Psychological Review, 56,* 51–65.

Heinemann, E.G., and Chase, S. (1975). Stimulus generalization. In W.K. Estes (Ed.), *Handbook of learning and cognitive processes* (pp. 305–349). Hillsdale, NJ: Erlbaum.

Hesketh, B. (1997). Dilemmas in training for transfer and retention. *Applied Psychology: An International Review, 46,* 317–339.

Hilgard, E.R., and Marquis, D.G. (1940). *Conditioning and learning.* New York: Appleton-Century.

Hinson, R.E. (1982). Effects of UCS preexposure on excitatory and inhibitory rabbit eyelid conditioning: an associative effect of conditioned contextual stimuli. *Journal of Experimental Psychology: Animal Behavior Processes, 8,* 49–61.

Hoehler, F.K., and Leonard, D.W. (1976). Double responding in classical nictitating membrane conditioning with single-CS dual-ISI pairing. *Pavlovian Journal of Biological Science, 11,* 180–190.

Hoehler, F.K., and Leonard, D.W. (1981). Motivated vs. associated role of the US in classical conditioning of the rabbit's nictitating membrane response. *Animal Learning & Behavior, 9,* 239–244.

Holland, P.C. (1983). Occasion-setting in Pavlovian feature positive discriminations. In M.L. Commons, R.J. Herrnstein, and A.R. Wagner (Eds.), *Quantitative analyses of behavior: discrimination processes* (pp. 182–206). New York: Ballinger.

Holland, P.C. (1992). Occasion setting in Pavlovian conditioning. In D.L. Medin (Ed.), *The psychology of learning and motivation* (pp. 69–125). San Diego: Academic Press.

Holland, P.C., Hamlin, P.A., and Parsons, J.P. (1997). Temporal specificity in serial feature-positive discrimination learning. *Journal of Experimental Psychology: Animal Behavior Processes, 23,* 95–109.

Holt, P.E., and Kehoe, E.J. (1985). Cross-modal transfer as a function of similarities between training tasks in classical conditioning of the rabbit. *Animal Learning & Behavior, 13,* 51–59.

Hull, C.L. (1943). *Principles of behavior.* New York: Appleton-Century-Crofts.

Hume, D. (1969). *A treatise of human nature.* New York: Penguin Books. (Original work published 1739.)

Hupka, R.B., Kwaterski, S.E., and Moore, J.W. (1970). Conditioned diminution of the UCR: differences between the human eyeblink and the rabbit nictitating membrane response. *Journal of Experimental Psychology, 83,* 45–51.

Ison, J.R., and Leonard, D.W. (1971). Effects of auditory stimuli on the amplitude of the nictitating membrane reflex of the rabbit (*Oryctolagus cuniculus*). *Journal of Comparative and Physiological Psychology, 75,* 157–164.

Kamin, L.J. (1968). "Attention-like" processes in classical conditioning. In M.R. Jones (Ed.), *Miami symposium on the prediction of behavior: aversive stimulation* (pp. 9–31). Miami: University of Miami Press.

Kandel, E.R., Abrams, T., Bernier, L., Carew, T.J., Hawkins, R.D., and Schwartz, J.H. (1983). Classical conditioning and sensitization share aspects of the same molecular cascade in *Aplysia. Cold Spring Harbor Symposia in Quantitative Biology, 48,* 821–830.

Kehoe, E.J. (1979). The role of CS-US contiguity in classical conditioning of the rabbit's nictitating membrane response to serial stimuli. *Learning and Motivation, 10,* 23–38.

Kehoe, E.J. (1982). Overshadowing and summation in compound stimulus conditioning of the rabbit's nictitating membrane response. *Journal of Experimental Psychology: Animal Behavior Processes, 8,* 313–328.

Kehoe, E.J. (1983). CS-US contiguity and CS intensity in conditioning of the rabbit's nictitating membrane response to serial compound stimuli. *Journal of Experimental Psychology: Animal Behavior Processes, 9,* 307–319.

Kehoe, E.J. (1986). Summation and configuration in conditioning of the rabbit's nictitating membrane response to compound stimuli. *Journal of Experimental Psychology: Animal Behavior Processes, 12,* 186–195.

Kehoe, E.J. (1988). A layered network model of associative learning: learning-to-learn and configuration. *Psychological Review, 95,* 411–433.

Kehoe, E.J. (1992). Versatility in conditioning: a layered network model. In D.S. Levine and S.J. Levin (Eds.), *Motivation, emotion and goal direction in neural networks* (pp. 63–90). Hillsdale, NJ: Erlbaum.

Kehoe, E.J. (1998). Can the whole be something other than the sum of its parts? In C.D.L. Wynne and J.E.R. Staddon (Eds.), *Models of action: mechanisms for adaptive behavior* (pp. 87–126). Mahwah, NJ: Erlbaum.

Kehoe, E.J., and Gormezano, I. (1980). Configuration and combination laws in conditioning with compound stimuli. *Psychological Bulletin, 87,* 351–378.

Kehoe, E.J., and Graham, P. (1988). Summation and configuration in negative patterning of the rabbit's conditioned nictitating membrane response. *Journal of Experimental Psychology: Animal Behavior Processes, 14,* 320–333.

Kehoe, E.J., and Holt, P.E. (1984). Transfer across CS-US intervals and sensory modalities in classical conditioning in the rabbit. *Animal Learning & Behavior, 12,* 122–128.

Kehoe, E.J., and Macrae, M. (1994). Classical conditioning of the rabbit nictitating membrane can be fast or slow: implications for Lennartz and Weinberger's (1992) two-factor theory. *Psychobiology, 22,* 1–4.

Kehoe, E.J., and Macrae, M. (1997). Savings in animal learning: implications for relapse and maintenance after therapy. *Behavior Therapy, 28,* 141–155.

Kehoe, E.J., and Morrow, L.D. (1984). Temporal dynamics of the rabbit's nictitating membrane response in serial compound conditioned stimuli. *Journal of Experimental Psychology: Animal Behavior Processes, 10,* 205–220.

Kehoe, E.J., and Napier, R.M. (1991). In the blink of an eye: real-time stimulus factors in delay and trace conditioning of the rabbit's nictitating membrane response. *Quarterly Journal of Experimental Psychology Comparative and Physiological Psychology, 43B,* 257–277.

Kehoe, E.J., and Schreurs, B.G. (1986). Compound-component differentiation as a function of CS-US interval and CS duration in the rabbit's nictitating membrane response. *Animal Learning & Behavior, 14,* 144–154.

Kehoe, E.J., and Weidemann, G. (1999a). Time course of conditioned responses in stimulus compounding in the rabbit nictitating membrane preparation. Sydney, Australia: University of New South Wales.

Kehoe, E.J., and Weidemann, G. (1999b). Within-stimulus competition in trace conditioning of the rabbit's nictitating membrane response. *Psychobiology, 27,* 72–84.

Kehoe, E.J., Gibbs, C.M., Garcia, E., and Gormezano, I. (1979). Associative transfer and stimulus selection in classical conditioning of the rabbit's nictitating membrane response to serial compound CSs. *Journal of Experimental Psychology: Animal Behavior Processes, 5,* 1–18.

Kehoe, E.J., Feyer, A., and Moses, J.L. (1981a). Second-order conditioning of the rabbit's nictitating membrane response as a function of the CS2-CS1 and CS1-US intervals. *Animal Learning & Behavior, 9,* 304–315.

Kehoe, E.J., Schreurs, B.G., and Amodei, N. (1981b). Blocking acquisition of the rabbit's nictitating membrane response to serial conditioned stimuli. *Learning and Motivation, 12,* 92–108.

Kehoe, E.J., Morrow, L.D., and Holt, P.E. (1984). General transfer across sensory modalities survives reductions in the original conditioned reflex in the rabbit. *Animal Learning & Behavior, 12,* 129–136.

Kehoe, E.J., Marshall-Goodell, B., and Gormezano, I. (1987a). Differential conditioning of the rabbit's nictitating membrane response to serial compound stimuli. *Journal of Experimental Psychology: Animal Behavior Processes, 13,* 17–30.

Kehoe, E.J., Schreurs, B.G., and Graham, P. (1987b). Temporal primacy overrides prior training in serial compound conditioning of the rabbit's nictitating membrane response. *Animal Learning & Behavior, 15,* 455–464.

Kehoe, E.J., Graham-Clarke, P., and Schreurs, B.G. (1989). Temporal patterns of the rabbit's nictitating membrane response to compound and component stimuli under mixed CS-US intervals. *Behavioral Neuroscience, 103,* 283–295.

Kehoe, E.J., Cool, V., and Gormezano, I. (1991). Trace conditioning of the rabbit's nictitating membrane response as a function of CS-US interstimulus interval and trials per session. *Learning and Motivation, 22,* 269–290.

Kehoe, E.J., Horne, P.S., and Horne, A.J. (1993). Discrimination learning using different CS-US intervals in classical conditioning of the rabbit's nictitating membrane response. *Psychobiology, 21,* 277–285.

Kehoe, E.J., Horne, A.J., Horne, P.S., and Macrae, M. (1994). Summation and configuration between and within sensory modalities in classical conditioning of the rabbit. *Animal Learning & Behavior, 22,* 19–26.

Kehoe, E.J., Horne, A.J., and Macrae, M. (1995a). Learning to learn: real-time features and a connectionist model. *Adaptive Behavior, 3,* 235–271.

Kehoe, E.J., Schreurs, B.G., Macrae, M., and Gormezano, I. (1995b). Effects of modulating tone frequency, intensity, and duration on the classically conditioned rabbit nictitating membrane response. *Psychobiology, 23,* 103–115.

Kehoe, E.J., Palmer, N., Weidemann, G., and Macrae, M. (2000). The effect of feature-target intervals in conditional discriminations on acquisition and expression of conditioned nictitating membrane and heart-rate responses in the rabbit. *Animal Learning & Behavior, 28,* 80–91.

Kiernan, M.J., Westbrook, R.F., and Cranney, J. (1995). Immediate shock, passive avoidance, and potentiated startle: implications for the unconditioned response to shock. *Animal Learning & Behavior, 23,* 22–30.

Kim, K.-S. (1986). Effects of context manipulation on latent inhibition: a study on the nature of context in classical conditioning. *Korean Journal of Psychology, 5,* 75–86.

Kinkaide, P.S. (1974). Stimulus selection in eyelid conditioning in the rabbit (*Oryctolagus cuniculus*). *Journal of Comparative and Physiological Psychology, 86,* 1132–1140.

Klopf, A.H. (1988). A neuronal model of classical conditioning. *Psychobiology, 16,* 85–125.

Konorski, J. (1967). *Integrative activity of the brain: an interdisciplinary approach.* Chicago: University of Chicago Press.

Kremer, E.F. (1978). The Rescorla-Wagner model: losses in associative strength in compound conditioned stimuli. *Journal of Experimental Psychology: Animal Behavior Processes, 4,* 22–36.

Lashley, K.S. (1916). The human salivary reflex and its use in psychology. *Psychological Review, 23,* 446–464.

Lennartz, R.C., and Weinberger, N.M. (1992). Analysis of response systems in Pavlovian conditioning reveals rapidly versus slowly acquired conditioned responses: support for two factors, implications for behavior and neurobiology. *Psychobiology, 20,* 93–119.

Leonard, D.W. (1975). Partial reinforcement effects in classical aversive conditioning in rabbits and human beings. *Journal of Comparative and Physiological Psychology, 88,* 596–608.

Leonard, D.W., and Theios, J. (1967). Classical conditioning in rabbits under prolonged single alternation conditions of reinforcement. *Journal of Comparative and Physiological Psychology, 64,* 273–276.

Leonard, D.W., Fischbein, L.C., and Monteau, J.E. (1972). The effects of interpolated US alone (USa) presentations on classical nictitating membrane conditioning in rabbit (*Oryctolagus cuniculus*). *Conditional Reflex, 7,* 107–114.

Levinthal, C.F. (1973). The CS-US interval function in rabbit nictitating membrane response conditioning: single vs multiple trials per conditioning session. *Learning and Motivation, 4,* 259–267.

Levinthal, C.F., and Papsdorf, J.P. (1970). The classically conditioned nictitating membrane response: the CS-US interval function with one trial per day. *Psychonomic Science, 21,* 296–297.

Levinthal, C.F., Tartell, R.H., Margolin, C.M., and Fishman, H. (1985). The CS-US interval (ISI) function in rabbit nictitating membrane response conditioning with very long intertrial intervals. *Animal Learning & Behavior, 13,* 228–232.

Levitan, L. (1975). Tests of the Rescorla-Wagner model of Pavlovian conditioning. *Bulletin of the Psychonomic Society, 6,* 265–268.

Liu, S., and Moore, J.W. (1969). Auditory differential conditioning of the rabbit nictitating membrane response: IV. Training based on stimulus offset and the effect of an intertrial tone. *Psychonomic Science, 15,* 128–129.

Lubow, R.E., and Moore, A.U. (1959). Latent inhibition: the effect of non-reinforced pre-exposure to the conditioned stimulus. *Journal of Comparative and Physiological Psychology, 52,* 415–419.

Mackintosh, N.J. (1975). A theory of attention: variation in the associability of stimuli with reinforcement. *Psychological Review, 82,* 276–298.

Macrae, M., and Kehoe, E.J. (1995). Transfer between conditional and discrete discriminations in conditioning of the rabbit nictitating membrane response. *Learning and Motivation, 26,* 380–402.

Macrae, M., and Kehoe, E.J. (1999). Savings after extinction in conditioning of the rabbit's nictitating membrane response. *Psychobiology, 27,* 85–94.

Maleske, R.T., and Frey, P.W. (1979). Blocking in eyelid conditioning: effect of changing the CS-US interval and introducing an intertrial stimulus. *Animal Learning & Behavior, 7,* 452–456.

Manning, A.A., Schneiderman, N., and Lordahl, D.S. (1969). Delay vs. trace heart rate classical discrimination conditioning in rabbits as a function of ISI. *Journal of Experimental Psychology, 80,* 225–230.

Marchant, H.G., III, and Moore, J.W. (1973). Blocking of the rabbit's conditioned nictitating membrane response in Kamin's two-stage paradigm. *Journal of Experimental Psychology, 101,* 155–158.

Marchant, H.G., Mis, F.W., and Moore, J.W. (1972). Conditioned inhibition of the rabbit's nictitating membrane response. *Journal of Experimental Psychology, 95,* 408–411.

Marshall Goodell, B., Kehoe, E.J., and Gormezano, I. (1992). Laws of the unconditioned reflex in the rabbit nictitating membrane preparation. *Psychobiology, 20,* 229–237.

Mauk, M.D., and Ruiz, B.P. (1992). Learning-dependent timing of Pavlovian eyelid responses: differential conditioning using multiple interstimulus intervals. *Behavioral Neuroscience, 106,* 666–681.

McGeoch, J.A. (1952). *The psychology of human learning,* 2nd ed. New York: Longmans, Green, & Co.

McNish, K.A., Betts, S.L., Brandon, S.E., and Wagner, A.R. (1997). Divergence of conditioned eyeblink and conditioned fear in backward Pavlovian training. *Animal Learning & Behavior, 25,* 43–52.

Meyer, D.R., and Miles, R.C. (1953). Intralist-interlist relations in verbal learning. *Journal of Experimental Psychology, 45,* 109–115.

Millenson, J.R., Kehoe, E.J., and Gormezano, I. (1977). Classical conditioning of the rabbit's nictitating membrane response under fixed and mixed CS-US intervals. *Learning and Motivation., 8,* 351–366.

Miller, R.R., and Matzel, L.D. (1989). Contingency and relative associative strength. In S.B. Klein and R.R. Mowrer (Eds.), *Contemporary learning theories: Pavlovian conditioning and the status of traditional learning theory* (pp. 61–84). Hillsdale, NJ: Erlbaum.

Minsky, M.L., and Papert, S. (1969). *Perceptrons: an introduction to computational geometry.* Cambridge, MA: MIT Press.

Mis, F., and Moore, J.W. (1973). Effect of preacquisition UCS exposure on classical conditioning of the rabbit's NMR response. *Learning and Motivation, 4,* 108–114.

Mis, F.W., Andrews, J.G., and Salafia, W.R. (1970). Conditioning of the rabbit nictitating membrane response: ISI by ITI interaction. *Psychonomic Science, 20,* 57–58.

Mitchell, D.S. (1974). Conditional responding and intertrial-interval variability in classical conditioning in the rabbit (*Oryctolagus cuniculus*). *Journal of Comparative and Physiological Psychology, 87,* 73–79.

Moore, J.W. (1972). Stimulus control: studies of auditory generalization in rabbits. In A.H. Black and W.F. Prokasy (Eds.), *Classical conditioning. II: Current research and theory* (pp. 206–230). New York: Appleton-Century-Crofts.

Moore, J.W., and Stickney, K.J. (1980). Formation of attentional-associative networks in real time: role of the hippocampus and implications for conditioning. *Physiological Psychology, 8,* 207–217.

Moore, J.W., Newman, F.L., and Glasgow, B. (1969). Intertrial cues as discriminative stimuli in human eyelid conditioning. *Journal of Experimental Psychology, 79,* 319–326.

Moore, J.W., Choi, J.-S. and Barto, A.G. (1996). TD Model of classical conditioning: response timing and cerebellar implementation. *Society for Neuroscience Abstracts, 22,* 1643.

Moscovitch, A., and Lolordo, V.M. (1968). role of safety in the Pavlovian backward fear conditioning procedure. *Journal of Comparative and Physiological Psychology, 66,* 673–678.

Napier, R.M., Macrae, M., and Kehoe, E.J. (1992). Rapid reacquisition in conditioning of the rabbit's nictitating membrane response. *Journal of Experimental Psychology Animal Behavior Processes, 18,* 182–192.

Nordholm, A.F., Lavond, D.G., and Thompson, R.F. (1991). Are eyeblink responses to tone in the decerebrate, decerebellate rabbit conditioned responses? *Behavioral Brain Research, 44,* 27–34.

Patterson, M.M. (1970). Classical conditioning of the rabbit's (*Oryctolagus cuniculus*) nictitating membrane response with fluctuating ISI and intracranial CS. *Journal of Comparative and Physiological Psychology, 72,* 193–202.

Pavlov, I.P. (1927). *Conditioned reflexes: an investigation of the physiological activity of the cerebral cortex* (trans. by G.V. Anrep). London: Oxford University Press.

Pearce, J.M. (1987). A model for stimulus generalization in Pavlovian conditioning. *Psychological Review, 94,* 61–73.

Pearce, J.M. (1994). Similarity and discrimination: a selective review and a connectionist model. *Psychological Review, 101,* 587–607.

Pearce, J.M., and Hall, G. (1980). A model for Pavlovian learning: variations in the effectiveness of conditioned but not of unconditioned stimuli. *Psychological Review, 87,* 532–552.

Pfautz, P.L., Donegan, N.H., and Wagner, A.R. (1978). Sensory preconditioning versus protection from habituation. *Journal of Experimental Psychology: Animal Behavior Processes, 4,* 286–295.

Plotkin, H.C., and Oakley, D.A. (1975). Backward conditioning in the rabbit (*Oryctolagus cuniculus*). *Journal of Comparative and Physiological Psychology, 88,* 586–590.

Port, R.L., Beggs, A.L., and Patterson, M.M. (1987). Hippocampal substrate of sensory associations. *Physiology and Behavior, 39,* 643–647.

Prokasy, W.F. (1965). Classical eyelid conditioning: experimenter operations, task demands, and response shaping. In W.F. Prokasy (Ed.), *Classical conditioning: a symposium* (pp. 208–225). New York: Appleton-Century-Crofts.

Prokasy, W.F., and Gormezano, I. (1979). The effect of US omission in classical aversive and appetitive conditioning of rabbits. *Animal Learning & Behavior, 7,* 80–88.

Randich, A., and LoLordo, V.M. (1979). Associative and nonassociative theories of the UCS preexposure phenomenon: implications for Pavlovian conditioning. *Psychological Bulletin, 86,* 523–548.

Razran, G. (1965). Empirical codifications and specific theoretical implications of compound-stimulus conditioning: perception. In W.F. Prokasy (Ed.), *Classical conditioning* (pp. 226–248). New York: Appleton-Century-Crofts.

Reiss, S., and Wagner, A.R. (1972). CS habituation produces a "latent inhibition effect" but no active "conditioned inhibition." *Learning and Motivation, 3,* 237–245.

Rescorla, R.A. (1967). Pavlovian conditioning and its proper control procedures. *Psychological Review, 74,* 71–80.

Rescorla, R.A. (1969). Pavlovian conditioned inhibition. *Psychological Bulletin, 72,* 77–94.

Rescorla, R.A. (1970). Reduction in the effectiveness of reinforcement after prior excitatory conditioning. *Learning and Motivation, 1,* 372–381.

Rescorla, R.A. (1973). Evidence for "unique stimulus" account of configural conditioning. *Journal of Comparative and Physiological Psychology, 85,* 331–338.

Rescorla, R.A. (1982). Effect of a stimulus intervening between CS and US in autoshaping. *Journal of Experimental Psychology: Animal Behavior Processes, 8,* 131–141.

Rescorla, R.A. (1997). Summation assessment of a configural theory. *Animal Learning & Behavior, 25,* 200–209.

Rescorla, R.A., & Wagner, A.R. (1972). A theory of Pavlovian conditioning: variations in the effectiveness of reinforcement and nonreinforcement. In A.H. Black and W.F. Prokasy (Eds.), *Classical conditioning. II* (pp. 64–99). New York: Appleton-Century-Crofts.

Robinson, G.B., Port, R.L., and Stillwell, E. (1993). Latent inhibition of the classically conditioned rabbit nictitating membrane response is unaffected by NMDA antagonist MK801. *Psychobiology, 21,* 120–124.

Rogers, R.F., and Steinmetz, J.E. (1998). Contextually based conditional discrimination of the rabbit eyeblink response. *Neurobiology of Learning and Memory, 69,* 307–319.

Ross, R.T., Scavio, M.J., Jr., and Erikson, K. (1979). Performance of the nictitating membrane CR following CS-US interval shifts. *Bulletin of the Psychonomic Society, 14,* 189–192.

Rudy, J.W., and Sutherland, R.J. (1995). Configural association theory and the hippocampal formation: an appraisal and reconfiguration. *Hippocampus, 5,* 375–389.

Rumelhart, D.E., Hinton, G.E., and Williams, R.J. (1986). Learning internal representations by error propagation. In D.E. Rumelhart, J.L. McClelland, and the PDP Research Group (Eds.), *Parallel distributed processing: explorations in the microstructures of cognition* (pp. 318–362). Cambridge, MA: MIT Press.

Saladin, M.E., and Tait, R.W. (1986). US preexposures retard excitatory and facilitate inhibitory conditioning of the rabbit's nictitating membrane response. *Animal Learning & Behavior, 4,* 121–132.

Saladin, M.E., Ten Have, W.N., Saper, Z.L., Labinsky, J.S., and Tait, R.W. (1989). Retardation of rabbit nictitating membrane conditioning following US preexposures depends on the distribution and number of US presentations. *Animal Learning & Behavior, 17,* 179–187.

Salafia, W.R., and Allan, A.M. (1982). Augmentation of latent inhibition by electrical stimulation of hippocampus. *Physiology & Behavior, 29,* 1125–1130.

Salafia, W.R., Mis, F.W., Terry, W.S., Bartosiak, R.S., and Daston, A.P. (1973). Conditioning of the nictitating membrane response of the rabbit (*Oryctolagus cuniculus*) as a function of length and degree of variation of intertrial interval. *Animal Learning & Behavior, 1*, 109–115.

Salafia, W.R., Daston, A.P., Bartosiak, R.S., Hurley, J., and Martino, L.J. (1974). Classical nictitating membrane conditioning in the rabbit (*Oryctolagus cuniculus*) as a function of unconditioned stimulus locus. *Journal of Comparative and Physiological Psychology, 86*, 628–636.

Salafia, W.R., Terry, W., and Daston, A.P. (1975). Conditioning of the rabbit (*Oryctolagus cuniculus*) nictitating membrane response as a function of trials per session, ISI, and ITI. *Bulletin of the Psychonomic Society, 6*, 505–508.

Salafia, W.R., Martino, L.J., Cloutman, K., and Romano, A.G. (1979). Unconditional-stimulus locus and interstimulus-interval shift in rabbit (*Oryctolagus cuniculus*) nictitating membrane conditioning. *Pavlovian Journal of Biological Science, 14*, 64–71.

Salafia, W.R., Lambert, R.W., Host, K.C., Chiaia, L.N., and Ramirez, J.J. (1980). Rabbit nictitating membrane conditioning: lower limit of the effective interstimulus interval. *Animal Learning & Behavior, 8*, 85–91.

Scavio, M.J., Jr. (1974). Classical-classical transfer: effects of prior aversive conditioning upon appetitive conditioning in rabbits (*Oryctolagus cuniculus*). *Journal of Comparative Physiological Psychology, 86*, 107–115.

Scavio, M.J., and Gormezano, I. (1974). CS intensity effects upon rabbit nictitating membrane conditioning, extinction, and generalization. *Pavlovian Journal of Biological Sciences, 9*, 25–34.

Scavio, M.J., and Gormezano, I. (1980). Classical-classical transfer: effects of prior appetitive conditioning upon aversive conditioning in rabbits. *Animal Learning & Behavior, 8*, 218–224.

Schafe, G.E., Sollars, S.I., and Bernstein, I.L. (1995). The CS-US interval and taste aversion learning: a brief look. *Behavioral Neuroscience, 109*, 799–802.

Schmajuk, N.A., and DiCarlo, J.J. (1992). Stimulus configuration, classical conditioning and hippocampal formation. *Psychological Review, 99*, 268–305.

Schmajuk, N.A., and Holland, P.C. (Eds.) (1998). Occasion setting: associative learning and cognition in animals. Washington, DC: American Psychological Association.

Schmajuk, N.A., and Moore, J.W. (1988). The hippocampus and the classically conditioned nictitating membrane response: a real-time attentional-associative model. *Psychobiology, 16*, 20–35.

Schmajuk, N.A., Lam, Y.-W., and Gray, J.A. (1996). Latent inhibition: a neural network approach. *Journal of Experimental Psychology: Animal Behavior Processes, 22*, 321–349.

Schmajuk, N.A., Lamoureux, J.A., and Holland, P.C. (1998). Occasion setting: a neural network approach. *Psychological Review, 105*, 3–32.

Schneiderman, N. (1966). Interstimulus interval function of the nictitating membrane response underlying trace versus delay conditioning. *Journal of Comparative and Physiological Psychology, 62*, 397–402.

Schneiderman, N. (1972). Response system divergencies in aversive classical conditioning. In A.H. Black & W.F. Prokasy (Eds.), *Classical conditioning: current theory and research* (pp. 341–376). New York: Appleton-Century-Crofts.

Schneiderman, N., McCabe, P.M., Haselton, J.R., Ellenberger, H.H., Jarrell, T.W., and Gentile, C.G. (1987). Neurobiological bases of conditioned bradycardia in rabbits. In I. Gormezano, W.F. Prokasy, and R.F. Thompson (Eds.), *Classical conditioning* (pp. 37–63) Hillsdale, NJ: Erlbaum.

Schreurs, B.G. (1989). Classical conditioning of model systems: a behavioral review. *Psychobiology, 17,* 145–155.

Schreurs, B.G. (1993). Long-term memory and extinction of the classically conditioned rabbit NM response. *Learning and Motivation, 24,* 293–302.

Schreurs, B.G. (1998). Long-term memory and extinction of rabbit nictitating membrane conditioning. *Learning and Motivation, 29,* 68–82.

Schreurs, B.G., and Alkon, D.L. (1990). US-US conditioning of the rabbit's nictitating membrane response: emergence of a conditioned response without alpha conditioning. *Psychobiology, 18,* 312–320.

Schreurs, B.G., and Gormezano, I. (1982). Classical conditioning of the rabbit's nictitating membrane response to CS compounds: effects of prior single stimulus conditioning. *Bulletin of the Psychonomic Society, 19,* 365–368.

Schreurs, B.G., and Kehoe, E.J. (1987). Cross-modal transfer as a function of initial training level in classical conditioning with the rabbit. *Animal Learning & Behavior, 15,* 47–54.

Schreurs, B.G., Kehoe, E.J., and Gormezano, I. (1993). Concurrent associative transfer and competition in serial conditioning of the rabbit's nictitating membrane response. *Learning and Motivation, 24,* 395–412.

Schreurs, B.G., Oh, M.M., Hirashima, C., and Alkon, D.L. (1995). Conditioning-specific modification of the rabbit's unconditioned nictitating membrane response. *Behavioral Neuroscience, 109,* 24–33.

Sechenov, I.M. (1863). *Reflexes of the brain.* Cambridge: MIT Press.

Sheafor, P.J. (1975). "Pseudoconditioned" jaw movements of the rabbit reflect associations conditioned to contextual background cues. *Journal of Experimental Psychology: Animal Behavior Processes, 104,* 245–260.

Siegel, S. (1969a). Effect of CS habituation on eyelid conditioning. *Journal of Comparative and Physiological Psychology, 68,* 245–248.

Siegel, S. (1969b). Generalization of latent inhibition. *Journal of Comparative and Physiological Psychology, 69,* 157–159.

Siegel, S., and Domjan, M. (1971). Backward conditioning as an inhibitory procedure. *Learning and Motivation, 2,* 1–11.

Skelton, R.W., Mauk, M.D., and Thompson, R.F. (1988). Cerebellar nucleus lesions dissociate alpha conditioning from alpha responses in rabbits. *Psychobiology, 16,* 126–134.

Smith, M.C. (1968). CS-US interval and US intensity in classical conditioning of the rabbit's nictitating membrane response. *Journal of Comparative and Physiological Psychology, 66,* 679–687.

Smith, M.C., Coleman, S.R., and Gormezano, I. (1969). Classical conditioning of the rabbit's nictitating membrane response at backward, simultaneous, and forward CS-US intervals. *Journal of Comparative and Physiological Psychology, 69,* 226–231.

Solomon, P.R. (1977). Role of hippocampus in blocking and conditioned inhibition of the rabbit's nictitating membrane response. *Journal of Comparative and Physiological Psychology, 91,* 407–417.

Solomon, P.R., and Moore, J.W. (1975). Latent inhibition and stimulus generalization of the classically conditioned nictitating membrane response in rabbits (*Oryctolagus cuniculus*) following dorsal hippocampal ablation. *Journal of Comparative and Physiological Psychology, 89,* 1192–1203.

Solomon, P.R., Brennan, G., and Moore, J.W. (1974). Latent inhibition of the rabbit's nictitating membrane response as a function of CS intensity. *Bulletin of Psychonomic Society, 4,* 445–448.

Sutherland, N.S., and Mackintosh, N.J. (1971). *Mechanisms of animal discrimination learning.* New York: Academic Press.

Sutherland, R.J., and Rudy, J.W. (1989). Configural association theory: the role of the hippocampal formation in learning, memory, and amnesia. *Psychobiology, 17,* 129–144.

Sutton, R.S., and Barto, A.G. (1981). Toward a modern theory of adaptive networks: expectation and prediction. *Psychological Review, 88,* 135–171.

Sutton, R.S., and Barto, A.G. (1990). Time-derivative models of Pavlovian reinforcement. In M. Gabriel & J.W. Moore (Eds.), *Learning and computational neuroscience* (pp. 497–537). Cambridge, MA: MIT Press.

Tait, R.W., and Saladin, M.E. (1986). Concurrent development of excitatory and inhibitory associations during backward conditioning. *Animal Learning & Behavior, 14,* 133–137.

Tait, R.W., Kehoe, E.J., and Gormezano, I. (1983). Effects of US duration on classical conditioning of the rabbit's nictitating membrane response. *Journal of Experimental Psychology: Animal Behavior Processes, 9,* 91–101.

Tait, R.W., Quesnel, L.J., and ten-Have, W.N. (1986). Classical-classical transfer: excitatory associations between "competing" motivational stimuli during classical conditioning of the rabbit. *Animal Learning & Behavior, 14,* 138–143.

Thomas, E., and Wagner, A.R. (1964). Partial reinforcement of the classically conditioned eyelid response in the rabbit. *Journal of Comparative and Physiological Psychology, 58,* 157–158.

Thompson, R.F., Donegan, N.H., Clark, G.A., Lavond, D.G., Lincoln, J.S., Madden, J., Mamounas, L.A., Mauk, M.D., and McCormick, D.A. (1987). Neuronal substrates of discrete, defensive conditioned reflexes, conditioned fear states, and their interactions in the rabbit. In I. Gormezano, W.F. Prokasy, and R.F. Thompson (Eds.), *Classical conditioning* (pp. 371–399). Hillsdale, NJ: Erlbaum.

Tolman, E.C. (1932). *Purposive behavior in animals and men.* New York: Century.

Wagner, A.R., and Brandon, S.E. (1989). Evolution of a structured connectionist model of Pavlovian conditioning (AESOP). In S.B. Klein and R.R. Mowrer (Eds.), *Contemporary learning theories: Pavlovian conditioning and the status of traditional learning theory* (pp. 149–190). Hillsdale, NJ: Erlbaum.

Weidemann, G., and Kehoe, E.J. (1997). Transfer and counterconditioning of conditional control in the rabbit nictitating membrane response. *Quarterly Journal of Experimental Psychology, 50B,* 295–316.

Weidemann, G., Georgilas, A., and Kehoe, E.J. (1999). Temporal specificity in patterning of the rabbit's conditioned nictitating membrane response. *Animal Learning & Behavior, 27,* 99–107.

Weisz, D., and Walts, C. (1990). Reflex facilitation of the rabbit nictitating membrane response by an auditory stimulus as a function of interstimulus interval. *Behavioral Neuroscience, 104,* 11–20.

Weisz, D.J., and LoTurco, J.J. (1988). Reflex facilitation of the nictitating membrane response remains after cerebellar lesions. *Behavioral Neuroscience, 102,* 203–209.

Weisz, D.J., and McInerny, J. (1990). An associative process maintains reflex facilitation of the unconditioned nictitating membrane response during the early stages of training. *Behavioral Neuroscience, 104,* 21–27.

Weisz, D.J., Harden, D.G., and Xiang, Z. (1992). Effects of amygdala lesions on reflex facilitation and conditioned response acquisition during nictitating membrane response conditioning in rabbit. *Behavioral Neuroscience, 106,* 262–273.

White, N. (1998). *Effects of reductions in US intensity on CR elimination in the rabbit nictitating membrane preparation.* Honours thesis. Sydney: University of New South Wales.

Zackheim, J., Myers, C., and Gluck, M. (1998). A temporally sensitive recurrent network model of occasion setting. In N. Schmajuk and P. Holland (Eds.), *Occasion setting: associative learning and cognition in animals* (pp. 319–342). Washington, DC: American Psychological Association.

7
Computational Theories of Classical Conditioning

SUSAN E. BRANDON, EDGAR H. VOGEL, AND ALLAN R. WAGNER

Computational theories of classical conditioning are theories whose propositions are stated as mathematical relationships. From such propositions one can compute, that is to say, deduce, presumed consequences for conditioned responding in circumstances addressed by the theory. This chapter is an attempt to roughly categorize and briefly summarize the major computational theories that have been developed to understand behavior in classical conditioning. It is organized around the essential ideas about conditioning that the various theories embrace. This chapter is not intended to provide a relative evaluation of the theories mentioned; to do so is considerably beyond the scope of what can be accomplished in the space available. Data are mentioned and judgments made, but only to provide understanding of the inspiration for the different models.

Computational theories that focus on classical conditioning are of recent origin, beginning with the formulation of Rescorla and Wagner (Rescorla & Wagner, 1972; Wagner & Rescorla, 1972). The computational theories of simple learning that preceded this effort were directed more to instrumental conditioning and selective learning than to classical conditioning, with some resultant ambiguity about how their propositions should be applied to classical conditioning. However, they provided an important background of computational machinery and theoretical ideas as a basis on which the subsequent models could build. In the section "Historical Background," we summarize some of the important legacies of such theorizing.

Of the modern computational theories, none addresses the full range of empirical phenomena and functional relationships that are known in classical conditioning. Different theories have focused on different phenomena, no doubt because their authors considered them to be critical to an adequate understanding of the conditioning process. Also, as expected, the sense of which are the remaining critical issues has changed. First-generation computational theories of classical conditioning were motivated by the attempt to deal with a set of phenomena involving "stimulus selection," discrimination learning, and generalization (Kamin, 1968; Wagner, 1969a). The section " Trial-Level Models" summarizes the models that originated in this literature and their recent extensions. A major concern since 1981 and the publication of the "time-derivative" model of Sutton and Barto (1981) and the SOP (Sometimes Opponent Process) model of Wagner and his colleagues (Donegan & Wagner, 1987; Mazur & Wagner, 1982; Wagner, 1981) has been with temporal phenomena, such as the effects upon conditioning of variation in the interval between conditioned stimulus (CS) and unconditioned stimulus (US). The section "Real-Time Models" summarizes the theoretical machinery that has been developed to handle such temporal phenomena. The most complex computational models of classical conditioning attempt to account for conditioned responses that appear to be controlled by separable associative systems. The last section, "Multiple Process Real-Time Models," briefly summarizes two such models.

All the computational theories of classical conditioning that we describe can be conceived as connectionist models in which classical conditioning is taken to involve the development of one or more associative links or "connections" between representations of the CSs and USs employed during training. In such theories, the major theoretical options are centered around three questions. How shall the CSs and USs be represented? How shall the links between stimulus representations be construed to change during conditioning? How do the measures of conditioned responding depend on the current values of the stimulus representations and their associative links? In each of the sections that follow, we attempt to point to the manner in which these issues have (or have not) been addressed.

Historical Background

The first well-developed quantitative learning theory was that offered by Clark Hull in a series of papers that began in the 1920s and culminated in *Principles of Behavior* (1943), *Essentials of Behavior* (1951), and *A Behavior System* (1952). What Hull offered, in effective collaboration with Kenneth Spence, was an example of a theoretical system that was previously unknown in psychology. He argued with a lively and impassioned voice for the development of quantitative theories, asserting that the value of a quantified behavior theory was in its rigor because it allowed the best test of an explanatory system, which is "to deduce correctly the results of experimental observations not yet made." He challenged competing theorists to exhibit similar rigor and come up with alternative predictions. If they could not, this would be an indication of "immaturity, possibly of inadequacy." If the competition could only make deductions based on essentially the same principles, but independently constructed, then this would be seen as supportive of the Hullian model. But if the "extremely interesting situation of parallelism would be presented," meaning that the same prediction is made on the basis of distinctly different principles, then this would be "mutually illuminating to all parties to the discussion" (Hull, 1930, p. 254).

Stimulus Representation

Breaking away from early structural and Gestalt psychology, both of which were dominated by attempts to characterize perception (Kohler, 1930; Titchener, 1896), the scientific analysis of learning focused on adaptive behavior and problem solving (Thorndike, 1898), with more emphasis on the nature of the response and the rules for its modification, than on the nature of the stimulus and rules for the formation of the *gnostic unit*. It is common, among current computational models of learning, to offer little specification of the nature of the stimulus representation other than to assume that there is some representation that can be elicited and thus serve as a retrievable link for associated stimuli or responses (Schmajuk, 1997; Sutton & Barto, 1981; Wagner, 1981), akin to earlier theories of information processing (Anderson & Bower, 1973) and conditioning (Konorski, 1967; Pavlov, 1927). However, it is possible to characterize broadly both early and current associative theories as assuming either a *molar* or a *componential* stimulus representation.

Hull (1937, 1943) borrowed from Pavlov (1927) the notion of the *molar stimulus trace*. This concept was an internal representation of the external stimulus, where the two were not isomorphic. Figure 7.1a is copied from Hull (1943), who himself borrowed it from Adrian (1928). It is a function that was generated by measuring the frequency of impulses emitted by the eye of an electric eel during continuous stimulation by a light. As can be seen, although the external stimulus might be a relatively constant event, the internal stimulus trace began at a low value and then rose rapidly to a comparatively high maximum, following which it fell off to some moderate level even as the external stimulus continued; when the external stimulus was terminated, the stimulus trace gradually decayed, but there was some activity even after stimulus offset (not shown in the figure), as if there was some "reverberation" in the nervous system as an after-discharge. Marks (1974), on the basis of extensive psychophysical experiments with human subjects, has reported a subjective "stimulus magnitude" function that is of the same form, that is, as involving periods of recruitment, adaptation, and decay. Marks' descriptive function is reproduced in Figure 7.1b.

In contrast to the essentially molar stimulus representation of Hull–Spence theory, in statistical learning theory (Atkinson & Estes, 1963; Estes, 1950, 1959b) the stimulus was assumed to be represented by a set of independently variable components or aspects of the total environment. At any target moment (identified as an experimental trial), only a sample (s) of elements from the total population was assumed to be effective. When experimental conditions involved repeated stimulation of an organism, each sample s was treated as a independent random sample from S.

This molecular view of the stimulus was anticipated by Konorski (1948), who himself borrowed it from Sherrington (1929), and is reproduced in Figure 7.2. It lends itself elegantly to conceptualizing many instances of stimulus generalization and discrimination. In the figure, A and B are two stimuli, each shown to activate its own sample of elements. Stimulus generalization is conceived of as following from shared elements, similar to Thorndike's notion of identical elements (Thorndike, 1932): having trained response to A, the same response is observed to B in proportion to the number of elements that are common to A and B. The proportion of common elements, conversely, serves as an index of the similarity of the two stimuli. The theoretical strategy is a powerful one and has been used in many subsequent models (Blough, 1975; Gluck & Bower, 1988; Wagner & Brandon, 2001; cf., Pearce, 1987, 1994).

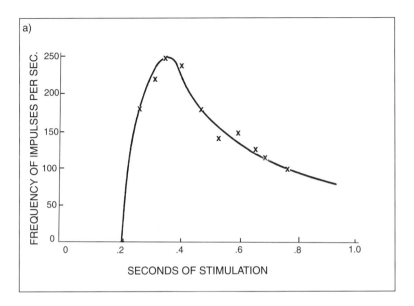

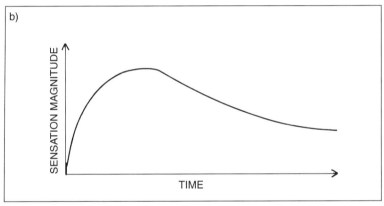

FIGURE 7.1. (**A**) Rise and gradual decay in the frequency of impulses emitted by the eye of an eel during continuous stimulation by a light of 830 meter-candles. (Adapted from Adrian, 1928, and copied from Hull, 1943, figure 4, p. 42.) (**B**) Human subjects reported sensation magnitude as a function of the presentation of stimuli from a variety of modalities across time. (Copied from Marks, 1974, figure 4.1, p. 100.)

Learning Rules

The Hull–Spence models were in the classic tradition of deterministic growth models: learning was represented by the growth of two hypothetical connections between stimulus representations and response representations, *habit strength* (sH_R) and *conditioned inhibition* (sI_R). The net excitatory tendency (sE_R) was an additive function of sH_R−sI_R. A simplified version of the presumed associative

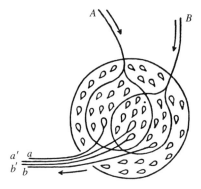

FIGURE 7.2. The "partial convergence of allied reflexes:" A, B two afferent nerve paths; a, a', b, b', efferent paths. (Taken from Sherrington, 1929, and copied from Konorski, 1948, figure 7, p. 110.)

structure and relevant connections is shown in Figure 7.3a. Adaptive behavior was acquired to the extent that stimulus representation came to be associated with response representation. In Hull's (1943) terms, the increment in habit strength on the occasion of "reinforcement" of that response was defined as proportional to the difference between a maximal value, M, and the strength of the habit held by the CS at the beginning of the trial:

$$\Delta sH_R = \theta(M - sH_R), \quad \text{where } sH_R = sH_R \times V_i, \tag{1}$$

where sH_R was the excitatory connection between stimulus and response representation, and V_i and θ were parameters for stimulus intensity and learning rate, respectively. The increment in sH_R was a linear function of sH_R, from which fact comes the term "linear operator model." The amount of increment in sH_R was further quantified by parameters for the amount and delay of reinforcement and for the temporal asynchrony of the antecedent stimulus and the response. To account for the decrement in performance that occurs when a designated response is not reinforced, Hull (1945) and Spence (1936) proposed the development of an antagonistic associative tendency, sI_R. Hull characterized it as an acquired tendency not to make the designated response to the stimulus, which varied with the amount of effort expended in the response. Spence characterized it as a similar tendency, the increment in which on a nonreinforced occasion was an increasing function of excitatory tendency, that is, $sI_R = \theta(sH_R - sI_R)$.

Statistical learning theory offered a probabilistic, rather than deterministic, account of learning (Bush & Moesteller, 1951; Estes, 1950, 1959b). Learning was presumed to reflect the changing proportion of the situational stimulus elements, S, that are conditioned to a designated response. It was assumed that, on each learning event, some sample, s, of the elements is selected and all of the elements in that sample will be "conditioned" to whatever response is reinforced. Thus, if p_i is the proportion of the elements of S (and of s) previously conditioned to a response, R_i, the increment in the proportion that would result from the reinforcement of that response would be:

$$\Delta p_i = \theta(1 - p_i), \tag{2a}$$

where the learning rate variable, θ, is equal to the proportion of the situational elements that are sampled on the trial, that is, Ns/NS. Following the same logic, if

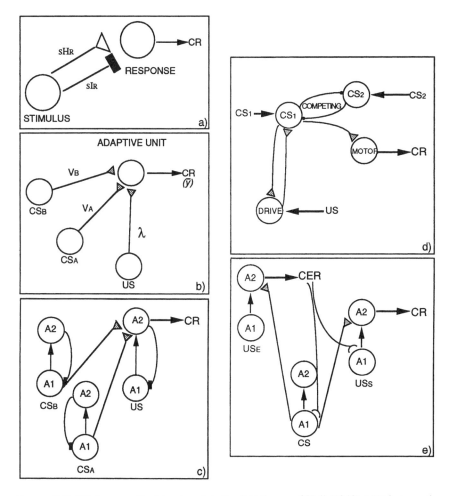

FIGURE 7.3. (**A**) A connectionist network in the S-R theory of Hull (1943). sH_R is an excitatory connection, and sI_R an inhibitory connection, between the *stimulus* representation and the *response* representation. The output of the response unit is assumed to be the CR. (**B**) The adaptive unit connectionist network that was used by Sutton and Barto (1981) and modified here for the Rescorla–Wagner model (Rescorla & Wagner, 1972; Wagner & Rescorla, 1972). Shown are representations for two CSs, *A* and *B,* and their connections with an adaptive unit that is also activated by a US according to the parameter λ. The output of the adaptive unit is assumed to be the CR. (**C**) A network rendition of SOP. Two CSs, *A* and *B,* and a US each are represented by two nodes, *A1* and *A2*. The associative learning that occurs when a CS is paired with a US in a forward manner consists of increases in connection strengths between the A1 nodes of the CS and the A2 nodes of the US. The model identifies the output of the US$_{A2}$ node as the CR. (**D**) A network rendition of the models of Grossberg and colleagues (Grossberg, 1982; Grossberg & Levine, 1987; Grossberg & Schmajuk, 1987). The interacting modules are a CS-sensory drive circuit, which amplifies and modifies the output of a CS-sensory motor circuit. Sensory-CS circuits compete with one another for memory capacity. (**E**) A network rendition of AESOP (Wagner & Brandon, 1989). The emotive aspect of the dual US representation is assumed to elicit a Conditioned Emotional Response (*CER*) as well as to modify the processing of concurrent stimuli. The sensory aspect of the dual US representation is assumed to elicit sensory/perceptual CRs, such as eyeblink closures.

some response other than R_i were reinforced, the conditioning of elements to that response would be at the cost of elements previously conditioned to R_i. That is, there would be a decrement:

$$\Delta p_i = \theta(p_i). \tag{2b}$$

Statistical learning theory and Hull–Spence theory provide two contrasting views of the nature of the association: for the probability theory of William Estes, the S–R connection reflected a single process, driven to a value of unity with reinforcement and a value of 0 with nonreinforcement. For the excitatory strength theory of Clark Hull and Kenneth Spence, the association reflected two opposing processes, excitation and inhibition. In this dual-process specification of sE_R, Hull and Spence were heavily influenced by Pavlov (1927), who was by training a physiologist and for whom inhibition and excitation were two fundamental processes of the nervous system. In his theoretical treatises, Pavlov distinguished *external inhibition,* which was some disruption of a CR caused by the action of an extraneous stimulus, from *internal inhibition,* which was like external inhibition in that it was antagonistic to the ongoing CR, but was different from external inhibition in that it developed only gradually, with training. It was internal inhibition that Hull and Spence incorporated into their view of the association via the construct sI_R.

The Hull–Spence notion that inhibition and excitation should be separately computed because they are distinguishable processes never has had a wealth of empirical support, although this assumption was common among the early strength learning models (Hull, 1943, 1951; Konorski, 1948; Spence, 1936, 1960). Spence (1937), however, used the assumption that there are differential generalization gradients of inhibition and excitation to explain the phenomenon of *transposition* (Kohler, 1938) in what remains even today as one of the most elegant derivations of modern psychology.

Performance Rules

All computational learning theorists have had some rule, or rules, whether implicit or explicit, about how learning translates into performance. It usually was assumed that variables other than S–R (stimulus–response) connection strength contributed to performance, either by increasing the strength or probability of a response, as *motivational* variables were likely to do, or by masking the learning that was otherwise evident, as *noise* factors were likely to do.

Hull (1950, 1952) and Spence (1936) proposed the working hypotheses that the final behavioral output, R, was a consequence of a multiplicative combination of the associative excitatory tendency for that response and the motivational state of the organism, M; that is, that $R = f(sE_R \times M)$. Thus, a conditioned response might be well learned but not exhibited under a low motivational state, and might show variation from occasion to occasion depending on a variety of motivational changes.

Hull (1945) and Spence (1960) further assumed that there were sources of processing noise, sO_R, which would influence the exhibition of an activated response. They summarized their noise as varying from moment to moment in a normally distributed fashion and as subtracting from the excitatory tendency. An important consequence of their assumption was that, although the rules for habit acquisition led to the expectation that sH_R and, hence, sE_R would grow as a negatively accelerated function of the number of training trials, some response measure should grow differently. For example, on the assumption that the probability of a response is determined by the probability that $sE_R - sO_R$ is above some threshold, L, one would expect an ogival-like function in which the periods of positive and negative acceleration would be more or less prominent, depending on parameters. Such tactics are necessary to rationalize the different forms that acquisition functions can exhibit in different situations (Spence, 1956).

An interesting question has been whether the latency of a learned response is a measure of associative strength. It could be taken to be such according to either stimulus sampling theory or Hull–Spence strength theory, via the reasoning that a response will occur after some number of successive stimulus samples (in stimulus sampling theory) or some number of successive fluctuations of oscillatory inhibition (in Hull–Spence theory) without a response, and that number will be smaller as p_i or the probability that ($sE_R - sO_R > L$) in each moment increases.

The "Overlap Problem"

Application of the linear operator rules of either Hull or Estes to the stochastic view of the stimulus offered in Figure 7.2 leads to what is called the "overlap problem" (Bush & Moesteller, 1955): that is, a stochastic conception of stimulus A and stimulus B, as rendered in Figure 7.2, where some degree of generalization between A and B is evident, cannot account for reports of errorless discrimination between two such stimuli (Robbins, 1970; Uhl, 1964). According to the stochastic view, if, after training with A, testing with a novel cue, B, elicits some of the same responding as did A, this is because A and B share some elements, as depicted in Figure 7.2. Subsequent discrimination training with A+, B−, will, with a linear operator rule for computing changes in associative strengths, increase the number of elements of A connected to the response and decrease the number of elements of B connected to the response so that the elements unique to A will have a net associative strength greater than 0 and the elements unique to B will have net associative strength equal to 0. However, those elements that are shared by A and B also will have a net associative strength greater than 0, due to their history of partial reinforcement. These then produce the response errors on B trials.

Pattern models (Estes, 1959b) were developed by statistical association theorists in part to contend with the overlap problem. The pattern model solution was to propose that conditioning accrued not to the individual elements in this instance, but to *gestalt* patterns that preempted control by the elements. Thus, the *configuration A* is distinguished from the *configuration B* as a function of the differential reinforcement of the two configural patterns, and the subject can discriminate perfectly between the two configurations. Application of the Hullian

(1935, 1937) *principle of afferent neural interaction,* which specified that the stimulus trace could vary as a result of "afferent neural interaction" with other stimulus traces that were impacting on the system at roughly the same time, so that the final trace was something partially different from the traces of which it was composed, also provides that the stimulus trace activated by the compound AX may differ from that activated by the compound BX. The Rescorla–Wagner learning rule offers a different solution to this problem, as we see next.

Trial-Level Models

Empirical Issues

The usefulness of these first-generation computational theories was substantively qualified by what have come to be known as demonstrations of "stimulus selection." One influential such demonstration was an experiment with pigeons reported by Jenkins and Harrison (1960). A group of birds was exposed to a manipulandum that consisted of a small plexiglas disk recessed in one wall of the experimental chamber. Whenever this "key" was illuminated with a white light, a tone was also present, and pecks at the key were reinforced (LT+). After responding to the light–tone compound was established, the birds were tested for what they had learned about the tone by presenting tones that varied in frequency systematically from the training frequency. In this *generalization test,* the pigeons responded virtually as much to the new tones as to the training tone. A second group of pigeons was trained in the same manner except that, following LT+ training, they were taught to discriminate between the presence versus the absence of the tone (LT+, L−). A subsequent generalization test with these birds produced a sharp generalization gradient: the highest level of responding was to the training tone, and response levels fell off in correspondence with the difference in frequencies between the test tone and the training tone.

The problem for the then-available theories posed by these results was to explain why the two generalization gradients differed.[1] Application of the linear operator rule such as that used in the Hull–Spence model to compute changes in connection weights (V) for the two groups results in the functions shown in Figure 7.4. The left-hand graph shows that LT+ training produced a similar level of conditioning to the light and the tone. The right-hand graph shows that discrimination training LT+, L− resulted in a smaller connection weight strength for L than for T. What is important is that the linear operator rule computes the same asymptotic level for learning about the tone (T) in the two training conditions. It appeared

---

[1] It was also a problem for these theories to explain why the second group of birds came to stop pecking at the light alone when the LT+, L− training was initiated. Hull (1952), for example, claimed that the equal numbers of reinforced and nonreinforced trials experienced with the light should leave that stimulus with a net associative strength of 0, but this was disingenuous because it was commonly acknowledged that a partial reinforcement schedule such as this was sufficient to establish and maintain high levels of responding.

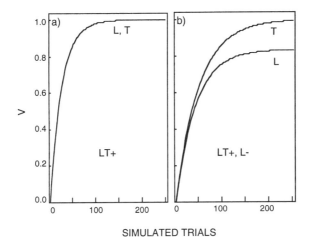

FIGURE 7.4. Simulations of a model applying the linear operator rule of Eq. 1 to training with a simple compound, LT+ (**A**) or discrimination training with LT+, L− (**B**).

necessary either to apply a different learning rule or to make some adjustment to a performance rule whereby the final response was a function of which of the available cues were most uniquely predictive of reinforcement.[2]

The Hull–Spence view, as well as that of statistical association theorists, was a *continuity* view: that all available or sampled cues contribute to discriminative performance and that all associative strengths are changed as a result of a single conditioning trial. The alternative view came to be known as the *noncontinuity* view, one version of which was that organisms learn only about stimuli to which they are "attending" (Lashley, 1942; Krechevsky, 1938) and that attention to a stimulus is a function of its predictiveness of reinforcement. It became increasingly apparent that important to this controversy was the argument that in an instrumental learning situation, where the reinforcement is contingent on the subject's making a particular response, it is difficult to determine whether a cue is effective because of its peculiar validity or the validity of the other cues also present, or because the subject's behavior brings it into contact with different cue-reinforcement contingencies (Wagner, 1969a, b). Obviously, this question is especially germane to experiments purporting to determine the putative stimuli controlling behavior. It was, thus, within the classical conditioning paradigm, where such subject variables are of much less consequence, that the theoretically informative experiments in stimulus selection subsequently were conducted. Recognition of this issue resulted in a large increase in the number of laboratories that primarily used classical conditioning techniques.

---

[2] It has been suggested that phenomena demonstrating apparent stimulus selection effects are not to be found in the literature on invertebrate learning, and thus that the continuity model still may be applicable to the neurophysiologically simpler animal (Couvillon & Bitterman, 1987, 1988; Couvillon et al., 1983).

Eventually, the most influential of these classical conditioning experiments were demonstrations of (1) *overshadowing,* reported first by Pavlov (1927; see also James & Wagner, 1980; Kamin, 1968; Mackintosh, 1975), which showed that if two cues are presented in compound and reinforced (AX+), one or either will be learned about less than if it were presented and reinforced alone (A+, X+); (2) *blocking* (Kamin, 1968; see also Wagner, 1969a), which occurred when prior training with one cue (A+) of a compound that is subsequently reinforced (AX+) impairs learning about the added cue (X); and (3) *relative validity effects* (Wagner et al., 1968; Wagner, 1969a, b), where the strength of conditioning to a CS (X) depends not only on how predictive that particular CS is of the reinforcer but also on whether that reinforcer is signaled by other events. When subjects were trained with a "correlated" problem AX+, BX− or with an "uncorrelated" problem AX±, BX±, there was less learning about the common cue X in the correlated condition.

The two primary theoretical responses to data such as these were to argue for an *attention* mechanism, whereby the effectiveness of the CS varies as a function of its validity (Lovejoy, 1965; Mackintosh, 1965; Sutherland, 1964; Pearce & Hall, 1980; Trabasso & Bower, 1968; Zeaman & House, 1963), or to argue for a "modified continuity theory," whereby the effectiveness of the US varies as a function of the extent to which it is predicted by the aggregate CSs that preceded it (Rescorla & Wagner, 1972; Wagner, 1969a–c, 1971; Wagner & Rescorla, 1972), in what has come to be known as the Rescorla–Wagner model.

Learning Rules

Theories of Variation in US Processing

The model of learning proposed by Rescorla and Wagner (1972; Wagner & Rescorla, 1972) was the first quantitative model that was explicitly directed at learning in instances of classical conditioning. Development of the model was tuned by the separate observations of Kamin (1968, 1969), Rescorla (1968), and Wagner (1969a,b), as well as by early reports of Egger and Miller on "information value and reliability of a stimulus" (1962), of Konorski on inhibition (1948), and of Pavlov on overshadowing (1927). What these experiments pointed to was that the effects of reinforcement and nonreinforcement on changes in associative strength for a particular stimulus appeared variable, depending on the existing associative strength not only of that stimulus, as was assumed in the linear operator models of Hull–Spence theory and of Estes (1950) and Bush and Mosteller (1951), *but also of other stimuli concurrently present.* This concept was expressed in the Rescorla–Wagner model by the learning equation:

$$\Delta V_i = \alpha_i \beta_j (\lambda_j - \Sigma V_i), \tag{3}$$

where ΣV_i = the sum of associative strengths of all stimulus elements present on that trial, α_i and β_j are learning rate parameters specific to the CS_i and US_j, respectively, and λ_j is the asymptotic strength that the US will support.

A network description of the Rescorla–Wagner model is shown in Figure 7.3b: when activation of an *adaptive unit* by a US (via λ) is different from the level of activation generated by concurrently activated, associated CSs (via ΣV), there is an adjustment in the strength of all CS-adaptive unit associations proportional to that difference, $(\lambda - \Sigma V)$. The model presents a formulation for associative strength, rather than response probability, more in line with the theories of Hull (1943) and Spence (1956) than of Estes (1950) or Bush and Mosteller (1951). However, unlike the Hull–Spence models, the Rescorla–Wagner model is a single-continuum theory; "excitation" is represented by values of $V > 0$, and "inhibition" by values of $V < 0$, but a single associative strength, V, for any CS is computed according to the trial-by-trial difference $(\lambda - \Sigma V_i)$. This model thus offers another view of the nature of the association: rather than separate excitatory and inhibitory processes, as was assumed by Hull–Spence theorists, and rather than replacement of one connection strength by another, as was assumed by statistical association theorists, this model assumes that single connections grow stronger or weaker in strength. Note that, as a result of training with reinforcement or nonreinforcement, the associative tendency accrued to a particular CS is assumed by the Rescorla–Wagner model to be independent of the history that established that particular connection weight, for all purposes of predicting the future efficacy of that CS, an assumption which has been described as an assertion of the *independence of path* of associative connection weights (Wagner & Rescorla, 1972).

As an illustration of how the model computes, consider overshadowing (Kamin, 1968; Pavlov, 1927). Application of the Rescorla–Wagner learning rule to the overshadowing experiment results in the prediction that learning about the individual CSs, A and B, is attenuated when the compound is reinforced, relative to the separate reinforcement of A+ and B+. Figure 7.5 shows the results of computer simulations of the Rescorla–Wagner model, assuming that $\lambda = 1.0$. As can be seen in Figure 7.5a and 7.5b, when each component A and B is assumed to have the same α value, after A+ and B+ training, V_A and V_B each approach λ, whereas after AB+ training, both V_A and V_B approach $\lambda/2$. Figures 7.5c and 7.5d show how the model computes when the presumed saliences of the components are different; here it was assumed that $\alpha_A = 0.10$, and $\alpha_B = 0.05$. In Figure 7.5c, when the cues have different saliences, the effect on acquisition with A+ and B+ is to make the cue with the lower salience approach asymptote more slowly, but the final V value for both still is λ. Figure 7.5d shows that overshadowing again is apparent after the compound training of two cues with different saliences, and that the cue with the lower salience reaches a lower asymptotic value than the cue with the higher salience.

Figure 7.5 illustrates one impact of "variation in US processing" on the fate of a single connection weight: that connection weight is less when the relevant CS is reinforced in compound than when it is reinforced alone; that is, the effectiveness of reinforcement is context dependent. The Rescorla–Wagner model assumes the same context dependency for nonreinforcement. The rule is that if $(\lambda - \Sigma V_i) > 0$, then weights increment, and if $(\lambda - \Sigma V_i) < 0$, then weights decrement. Thus, the

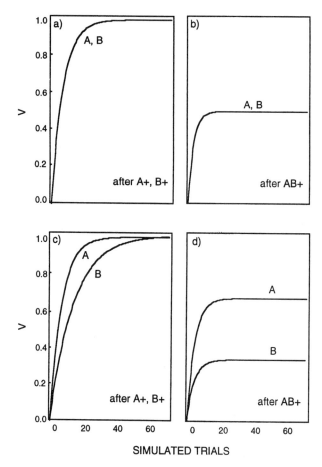

FIGURE 7.5. Simulations of the Rescorla–Wagner model for overshadowing. (**A**) Computed Vs across simulated trials for A and B after training with each CS alone, i.e., $A+, B+$, where the α values for A and B were equal (0.10). (**B**) Computed Vs for compound training, i.e., AB+ with the same α values. (**C**) Computed Vs for $A+, B+$, where $\alpha_A = 0.10$ and $\alpha_B = 0.05$. (**D**) Computed Vs for AB+, where again, $\alpha_A = 0.10$ and $\alpha_B = 0.05$.

model captured one of the most important characterizations about inhibition that has been made since Pavlov, which is that by all known tests, inhibition to a CS accrues to the extent that the CS is nonreinforced in the context of otherwise excitatory cues (Fowler et al., 1985; Wagner & Rescorla, 1972).

The "Overlap Problem" Solved

The notion that inhibition accrues if $(\lambda - \Sigma V_i) < 0$ allows the Rescorla–Wagner model to solve the overlap problem AX+, BX− that was referred to earlier (see Figure 7.2), without assumptions about configural or pattern cues. Recall that, for the linear operator model, the problem cannot be solved without errors because, al-

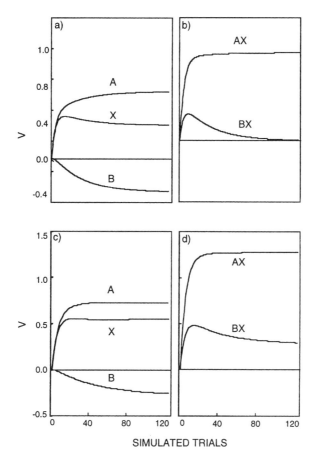

FIGURE 7.6. Simulations of two versions of the Rescorla–Wagner model for an AX+, BX– discrimination problem. (**A**) and (**B**) Simulations with the unmodified model (Eq. 3). (**C**) and (**D**) Simulations with the modified "detection" Rescorla–Wagner model (Eq. 13). The first of each pair shows the associative strengths (V) for each element A, B, and X; the second shows the same for the compounds, AX and BX, across training trials.

though $V_B = 0$, the associative strength of X elements, which are greater than 0 due to their history of partial reinforcement, adds to the response tendency of B on BX trials and produces errorful performance. The Rescorla–Wagner model solution to this problem is shown in Figure 7.6. Figure 7.6a shows the connection weights for the individual cues A, X, and B and 7.6b shows the assumed connection weights for the compounds AX and BX. At asymptote, V_A approaches λ; V_X is excitatory, but less so than V_A, and V_B is as inhibitory as V_X is excitatory. Assuming that $V_{BX} = V_B + V_X$, the problem of the excitatory strength of the common cues, X, thus is dealt with by the fact that the elements of B become sufficiently inhibitory that they result in a prediction of zero responding to the BX compound.

The Rescorla–Wagner model has enjoyed considerable success regarding the known phenomena of classical conditioning, including stimulus selection effects.

It also has appealed to investigators outside the area of animal learning. As noted by Sutton and Barto (1981), the learning rule specified by Rescorla and Wagner was essentially the same as the least-means-squares (LMS) algorithm of adaptive network theory (Widrow & Hoff, 1960), also known as the "delta rule." Parker (1985) showed how backpropagation LMS could be implemented using elementary Rescorla–Wagner/LMS components, and Rumelhart and McClelland (1986) demonstrated that the backpropagation rule for training multilayer networks is a generalization of the Rescorla–Wagner LMS rule (Gluck et al., 1990; see also Schmajuk et al., 1998). Modifications and elaborations of the Rescorla–Wagner rule are central to later trial-level (Blough, 1975; Chiang, 1993; Daly & Daly, 1982; Frey & Sears, 1978; Mackintosh, 1975; Pearce, 1987, 1994; Pearce & Hall, 1980; Wagner, 1978) and real-time (Desmond, 1990; Gluck et al., 1990; Mauk & Donegan, 1997; Moore & Stickney, 1980, 1982; Schmajuk, 1997; Schmajuk & DiCarlo, 1992; Schmajuk & Moore, 1985, 1988; Wagner & Brandon, 2001) models in animal learning, as well as to theorists in human cognition (Allan, 1993; Gluck & Bower, 1988; Rudy, 1974; Shanks, 1993; Van Hamme & Wasserman, 1994).

Theories of Variations in CS Processing

The characteristic assumption of attention theories of stimulus selection effects was that the degree of consistency with which a CS was associated with reinforcement and nonreinforcement determined the effectiveness of that CS in controlling behavior, that is, its capacity to elicit a previously learned response or to acquire a new response. The two-stage attention theory proposed by Sutherland (1964) and Mackintosh (1965), based largely on discrimination data in instrumental conditioning, held that the subject first observes or attends to a set of stimuli on each trial, and then makes an overt response that is determined only by those stimuli attended to.

Mackintosh (1975) offered a model of Pavlovian conditioning, which represented both assumptions of two-stage attention theory. The equations specified that α_i, a stimulus-specific learning rate parameter, may vary not only with the intensity or salience of a stimulus, but also as a result of experience with that stimulus's correlation with reinforcement. Specifically, the α_i of a cue was proposed to increase when the cue predicts an otherwise unexpected reinforcer, and to decrease if it signals no change in reinforcement from the level expected on the basis of other cues present. Salience then affects how much is learned about that stimulus according to a linear operator equation. That is:

$$\Delta\alpha_i > 0 \quad \text{if } |\lambda - V_i| < |\lambda - \Sigma V_j|, \tag{4}$$

where the change in α_i is proportional to the discrepancy between $|\lambda - V_i|$ and $|\lambda - \Sigma V_j|$, V_i is the current associative strength of a CS, ΣV_j is the current associative strength of all other cues available on that trial, and λ is the asymptotic associative strength conditionable by that reinforcer (having a value > 0 on reinforced trials and ≤ 0 on nonreinforced trials). Conversely,

$$\Delta\alpha_i < 0 \quad \text{if } |\lambda - V_i| \geq |\lambda - \Sigma V_j|, \tag{5}$$

where the change in α_i again is proportional to the discrepancy between $|\lambda - V_i|$ and $|\lambda - \Sigma V_j|$. Change in associative strength is then computed by the formula

$$\Delta V_i = \alpha_i \theta (\lambda - V_i), \tag{6}$$

where $0 \leq \alpha_i \leq 1$, and θ is a learning rate parameter.

The Mackintosh model assumed that the starting value for α, where $0 \leq \alpha \leq 1$, was positively correlated with the intensity and salience of the stimulus, and was determined by the extent to which that CS may have some associative strength via some generalization from another CS, as a function of their similarity. Mackintosh also asserted that the assumption that $\Delta V_i > 0$ for reinforcement, and $\Delta V_i \leq 0$ on nonreinforced trials, that is, that both reinforcement and nonreinforcement result in changes in the single variable, V, could as well be exchanged with an assumption that the omission of an expected reinforcer resulted in the increment in some specific inhibitory process, which grew to some asymptote. Then, V would represent a net associative strength and be some function of the difference between excitation and inhibition. However, there were no special assumptions made about inhibition that would make one construe it as under different parametric control than excitation, so that the "independence-of-path" assertion held.

This model was applied to the same kinds of stimulus selection problems as was the Rescorla–Wagner model. Consider again overshadowing (Pavlov, 1927). The model computes that, for training with AB+ versus A+, B+, if the salience of A and B are assumed to be equal (and relatively low) at the outset of training, α_A will increase faster if reinforcement is signaled by A alone rather than by the compound. This is true because in single-cue training, where V_A is the only predictor of the US, $|\lambda - V_A| < |\lambda - \Sigma V_i|$, where V_i are all cues present on a trial other than A, so that α_A increases. In the instance of compound training, this inequality cannot apply, as it must be assumed that $|\lambda - V_A| \geq |\lambda - V_B|$.

Problematic for this model was evidence that a less valid stimulus will not only fail to gain associative strength, as the model predicts, but also may lose the apparent associative strength that it had acquired as a result of an earlier history of reinforcement, under conditions of no change in reinforcement contingency for that stimulus. For example, Wagner et al. (1968), investigating performance in the correlated AX+, BX− and uncorrelated AX±, BX± problems referred to earlier, showed there was less control of response by X in the correlated condition even when the subjects had been trained first in the uncorrelated condition. As noted (Mackintosh, 1975), the model predicts no change in V_x in the switch from uncorrelated to correlated training; to accomodate these data within the model, it would be necessary to assume that α affects both learning and performance.

Theories of Variations in CS and US Processing

The observation that CS preexposure not only impairs CS capacity to acquire excitatory (Lubow & Moore, 1959) but also inhibitory response tendencies (Reiss &

Wagner, 1972) prompted Wagner and Rescorla (1972) to suggest that, under some conditions, it may be necessary to specify that α, the parameter for CS salience, is modified by learning. Wagner (1978) suggested this adjustment:

$$\Delta V_i = \alpha_i (l - \Sigma v) \beta_j (\lambda_j - \Sigma V_i), \tag{7}$$

where the additional terms, $(l - \Sigma v)$, that modify α, represent the asymptotic conditioning that can be supported by the CS (l) and the aggregate prediction of the CS by preceding stimuli (Σv). Many of the notions of this *priming model* were subsumed in a subsequently developed, real-time model, SOP (Donegan & Wagner, 1987; Mazur & Wagner, 1982; Wagner, 1981), that is described in a later section. However, the appeal of allowing for changes in both CS and US effectiveness, as suggested by Wagner (1978), was at the heart of the model proposed by Pearce and Hall (1980).

Pearce and Hall (1980) proposed a theory where $(\lambda - \Sigma V_i)$ again was used for determination of CS associability, but the Mackintosh (1975) rule is reversed, so that a stimulus is likely to be processed to the extent that it is not an accurate predictor of its consequences. Their proposal was that the current *associability* of a CS, α_i, could be computed with the formula:

$$\alpha_i = |\lambda - \Sigma V_i|, \tag{8}$$

where λ was the reinforcement parameter, ΣV_i was the sum of the associative strengths of all stimuli present on the preceding CS occurrence, and $0 \leq \alpha \leq 1$, and $0 \leq \lambda \leq 1$.

Excitatory conditioning accrued when $\lambda \geq \Sigma V$, and depended not only on the associability of the CS, but also on its intensity and the intensity of the US, according to the equation

$$\Delta V_i^+ = S_i \alpha_i \lambda_j, \tag{9}$$

where S was the parameter of CS intensity, and $0 \leq S_i \leq 1$. "Inhibitory" conditioning accrued when $\lambda_j < \Sigma V_i$, according to the delta rule:

$$\Delta V_I^- = S_i \alpha_i (\Sigma V_i - \lambda_j). \tag{10}$$

Note that this formulation avoided the prediction of the Rescorla–Wagner model that there is no extinction of a conditioned inhibitor via nonreinforced presentations of the inhibitory CS, which had not received empirical support (Zimmer-Hart & Rescorla, 1974).

It was assumed that net associative strength was determined by the sum of V_i^+ and V_I^-; however, whereas V^+ represented the connection strength between a CS and a US, V^- represented connection strength between a CS representation and the representation of no US (no-US). The notion of different stimulus representations for excitation and inhibition was borrowed from Konorski (1967) and developed further by Rescorla (1979), but it is relatively rare among current quantitative models. Pearce and Hall argued that the distinction allows for the observation that, in some instances, conditioned inhibitors appear to elicit responses rather than just to antagonize responses. For example, pigeons are observed to withdraw from a keylight that has been established as a conditioned inhibitor, whereas they will approach a keylight that has been established as a conditioned excitor (Hearst & Franklin, 1977; Wasserman et al., 1974). However, it is equally possible to understand the pigeons' withdrawal in terms of a conditioned emotional state that interferes with an approach CR (Amsel, 1958; Denny, 1971), or to assume that an antagonistic CR is the outcome of a negative V, as with Rescorla and Wagner (1972).

For the Pearce–Hall model, overshadowing that is, the diminished conditioning to a cue trained in compound versus alone (Pavlov, 1927), requires a relative decrease in α over training. Thus, like models of Rescorla–Wagner and Mackintosh, there should be no "one-trial overshadowing." Evidence of just such an effect (James & Wagner, 1980) has encouraged theorists since the 1980s to incorporate some mechanism whereby processing of a stimulus might be limited by interference from the processing of another stimulus (Wagner, 1978, 1981), in a manner reminiscent of attentional theories (Sutherland & Mackintosh, 1971) and distributed capacity theories (Revusky, 1971).

Stimulus Representation

Rescorla–Wagner Model with Unique Cues

As already described, application of the Rescorla–Wagner learning rule to a set theoretic view of stimulus representation, as pictured in Figure 7.2, allows one to deduce eventual errorless discrimination between two stimuli even though, before training, there was some apparent generalization between the discriminanda (Gluck, 1992; Rescorla & Wagner, 1972; Wagner & Brandon, 2001). That is, the model solves the "overlap problem."

The simple linear rule, that $V_{BX} = V_B + V_X$, was successfully applied to most of the stimulus selection problems involving compound stimuli to which the Rescorla–Wagner rule initially was addressed, such as overshadowing, blocking, and relative stimulus validity effects. This rule, however, cannot work for nonlinear representation problems. For example, in the negative patterning experiment of Woodbury (1943), dogs learned to respond to each of two stimuli separately, A+, B+, but not to their joint occurrence, AB− (see also Bellingham et al., 1985; Kehoe & Graham, 1988; Pavlov, 1927; Rescorla, 1972, 1973; Whitlow & Wagner, 1972). A similar challenge has been that subjects learn to respond appropriately in a problem that is a variation on the negative patterning problem, referred to as a *biconditional discrimination:* AX+, BX−, AY−, BY+ (Saavedra, 1975). In this problem, no one element predicts reinforcement or nonreinforcement; rather, the conjunction of stimuli alone is predictive.

Rather than abandon an elementistic view of the stimulus, Wagner and Rescorla (1972; see also Rescorla, 1972, 1973) borrowed from Spence (1960) and made the nonlinear problem a linear one by postulating that there is a *unique cue* that is created only by the joint occurrence of two elemental inputs. This is an approach that we refer to as an "added element" approach (Wagner & Brandon, 2001). So, for example, in the negative patterning problem, when the subject is asked to discriminate among A+, X+, AX−, they do so via $\bar{a}\bar{x}$, a unique cue formed by the conjunction of AX. It is assumed that $V_{AX} = V_A + V_X + V_{\bar{a}\bar{x}}$. Application of the Rescorla–Wagner learning rule results in V_A and V_X approaching λ, and the added, configured $V_{\bar{a}\bar{x}}$ approaching −2λ. Summation of the appropriate Vs results in substantial responding to A and X and virtually none to the compound AX. For a biconditional discrimination, there would be four added configured cues, $\bar{a}\bar{x}$, $\bar{b}\bar{x}$, $\bar{a}\bar{y}$, and $\bar{b}\bar{y}$. Objections have been raised to this strategy on the basis that with systems with more than two stimulus inputs, it would be necessary to postulate added configured cues for each possible combination, which could create an unwieldy proliferation of special input units (Barto et al., 1982; Kehoe, 1988; Kehoe & Gormezano, 1980; Rumelhart et al., 1986); however, given the extraordinary information processing capacity of the nervous system of even infrahuman organisms, and the fact that such configured elements are required only to the level of configuration problems that prove solvable, this is perhaps an inconsequential objection.

A Configural Model

Using the Rescorla–Wagner rule for increments and decrements in associative strength, together with a partly pattern, partly elementistic view of stimulus representation, Pearce (1987, 1994; Young & Pearce, 1984) offered a model of discrimination and generalization in classical conditioning that designated a *configural* representation for all CSs; that is, the configuration is considered as the basic unit of conditioning. This mixed model approach, as noted earlier, has considerable precedence in analysis of both infrahuman and human sensory processing and learning (Atkinson & Estes, 1963; Bellingham et al., 1985; Estes & Hopkins, 1961; Gulliksen & Wolfle, 1938; Medin, 1975).

Interestingly, the Pearce model considers the configuration as the basic associative unit, but describes the similarity of that unit to other units in terms of common elements.[3] That is, the similarity, $_AS_B$, between any compound A and other compound B, is asssumed to be given by the product of the proportion of elements of A that are common to the two compounds and the proportion of the elements of B that are common to the two compounds. In this model, the associative strength of a configuration, V_C, is given as the sum of the associative strength directly con-

---

[3] This approach prompted a conceptualization of the Pearce model by Wagner and Brandon (2001; see also Baçhekapili, 1997) as an elementistic model where some elements, in instances of patterning, are "inhibited" by other elements. This view computes virtually the same predictions as the configural rules of the original Pearce model.

ditioned to that configuration E_C and the associative strength that generalizes to the configuration from all other configurations e_C, according to the formula

$$e_C = \Sigma_j {}_jS_C E_j, \qquad (11)$$

where E_j is the direct associative strength of configurations j, and ${}_jS_C$ is the degree of similarity between the configurations C and j. The change in direct associative strength of a configuration C, E_C, in any trial then is given by the equation

$$\Delta E_C = \beta_j(\lambda_j - [E_C + e_C]), \qquad (12)$$

where β_j and λ_j are reinforcement parameters.

Pearce has reported a variety of experimental results that are more consistent with this model (Pearce, 1987, 1994; Pearce & Redhead, 1993; Young & Pearce, 1984) than the Rescorla–Wagner conception. However, the picture is complicated by the fact that the findings appear to depend on the stimuli used (Myers et al., 2001). For example, Pearce predicts that a feature negative discrimination AX−, X+ should be solved more easily than should the same problem with an additional common cue, AXY−, XY+, whereas the Rescorla–Wagner model makes the opposite prediction. Experiments using combinations of visual stimuli (Pearce & Redhead, 1993) support Pearce; experiments using combinations of stimuli from different modalities support the Rescorla–Wagner model (Bahçekapili, 1997).

Rescorla–Wagner Model with Replaceable Elements

Wagner and Brandon (2001) recently investigated a model that uses the Rescorla–Wagner learning rule and a set theoretic view of stimulus representation (see Figure 7.2). The model can be conceptualized as a way of representing Hull's principle of afferent neural interaction in computational form (Hull, 1943, 1945, 1952; see also Bellingham et al., 1985; Kehoe, 1988). As depicted in Figure 7.7a, the *replaceable elements* model specifies an elementistic view of stimulus representation in which the representation consists of some elements that are "context independent" and other elements that are "context dependent." When two CSs, A and X, are presented together in compound, it is assumed that configural elements α_X and x_a are formed, which are unique to the conjunction of the two, similar to the added elements notion, but that replace specified context-dependent elements α_0 and x_0, respectively; that is, stimulus compounding results in replacement of the context-dependent elements that are intrinsic to the single stimulus representation by context-dependent elements which are unique to the conjunction of the stimuli. Nonlinear problems, such as negative patterning and biconditional discriminations, are solved on the basis of the context-dependent elements that are unique to the occasions of reinforcement and nonreinforcement. Brandon et al. (2000b) present some generalization data that favor this conception.

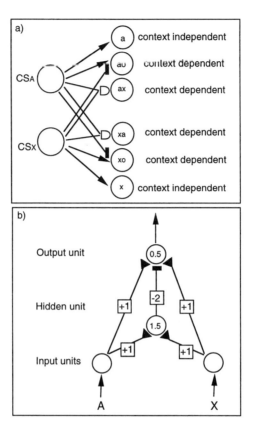

FIGURE 7.7. (**A**) Stimulus representation in the replaceable elements model of Wagner and Brandon (2001). Presentation of a compound AX activates context-independent and context-dependent elements; some of the context elements are replaced when A and X are presented together versus when they are presented alone. Connections are excitatory (*arrows*), inhibitory (*rectangles*), or conjunctive (*and gates*). (**B**) The hidden unit network model of Rumelhart et al., (1986). Two input units activate an output unit, directly and via activation of a hidden unit. All connections are excitatory except that of the hidden unit to the output unit, which is inhibitory. The weights of the connections, shown in the *boxes*, are assumed to be those that allow the network to solve a negative patterning, or XOR, problem: activation of either input unit alone was followed by reinforcement, and activation of both together was not. The *numbers inside the circles* represent the threshold of the units.

A Layered Network

An alternative and elegant solution to the problem of stimulus patterning comes from treatments of stimulus representation in human cognition and artificial intelligence (Anderson, 1985; Gardner, 1985; Rumelhart et al., 1986). Stimuli are represented by a *layered network* in a parallel distributed processing system (Anderson, 1973; Barto et al., 1982; Grossberg, 1970, 1976; Rumelhart & McClelland,

1986). This conception of stimulus representation differs from those that have been mentioned in supposing that there are hidden units, interspersed between input and output units, with the characteristics that (1) each hidden unit is connected to all input units, that is, there is none that is "hard wired" to represent only one stimulus or one stimulus conjunction; and (2) the input-hidden unit connections are modifiable, that is, are subject to strengthening and weakening according to specified learning rules.

Figure 7.7b is a hidden unit model that solves a negative patterning (A+, X+, AX−) problem (Rumelhart et al., 1986). The numbers on the connections represent the strengths of the connections among the units. The numbers inside the circles represent the thresholds of the units. The value of .5 for the output unit means that it will turn on only when it gets a net positive input greater than .5, which will occur when either A or X is on. The value of +1.5 for the threshold of the hidden unit means that it will be turned on only when both input units are on, but the weight of −2.0 from the hidden unit to the output unit ensures that the output unit will not come on when both input units are on (on an instance of AX). In general, for this kind of network to solve negative patterning problems, it is necessary to assume a differential threshold for activation of hidden units versus input units (Rumelhart et al., 1986).

A New Version of the "Overlap Problem"

Consider again the set theoretic view of the stimulus, as depicted in Figure 7.2. It is in the spirit of this model to expect that the greater the number of common elements, that is, the greater the tendency to generalize between A and B, the more difficult should be the discrimination. In an extreme case, one might suppose that if the number of common elements substantially outweighs the number of unique elements, the problem would be virtually insoluble. However, the delta rule applied here computes asymptotic values of V_A and V_B such that V_B is sufficiently negative that, despite the presence of a number of unique A and B elements relative to a large number of X elements, an errorless discrimination eventually will be achieved. We might consider this as a new version of the "overlap problem."

One way to avoid this intuitively unattractive prediction (although there are no data that we know of to disprove it), is to let the parameter α of the Rescorla–Wagner model represent a variable probability that a CS component will be "detected" on a trial and to allow a cue to influence ΣV and be changed in associative strength only on occasions on which it is detected. In this case, it can be shown (Wagner, 1972) that ΔEV_i, the *expected* change in V_i, will be

$$\Delta EV_i = \alpha_i \beta_j [\lambda_j - (V_i + \Sigma \alpha_k V_k)], \tag{13}$$

where $k \neq i$, and the parameters are the same as those already specified for Eq. 3, except that α_i expresses the independent probabilities of detection of the several cues of a compound. A sample derivation from this "detection model" is shown in Figure 7.6c and 7.6d for the AX+, BX− problem. Essentially, the model assumes in this case that AX+, A+, X+, BX−, B−, and X− events occur with various probabilities, and, as can be seen, predicts that the problem will not be solved absolutely;

that is, at asymptote, there is a persisting nonzero tendency to BX. There may be other instances in which this or similar modifications of the Rescorla–Wagner rule can be useful, as is discussed later (see also Wagner & Brandon, 2001).[4]

Real-Time Models

General Model Types

The 1980s saw the advent of *real-time* models of Pavlovian conditioning. These models were motivated primarily by the need to account for temporal phenomena that were not addressed by trial-level models, such as (1) the *temporal dependence of extratrial stimulus effects,* where the processing of a CS or a US is changed by prior presentation of the same or an associated stimulus (Davis, 1970; Kimble & Ost, 1961; Terry & Wagner, 1975; Whitlow, 1975); (2) the observation that in many classical conditioning preparations, the CR comes to occur with maximum amplitude or likelihood at that time within the CS presentation when the US regularly is presented, the most prominent of the so-called *CR timing* effects (Millenson et al., 1977); and (3) the dependence of CR acquisition on the temporal relationship of the paired CS and US, that is, on what is called the *interstimulus interval (ISI) function* (Frey & Ross, 1968; Razran, 1957; Schneiderman, 1966; Schneiderman & Gormezano, 1964; Smith et al., 1969). The subsequent further development of some of these real-time models was motivated by the need to account for *occasion setting* in Pavlovian conditioning, where the CRs in a sequential feature positive (A−X+, X−) or feature negative (A−X−, X+) problem appear to be under the control of the temporally distant feature, A, even though they are elicited only by the temporally proximal target, X (Holland, 1983, 1985; Jenkins, 1985; Rescorla, 1985; Ross & Holland, 1981).

In trial-level models, the unit of calculation for a change in associative strength or a measure of performance is the trial. In contrast, real-time models calculate "moment-by-moment." Rather than specifying the calculation of associative strength at the beginning or end of a trial, real-time models provide a running account of associative strength within the trial, with an incumbent requirement of specifying which intratrial events are critical for associative change. Those real-time models that were proposed by learning theorists, some of which we describe here, were challenged and enhanced by neural network models that came from adaptive system theory (Grossberg, 1975; Klopf, 1988; Sutton & Barto, 1981) and by connectionist models that had focused largely on phenomena of human cognitive processing (see Rumelhart & McClelland, 1986).

There are several types of real-time models, reflecting differences in their individual origins. The first published real-time model was that of Moore and Stick-

---

[4] It was assumed here that $\alpha = 0.50$. We should note here that this model makes some interesting predictions that await evaluation, such as that blocking (Kamin, 1968) and the relatively validity effect (Wagner et al., 1968) are transient.

ney (1980), who offered a real-time, computational version of Mackintosh's attentional model of classical conditioning (1975). Wagner (1981) presented a model intended to deduce some of the features of the Rescorla–Wagner (1972) model within a real-time framework. Schmajuk and Moore (1985, 1988) offered a real-time rendition of the model of Pearce and Hall (1980), extended by parameters that allowed for a detailed description of many of the temporal and topographical characteristics of the rabbit nictitating membrane (NM) conditioned response. Later models likewise have tried to offer a precise description of the temporal characteristics of this same CR (Desmond & Moore, 1988; Moore, 1991; Moore et al., 1998).

In some real-time models, the rules for learning and performance are like those for trial-level models, that is, both are a function of activity in CS and adaptive unit representations (cf. Figure 7.3b), but computations are made on a moment-by-moment basis (Desmond, 1990; Gluck et al., 1990; Schmajuk, 1997; Schmajuk & DiCarlo, 1992; Wagner, 1981; Wagner & Brandon, 2001). These *momentary* models can be distinguished from those that follow the prototype of Sutton and Barto (1981) that computes learning and performance only when there is a change in adaptive unit activity, that is, according to *time-derivative* learning rules (Klopf, 1988; Moore et al., 1986; Sutton & Barto, 1990).

In the present section, we describe some real-time models in terms of the issues already deemed important in formal models of animal learning, specifically, conceptualizations of the stimulus representation (and corresponding stimulus trace), and statements of the rules for changes in connection strength. We begin with a brief description of the data that were the primary impetus for the development of real-time models. Then we show how it is possible, using an appropriate stimulus representation trace, to derive several of the temporal effects already listed: the temporal sensitivity of CS and US priming effects, CR timing, and occasion setting. In large part, these derivations are independent of the particular learning rules that the model employs and are almost entirely dependent on the specification of the stimulus representation. Then, we show how an explanation of the nonmonotonic ISI function appears to follow from a more limited range of stimulus representation and learning rules.

Empirical Real-Time Issues

Priming Effects

Presentation of an instance of a CS or US prior to a CS–US pairing, or prior to a measurement of responding to these respective stimuli, has been reported to have a decremental effect on conditioning and performance (Best & Gemberling, 1977; Davis, 1970; Donegan, 1981; Kalat & Rozin, 1973; Pfautz & Wagner, 1976; Terry, 1976; Terry & Wagner, 1975; Wagner, 1976, 1979; Wagner et al., 1973; Whitlow, 1975). These CS and US *priming* effects are dependent upon the temporal interval between the priming and the target (test) stimulus. The data shown in Figure 7.8a provide one example of priming, taken from a study of the conditioned diminution of an unconditioned eyelid closure response in rabbits by

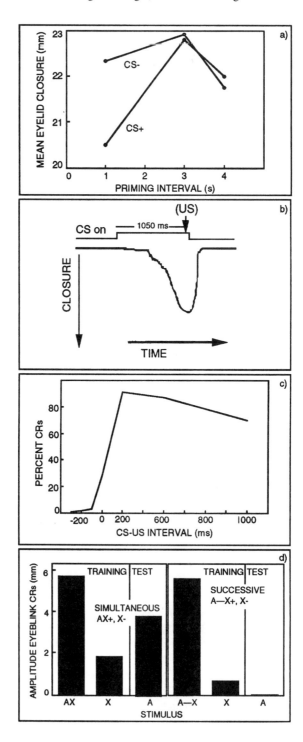

Donegan (1981). Animals received CS–US pairings sufficient for that CS to elicit an eyelid closure CR, and then were assessed for decrements in UR amplitudes when the paraorbital shock US was preceded by the CS that previously had been paired with that same US (CS+), or by a CS that had not been paired with that US (CS−). The figure illustrates that the decremental effect of the preceding CS on the eyeblink UR was associative; that is, it occurred with CS+ and not with CS−, and that it was time dependent, evident when the priming stimulus occurred 1 s before US presentation and not when it occurred at 3 or 4 s before US presentation.

CR Timing

One of the characteristics of a Pavlovian CR is that, in many conditioning situations, the well-trained response occurs with some latency after CS onset, at a time that approximates the CS–US training interval. This observation has been made for the rabbit's nictitating membrane (Coleman & Gormezano, 1971; Smith, 1968), eyelid closure (Mauk & Ruiz, 1991) and heart rate CR (Wilson, 1969), and for human eyelid (Martin & Levey, 1965) and galvanic skin (Kimmel, 1965) CRs. Figure 7.8b is a trace of a well-trained conditioned eyelid closure response generated by a rabbit in our laboratory: in training, the CS was a 1050-ms tone, and the US was a 50-ms paraorbital shock that was applied at a time (indicated by the arrow) so that it terminated with the CS. The trace in the figure is taken from a test trial in which the CS was presented in the absence of the US. As is evident, the CR was not initiated with the onset of the CS, but was apparent some 500 ms later, and reached its peak when the US usually was delivered.

The question that has been posed about CR timing is whether it can be derived on the basis of associative strength; that is, can variation in CR timing be understood in terms of the notion that as the CS–US interval is longer in training, the CS–US association is weaker, and as the association is weaker, the CR latency is longer? Alternatively, is a "shaping" analysis demanded (Boneau, 1958; Logan, 1956), where the temporal energetic aspects of behavior, in this case, the CR latency, are part of what is learned? The strength hypothesis is encouraged by observations that (1) onset latency has been shown to decrease, that is, move toward the onset of the CS, over extended training, and this decrease in latency is concomitant with an increase in peak amplitude (Ebel & Prokasy, 1963), and

---

FIGURE 7.8. (**A**) Mean amplitude eyelid closure in the rabbit when a paraorbital shock US was preceded by a *CS+* or a *CS−* at three different temporal priming intervals. (Redrawn from Donegan 1981.) (**B**)The polygraph record generated by the closure of a rabbit's eyelid upon presentation of a CS. The US would have occurred at the point indicated by the *arrow;* this record was taken from a nonreinforced test trial. (**C**) Mean percentage rabbit nictitating membrane closure CRs after forward delay conditioning with CS-US intervals of 0, 80, 120, 200, 400, 800, 1800, and 3000 ms. (Redrawn from Gormezano et al. 1983.) (**D**) Mean amplitude of eyelid closure CRs after training with simultaneous AX+, X− (*left graph*) and successive A—X+, X− (*right graph*) training, and in tests with A alone following such training, in each condition.

(2) shifts in ISI from a shorter to a longer duration result in increases in CR onset latency that correspond to a concomitant decrease in peak amplitude (Coleman & Gormezano, 1971; Ebel & Prokasy, 1963; Pavlov, 1927; Prokasy & Papsdorf, 1965). The shaping hypothesis is supported by reports that (1) a shift in ISI, from longer to shorter or shorter to longer, after acquisition of a CR at the first training interval, produces a CR that peaks at the point of the shifted ISI, where the shift is not gradual but appears to be a loss of one latency and the relatively discrete acquisition of another (Coleman & Gormezano, 1971; Pavlov, 1927), and even more importantly, that (2) at least one Pavlovian CR, that of the rabbit's nictitating membrane, exhibits two peaks when a single CS is trained with a mixture of a short and a long CS–US interval, the two peaks corresponding to the two CS–US training intervals (Hoehler & Leonard, 1976; Millenson et al., 1977).

The ISI Function

Another issue important to the development of real-time models has been what appears to be a nonmonotonic relationship between acquisition of a CR and the CS–US interval. The empirical challenge is exemplified in the function drawn in Figure 7.8c (from Gormezano et al., 1983), taken from studies *of delay conditioning* in rabbit eyeblink conditioning, where the CS duration extends to the moment of application of the US. What is apparent is that very short and very long CS–US intervals show poor acquisition in terms of CR frequency, relative to that seen with intermediate intervals. Essentially the same function has been observed for *trace conditioning,* where the CS is of a constant duration and may terminate before the presentation of the US (Smith et al., 1969).

The theoretical import of this inverted U-shaped function is difficult to assess, for several reasons. First, the particular nonmonotonic ISI function described most often and, certainly, that to which real-time models of conditioning have most frequently been addressed (Gluck et al., 1990; Mauk & Donegan, 1997; Moore et al., 1986), is that generated for the rabbit eyelid closure or NM CR under conditions of training when there is a relatively high density of training trials in multiple training sessions. However, the function does not always obtain even with this response and under these conditions (Schneiderman, 1966); nor does it obtain under different conditions, such as one trial a day or where a discriminative response is required. In these cases, the function shifts to the right, so that longer intervals are optimal (Levinthal, 1973; Levinthal & Papsdorf, 1970; Levinthal et al., 1985). The function also shifts to the right when different responses are measured, such as the Galvanic Skin Response (GSR) in humans (Jones, 1962). The ISI function for the Conditioned Emotional Response (CER) conditioning appears to be monotonic, with the best conditioning at shorter intervals (Kamin, 1965; cf. see Yeo, 1974).

Second, it may be that associative weight changes are greatest with simultaneity at a cellular level (Kelso et al., 1986), but not when assessments are made at a behavioral level, perhaps because there are differential rates of processing of the CS and the US or differential latencies with which the CR and UR can be generated (Hull, 1943, 1952; Jones, 1962). Third, it also may be that the greater stimulus generalization decrement that must obtain for simultaneous CS–US training and subsequent CS-alone assessment, relative to a forward CS–US interval training and subsequent CS-alone assessment, degrades the apparent efficacy of the simultaneous arrangement only for performance and not for learning (Rescorla, 1980). Finally, and importantly, with some exceptions (Bitterman, 1964; Patterson, 1970; Smith et al., 1969), experiments that attest to this nonmonotonic function are almost universally experiments in which subjects are trained and tested with different durations of CSs, so that it is unclear what is caused by differences in learning versus differences in performance. It might be, for example, that a short CS–US interval looks ineffective because short CSs allow little time to produce a CR. Virtually all reports of real-time model simulations of ISI intervals, so far as we are able to assess, are confounded assessments.

Occasion Setting

Originally described by Skinner and investigated in operant conditioning situations (1938), occasion setting has come under more recent scrutiny in classical conditioning preparations, notably those of appetitive conditioning in rats (Holland & Ross, 1981; Holland, 1983, 1985) and pigeons (Jenkins, 1985; Rescorla, 1985). There has been further investigation of this phenomenon with other behaviors, including the rabbit eyeblink and NM response (Brandon & Wagner, 1991; Wagner & Brandon, 1989; Kehoe et al., 1987; Macrae & Kehoe, 1995; Wiedemann & Kehoe, 1997). What is interesting about occasion setting is that the basic design incorporates a temporal change in an otherwise familiar problem, that is, a discrimination task of the sort AX+, BX−. Extensive investigations have shown that training with this problem, where the compounds are simultaneous, produces a significant excitatory response tendency for A and a significant inhibitory tendency for B, whereas there is relatively little excitatory tendency for X (Wagner, 1976; Wagner et al., 1968). However, when the compound is sequential as in occasion setting, that is, A—X+, B—X−, a different pattern of behavior occurs. Now, the CR is produced only when X occurs and neither A nor B appears to have significant direct response-eliciting or response-inhibiting properties. However, behavioral control by A and B is evident from the fact that the CR to X occurs only when X follows A and not when it follows B; that is, A and B are said to *set the occasion* for response to X.

Figure 7.8d depicts data collected in our laboratory, using the rabbit eyeblink CR as a measure of occasion setting. Shown on the left are the results of training with a simultaneous, AX+, X− arrangement of cues, and on the right, the results of training with a successive arrangement. In the former, both A and X were 1050 ms in duration; in the latter, A was 10,050 ms in duration and X was 1050 ms in duration, delivered 9000 ms after the onset of A. The US was a paraorbital shock applied in the last 50 ms of compound stimulus presentation. The discriminative performance after training with both arrangements can be seen: there is a greater frequency of eyelid closure to X within A than to X alone. When the rabbits were well trained on the discrimination, they were tested with cue A alone. The data presented on the right side of each graph show that although there was substantial responding to A after simultaneous training, there was no responding to A after successive training.

Stimulus Representation

Molar Trace Models

For many real-time models, the stimulus representation is a molar trace, similar to that proposed by Pavlov (1927) and borrowed by Hull (1943, 1952). Schmajuk and Moore (1988) specified a short-term stimulus trace that they called τ, changes in which were defined by the equation:

$$\Delta \tau_i = k1[CS_i \max - \tau_i], \tag{14}$$

where CS_imax is the maximum intensity of the trace recruited by CS_i (= 0 when the CS terminates), and k1 is a constant such that $0 < k1 \leq 1$ (see Grossberg, 1975, for an earlier use of the same equation). One version of this stimulus trace is shown in Figure 7.9a. Essentially the same function was used in later models developed by Schmajuk and colleagues (Schmajuk & DiCarlo, 1992; Schmajuk, 1997; Schmajuk et al., 1996); in some versions, the passive decay of the trace was delayed by an additional term (Schmajuk & Buhusi, 1997; Schmajuk et al., 1998), but this does not change the overall shape of the trace.

The model of stimulus configuration in classical conditioning proposed by Schmajuk and DiCarlo (1992) is a hidden unit model that takes advantage of the ability of such models, as already described (Rumelhart et al., 1986), to solve problems that otherwise can be solved only via a nonlinear representation of the eliciting stimuli (see also Kehoe, 1988). The stimulus representation assumed by this model is similar to that shown in Figure 7.7b. For Schmajuk and DiCarlo, as for Rumelhart et al. (1986), there are direct input–output connections, as well as indirect input–hidden-unit–output connections. It is assumed that there is some threshold function for hidden-unit activation that makes the hidden units harder to activate than input units to some effective level. The molar trace activity presented to each unit follows from the momentary stimulus and the connection weights.

The representation of the molar stimulus trace in the SOP model (Wagner, 1981) is not arbitrary, but is determined by the underlying stochastic assumptions of the model. A network characterization of SOP (Wagner, 1992) is shown on the right side of Figure 7.9b. The model assumes that each stimulus representational unit consists of two nodes, referred to as A1 and A2, each with a large but finite set of elements that may be active or not. Presentation of a stimulus results in the activation of some proportion of eligible inactive elements in the A1 node (according to a parameter *p1*), in each moment that the initiating stimulus is present. A1 elemental activity is followed by activation of inactive elements in the secondary, A2 node (according to a parameter pd1), which recurrently inhibit corre-

7. Computational Theories of Classical Conditioning   261

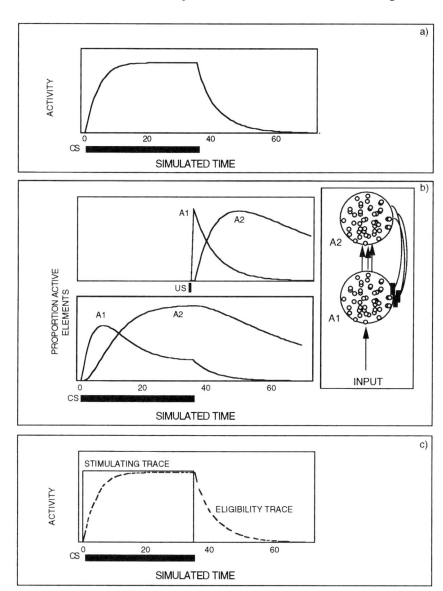

FIGURE 7.9. Stimulus traces generated by three real-time models of classical conditioning. (**A**) CS trace for the model of Schmajuk and Moore (1988; see also Schmajuk & DiCarlo, 1991). (**B**) US and CS traces used in the SOP model of Wagner (1981; see also Mazur & Wagner, 1982), along with the network rendition of stimulus representation that is used to generate the stimulus traces. (**C**) Two traces used in the model of Sutton and Barto (1981), the stimulating trace and the eligibility trace. For all three panels, the duration of CS presentation is indicated by the *black bars*.

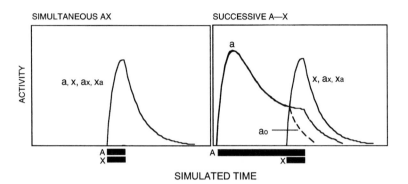

FIGURE 7.10. Elemental CS activity as predicted by the real-time replaceable elements model (Wagner & Brandon, 2001). Simulations are of CS representations generated as a function of simultaneous AX (*left graph*) and successive A–X (*right graph*) stimulus arrangements. See text for details.

sponding elements in A1, during which the A1 elements are refractory to activation. Elements of A2 eventually decay from activation, releasing the inhibited A1 elements (according to a parameter pd2). Application of these parameters to stimulus presentations that are short (such as USs typically are) or relatively longer (such as companion CSs) leads to stimulus traces that have the form shown in Figure 7.9b. Shown for both US and CS traces are the proportion of A1 and A2 elements that are active across time. The assumptions of this model about the relationship between A1 and A2 result in a primary (A1) "CS" trace that differs from that used in the Schmajuk models, and is similar to the form that was proposed by Hull and was found in psychophysical measurements by Marks (1974), as depicted in Figure 7.1b: a period of recruitment is followed by some period of adaptation, before a subsequent decay when the stimulus is removed.

Brandon and Wagner (1998) translated their previously described, replaced elements conception of stimulus compounding into a real-time formulation within the framework of the SOP trace. Consider a compound made up of two stimuli, A and X, in either a simultaneous arrangement, AX, or a successive arrangement, A—X, as indicated in Figure 7.10. The context-independent elements of the representation of A and X, designated as "a" and "x," respectively, are presumed to follow a characteristics SOP course, with the activation parameters $p_{1,a}$ and $p_{1,x}$, having nonzero values in any moment in which the stimulus is present. In contrast, the context-dependent elements of A-alone, "a_o," or A in the context of X, "a_x," and the context-dependent elements of X-alone, "x_o," or X in the context of A, "x_a," are assumed to have activation parameters that follow the rules

$$\begin{aligned} p_{1,ao} &= k \quad \text{when } (p_{1,a} > 0 \text{ and } p_{A1,x} < T), \quad \text{else } 0, \\ p_{1,ax} &= k \quad \text{when } (p_{1,a} > 0 \text{ and } p_{A1,x} \geq T), \quad \text{else } 0, \\ p_{1,xo} &= k \quad \text{when } (p_{1,x} > 0 \text{ and } p_{A1,a} < T), \quad \text{else } 0, \\ p_{1,xa} &= k \quad \text{when } (p_{1,x} > 0 \text{ and } p_{A1,a} \geq T), \quad \text{else } 0, \end{aligned} \qquad (15)$$

where $k \leq 1$, T is a threshold value, and p_{A1} refers to the proportion of elements of that stimulus that are in the A1 state. Figure 7.10 depicts the consequent course of the several representations in the two example temporal arrangements. Note that the extent to which the elemental representations are concurrent with each other is different: the context-dependent cues a_x and x_a are more concurrent with the context-independent cues, a and x, in the simultaneous case than in the successive case, where the trace is relatively degraded by the time the X stimulus occurs and the context-dependent elements, a_x and x_a, might be generated.

Most molar-trace models assume an isomorphism between the theoretical representational activity that is generated by the eliciting stimulus and the availability of the connection weights for that representational entity being modified by learning. An early version of the *adaptive unit model* proposed by Sutton and Barto (1981) differs in that it proposed a distinction between the *stimulating* trace, that is, the function that describes the change in stimulus output across time, and the *eligibility* trace, which determines whether the stimulus entity is available for modification of connection weights (see also Klopf, 1972). Figure 7.9c shows the relationship that is specified between the two: the stimulating trace is assumed to be rectilinear across time, whereas the eligibility trace has a form that shows some recruitment and gradual decay with the offset of the stimulating trace. The eligibility trace differs from the stimulating trace not only in its time course, but because the two may be affected differently by a host of variables, such as attention, stimulus salience, generalization, contrast, and previous learning (Barto & Sutton, 1982; Grossberg & Levine, 1987; Klopf, 1972, 1988, 1989).

Componential Trace Models

Several real-time models of classical conditioning have adopted a componential stimulus representation to allow for an associative strength analysis of CR timing. One such theoretical device was used by Pavlov (1927); it assumes that stimulus onset initiates a *serial* representation of the CS, where each successive element acquires its own associative strength as a function of its temporal relation to the US. If it is assumed that elements close to US delivery are more excitatory than those elements that more distantly precede the US, one can, with specifiable assumptions (Vogel, 2001) allow for *inhibition of delay,* which is characterized as an increase in the onset latency of a CR with increased training (Pavlov, 1927). Another theoretical strategy has been to postulate that CS onset generates a *pattern* of stimulation among elemental components of the CS that differs as a function of CS duration, and again assume that the juxtaposition of elements present at the moment of US delivery are those that become excitatory, whereas others may become inhibitory (Desmond, 1990; Gluck et al., 1990).

The serial CS hypothesis is evident in the models of CR timing that have been proposed by Moore and colleagues, who conceive of the CS as a *tapped delay-line,* where each component plays a role in both the learning and the production of the CR (Desmond, 1990; Desmond & Moore, 1988; Moore, 1991; Moore et al., 1989, 1998;

see also Tieu et al., 1999). A simplified picture of the aggregate output of a tap delay-line model (Desmond & Moore, 1988), with six eligibility trace elements, is shown in Figure 7.11a.[5] It is proposed that each stimulus process consists of the sequential and overlapping activation of elements within an array of elements, where each element is identified in terms of the CS that activates it ($i = 1, \ldots, n$), the trace process to which it belongs ($j = 1$ if onset, and $j = 0$ if offset), and a unique element number ($k = 1, \ldots, N$). Once an activation process is initiated, by either the onset or the offset of a CS, a new element is recruited at every time step. These x_{ijk} elements make modifiable connections with an adaptive unit (Figure 7.3b). There are multiple eligibility traces, one for each V_{ijk} connection that make up the overall CS eligibility.

In support of the serial hypothesis, there is considerable evidence that explicit, serially presented stimuli can gain discriminative control of the timing of CRs (Boyd & Lewis, 1976; Dubin & Levis, 1973; Gaioni, 1982; Holland & Ross, 1981; Kehoe & Morrow, 1984; Kehoe & Napier, 1991; Newlin & LoLordo, 1976). However, for the rabbit NM response, it appears that, for relatively short CS–US intervals, even the earliest portion of the CS acquires considerable tendency to evoke a CR: when a CS that consisted of four 50-ms pulses, each of a different stimulus type, was trained with a 400-ms CS–US interval, the first 50-ms pulse was sufficient to evoke a substantial and relatively well-timed CR (Kehoe & Napier, 1991). For this response system, it may be that inhibition of delay is a phenomenon that can develop only with longer CS–US intervals (Millenson et al., 1977; Vogel, 2001).

Stimulus representation in the model of Desmond (1990; see also Desmond et al., 1986; Desmond & Moore, 1988; Moore et al., 1989) is a two-dimensional array of elements, where it is assumed that each element of the array may activate not only the element in front of it (i.e., in the next column but in the same row) but also elements in adjacent rows. The CS is viewed as initiating a propagation of activation of elements throughout the array: elements show convergence of inputs and divergence of outputs (spatial summation), and the time delay for recruitment of any particular element varies as a function of the number of inputs (temporal summation). By assuming properties of spatial and temporal summation of activation, along with a property of decaying propagation, the number of elements recruited to activity in the planar array is not constant but varies as a function of stimulus duration. The conception produces a stimulus representation in which the number of active elements first increases and then decreases, that is, is an inverted U-shaped function of stimulus duration.

The stimulus representation offered by Gluck et al. (1990), in an account of cerebellar processing in classical conditioning and adaptation of the vestibulo-ocular reflex, is one in which the CS assumes a complex pattern that is unvarying across trials, so that unique portions of the pattern can acquire excitatory or inhibitory associative strength, given a consistent CS–US interval in training. Spe-

---

[5] In Figure 7.10a, 50 eligibility trace elements were used to generate the aggregate trace, although the figure shows the activity of only 6.

cifically, these authors suggested that the CS and US each generate a collection of sine wave signals that vary in phase (0, $\pi/2$, π, $3\pi/2$) and frequency (from 1 to 30 cycles per trial). Each training trial consists of a sweep through discrete time steps, taking the current value of each of the CS-generated waveforms as the inputs and the current value of the US as the desired CR output, and then applying a delta rule algorithm to update the connection weights. A simplified version of such a stimulus representation is shown in Figure 7.11b.[6]

Grossberg and Schmajuk (1987, 1989) proposed a complex pattern–CS trace that is derived from two processes that are elicited with CS onset: an activity process and a habituation process. The model assumes that a CS activates a population of cells whose members react and habituate at different rates, which generates a *gated signal spectrum*. Examples of these gated signals are shown in Figure 7.11c.[7]

We have been investigating (Vogel, 2001; Wagner & Brandon, 2001) a molecular version of the SOP trace described by Wagner (1981), predicated on the added assumption that there is some trial-to-trial consistency in the temporal activity course of the individual elements that contribute to the overall trace. The function shown in Figure 7.11d is a raster plot depicting the momentary activity over successive moments of time of each of a large number of SOP elements representing a CS as shown. It was generated from the assumption that any inactive, nonrefractory element had a constant probability, $p1$, of being activated in any moment of the CS, that any active element had a constant probability, $pd1$, of decaying in any moment into a refractory state, and that any refractory element had a constant probability, $pd2$, of coming back in any moment to the inactive state from which it might again be activated. The overlying function represents the summed number of active elements. The proposal is that, although the summary function carries no possibilities for temporal discrimination, an assumed trial-to-trial consistency in the temporal activity of the individual elements can do so.

Learning Rules

Momentary Models

The Rescorla–Wagner model accounted for associative change at the level of the trial and acted, essentially, as if all stimuli present on a trial (CSs and USs), were punctate and concurrent. With reference to Figure 7.3b, changes in associative strength of a CS were computed according to the discrepancy between the activation of the adaptive unit by the US and its concurrent activation by associated

---

[6] In Figure 7.10b, four sine waves were generated, assuming an amplitude of $2\pi*$(time) for each, and a period of 12, 24, 48, or 96. A single trial was modeled as 100 discrete time steps with the CS starting at step 1 and the US occurring between steps 49 and 51, inclusively.
[7] In Figure 7.11c, it was assumed that I = 1, 2, 5, 10, 20, 35, and 50, where I = internal input generated by the CS to the timing circuit. Each trace was generated with a simulation that was run for 1600 time steps, where it was assumed that the CS came on at the first moment and stayed on for the duration of the simulation.

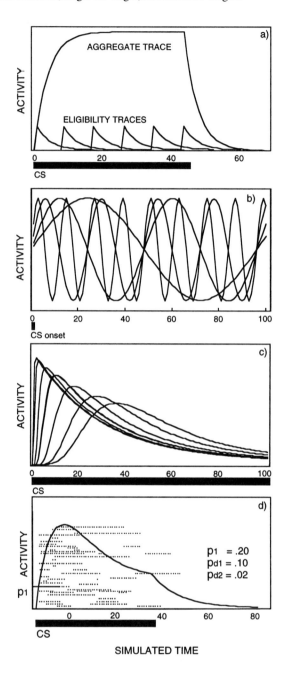

CSs. Increments in associative strength were assumed to occur when the activation of the adaptive unit by the US was greater than its activation by associated CSs, decrements in associative strength were assumed to occur when the activation by the adaptive unit was less than its activation by associated CSs, and no change was computed when the CS and the US activated the adaptive unit equally. A simple real-time application of this rule is to represent the continuous variable of time by discrete *moments,* and to compute the same discrepancies in each theoretical moment according to the rule:

$$\Delta V_i = \Sigma[\tau_i \beta_j (\lambda_j - \Sigma \tau_i V_i)], \qquad (16)$$

where β_j and λ_j assume different values for those moments when the US is present versus absent, and τ_i represents the stimulus trace, which may be of various sorts (we borrow the term from Schmajuk & Moore, 1988).

Figure 7.12 is a simulation of the Rescorla–Wagner model using the rule expressed, assuming a 10-moment binary (on/off) CS that coterminates with a 1-moment binary US ($\beta^+ = 0.10$, $\beta^- = 0.02$; $\lambda^+ = 1.00$; $\lambda^- = 0.00$, $\tau_i = 1.00$ when the CS is on and 0 when it is off). The main graph depicts the change in associative strength on a trial-by-trial basis, for 50 trials of acquisition (CS+US) followed by 50 trials of extinction (CS alone). The smaller graphs depict the changes in associative strength in the first trial of acquisition (a), at the 50th trial, at which time V was asymptotic (b), and on trial 51, the first extinction trial (c), across simulation moments. As is evident, on the first trial there is no change in V until the moment of application of the US, at which time $(\lambda_j - \Sigma V_i) > 0$, so that there is a net increment in V for that trial. However, at asymptote, the onset of the CS results in a decrement in V across the extent of the CS, as, before the US, $(\lambda_j - \Sigma V_i) < 0$; there is an increment in V with application of the US, but the net effect is no change in V for that trial. When the US is omitted, in extinction, the pre-US extinction effectively results in a net decrement in V. The assumption that $\beta^+ > \beta^-$ is in line with the original Rescorla–Wagner model (1972), as well as with earlier models (e.g., Bush & Mosteller, 1955), and allows for the longer period of CS nonreinforcement to be balanced by a shorter period of CS reinforcement on acquisition trials (see also Schmajuk & Moore, 1985).

FIGURE 7.11. Some stimulus traces used by four computational models of CR timing. The first (**A**) shows the eligibility traces and an aggregate output trace as proposed by the tapped delay-line model of Moore et al. (1986). Although only 6 eligibility traces are shown, the aggregate trace was made assuming 50 such traces, each generated in successive moments for as along as the CS was present, as indicated by the *black bar.* The second trace (**B**) is that used by Gluck et al. (1990); the third (**C**) is that used by Grossberg and Schmajuk (1987); and the fourth (**D**) is that used by the current authors (see also Vogel, 2001). The parameters for the simulations shown in **B** and **C** are briefly described in footnotes 6 and 7.

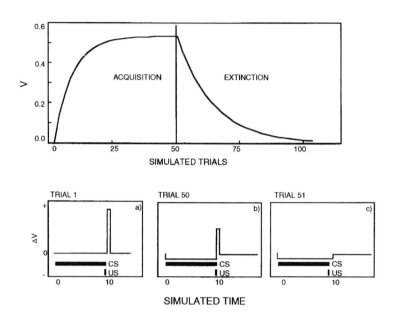

FIGURE 7.12. Simulations using a real-time version of the Rescorla–Wagner model, as specified in Eq. 16. The *top graph* shows V in acquisition and extinction for 50 simulated trials with each; the *bottom graphs* (**A–C**) show V across simulations moments for the first (**A**) and 50th (**B**) acquisition trial and the first (**C**) extinction trial. CS and US durations are indicated by the *black bars*.

Figure 7.13 was drawn to correspond to the simulations of the trial-level Rescorla–Wagner model that were depicted in Figure 7.5. Figure 7.13a and 7.13b shows simulated V values for the real-time Rescorla–Wagner model for instances of overshadowing, that is, training with an AB+ compound or the components alone, A+, B+. Each CS was a 10-moment binary signal and the US was a 1-moment binary signal that coterminated with the CS. When the salience of the two components was the same, training the AB+ compound resulted in an overshadowing effect, that is, each component trained alone reached a higher asymptote (Figure 7.13a) than did either when trained in compound (Figure 7.13b).

Figure 7.13c shows the outcome of a simulation that, as is shown next, is relevant to derivations of CR timing and ISI functions. The simulation uses two CSs, one that changes across time and one that does not: component B is less salient ($\alpha_B = 0.05$) than component A ($\alpha_A = 0.10$), over half the time period of stimulus presentation. Note that stimulus B, which has a smaller trace pre-US, shows a higher level of associative strength at asymptote that does stimulus A, because the smaller trace produced less decrement during the nonreinforced period. This inverse relationship between α (or an equivalent representation of CS salience) and asymptotic V appears to be typical of time-derivative models as well (see Klopf, 1988, p. 96, figure 7). Figure 7.13d shows that when the same two components are trained in compound, the sometimes smaller trace cue, B, acquires a positive as-

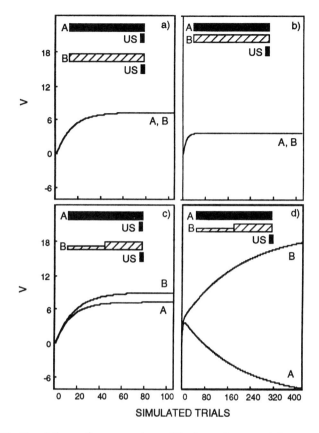

FIGURE 7.13. Simulations of compound conditioning using a real-time version of the Rescorla–Wagner model, as specified in Eq. 16, with rectilinear CS traces. (**A**) shows V_is computed across simulated trials for two CSs, *A* and *B*, each with the same salience values and durations (indicated by the *black* and *striped bars*). Each was separately paired with the US, A+, B+. (**B**) The same two CSs were trained in compound, AB+. The same conditions obtained for (**C**) and (**D**) as for (**A**) and (**B**), respectively, except that the salience of B was half that of A for the first half of B's presentation.

sociative value whereas the sometimes larger trace, A, acquires a negative associative value, a prediction of the model that so far as we know has not been tested.

The relationship between the strength of the assumed stimulus trace and the acquisition of associative strength is shown in Figures 7.14a and 7.14b for curvilinear stimulus traces generated by short-duration CSs that are characteristic of those used in real-time models of classical conditioning. The point of delivery of the US is shown by the black bar in the inset CS traces. Figure 7.14a shows the acquisition of associative strength for stimulus A and B when each cue is trained alone and where A is assumed to have a higher trace value than B (that is, trace A shows a faster recruitment and a higher peak), as depicted in the inset. As in Figure 7.13a, the cue with the larger trace, A, reaches a lower asymptotic level of associative

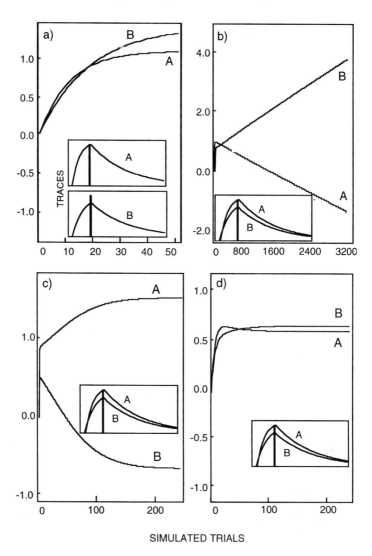

FIGURE 7.14. (**A**) and (**B**) Simulations using a real-time version of the Rescorla–Wagner model, as specified in Eq. 16, with curvilinear CS traces. (**A**) V_is computed across simulated trials for two CSs, *A* and *B*, where the stimulus trace for A was larger (showed faster recruitment and a higher asymptote) than that for B, as shown in the *inset graphs*. (**B**) The same two CSs were trained in compound, AB+. (**C**) and (**D**) Simulations of compound conditioning with a compound CS in which the two elements of the compound are formed by disparate stimulus traces (*insets*). (**C**) Simulations using the CS trace function (Eq. 14) and learning rule (Eq. 17) of Schmajuk (1997). (**D**) Simulations using the same CS trace function and a differently constrained, real-time version of the Rescorla–Wagner learning rule (Eq. 18).

strength than the cue with the smaller trace, B, although A shows a faster rate of acquisition so that, early in training, it appears to be acquiring more strength than B. Figure 7.14b shows the two are trained together in compound. As may be seen, the functions reach asymptote very slowly, with B becoming increasingly positive and A increasingly negative.[8] The functions in Figure 7.14a are germane to some of the questions that have been posed about CR timing and the ISI function, as we show. The functions in Figure 7.14b, illustrate the fact that, in the instance of compound cues, the calculated component Vs may reach asymptote at substantially different rates, depending upon the manner in which the stimulus traces assumed for each component are designated (Blough, 1975).

Equation 16 is a close approximation to another real-time learning rule, that used by Schmajuk and DiCarlo (1992; see also Schmajuk, 1997; Schmajuk et al., 1996, 1998), rewritten here to show the similarity:

$$\Delta V_i = [\tau_i \beta (\lambda_j - \Sigma \tau_i V_i)][|1 - V_i|], \qquad (17)$$

where τ_i describes the molar stimulus trace, $\lambda = 1$ if the US is present and $\lambda = 0$ if the US is absent, and β is a constant.

The effect of the additional term, $|1-V_i|$, is to make the model unable to compute changes in ΔV in those instances where $V_i = 1$, and although such values are unlikely given the usual parameter values selected, the term also ensures an inverse relationship between asymptotic level of V_i and rate of extinction, about which there are few data for Pavlovian CRs. Our simulations of this model show that the additional term has a relatively stabilizing effect on the outcome of training a compound such as that depicted in Figure 7.14, that is, one that consists of two unequal traces. Shown in Figure 7.14c are simulations of training with the compound AB as indicated in the inset graph, using the learning rule of Eq. 17.[9] With learning rate parameters comparable to those used for the simulations in Figure 7.14b, the associative weights for the elements of the compound approach a stable asymptote relatively quickly, and they do so with the associative strength for the larger trace (A) stabilizing at a positive value, and that for the smaller trace (B) stabilizing at a negative value.

Figure 7.14d shows simulations of training with the same AB compound, using a different constraint than that used by Schmajuk and colleagues (Schmajuk, 1992; Schmajuk & DiCarlo, 1992; Schmajuk et al., 1996, 1998) but the same Rescorla–Wagner learning rule. It was assumed that the associative strength of any CS was limited to the value Λ, but otherwise accrued according to the rule:

$$\Delta V_i = \tau_i \beta (\lambda_j - \Sigma \tau_i V_i), \qquad (18)$$

---

[8] Using relatively large learning rate parameters, it often is possible to evaluate whether these functions are approaching asymptote only by ascertaining that the slopes for each function are decreasing, with increasing simulated trials.
[9] For this and all subsequent simulations of models other than our own, we offer in advance whatever apology might be deemed necessary to the respective models' authors for any errors of simulations that might have occurred, and we take full responsibility for such.

where it is assumed that $V_i \leq |\Lambda| < \lambda$, and where the parameters are as for Eq. 16, that is, $\beta^+ > \beta^-$.[10] As seen in Figure 7.14d, this kind of limitation on the individual associative weights for A and B results in both weights assuming positive asymptotic values relatively quickly, and, consistent with the unconstrained model, the asymptotic weight for the cue which generates the smaller trace (B) is larger than that for the cue which generates the larger trace (A). One attribute of the use of this constrained rule is that, in instances of conditioning with stimuli that are assumed to have disparate traces (such as those shown in the figure), one can be certain that the calculated associative strengths are stable after a number of trials that is comparable to the number needed in the instance of training with single cues. Another advantage is that the constrained rule allows for the deduction of some of the temporal phenomena with which we are concerned here, primarily, the ISI function, as is shown later.

The algebraic difference term in the Rescorla–Wagner model is paralleled in another real-time model, SOP (Wagner, 1981; see also Donegan & Wagner, 1987; Mazur & Wagner, 1982), by the following rules for computation of moment-by-moment increases in the excitatory and inhibitory connections between any two stimuli (depicted here as between as CS and a following US), respectively:

$$\Delta V_i^+ = L^+ \Sigma (p_{A1,CSi} \times p_{A1,USj}) \quad \text{for excitatory learning,}$$
$$\Delta V_i^- = L^- \Sigma (p_{A1,CSi} \times p_{A2,USj}) \quad \text{for inhibitory learning,}$$
$$\Delta V_i = \Delta V_i^+ - \Delta V_i^+, \tag{19}$$

where $p_{A1,CSi}$ is the proportion of the total number of primary CS elements that are active following stimulus presentation, and $p_{A1,USj}$ and $p_{A2,USj}$ are the proportion of primary and secondary US elements, respectively, that are active upon presentation of the US. These rules are meaningful with respect to the SOP assumptions about stimulus representation, shown previously in terms of the network model in Figure 7.9b (Wagner, 1992), and were rationalized in large part on the basis of priming data.

As shown in Figure 7.3c, learning is represented in SOP by associations between A1 nodes and A2 nodes; in this case, between the nodes CSA_{A1} and US_{A2}, and CSB_{A1} and US_{A2}. The rules for associative change are simple: contiguous A1/A1 activation results in excitatory learning, and contiguous A1/A2 activation results in inhibitory learning. Importantly, although SOP postulates that the strengthened connection is that between the primary elements of a CS and the secondary elements of a US, there is no assumption that excitatory processes are qualitatively different from inhibitory processes, so that the model can as well be stated in terms of a single learning rule, with variance in V along a single continuum that includes negative and positive values, such that

$$\Delta V_i = \Sigma p_{A1,CSi} (L^+ p_{A1,USj} - L^- p_{A2,USj}). \tag{20}$$

---

[10] The values for β^+ and β^- were 0.1 and 0.02, respectively, and $\Lambda = .90$.

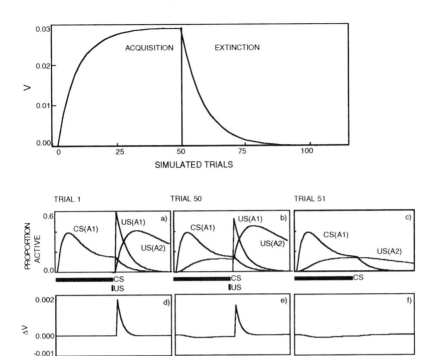

FIGURE 7.15. Simulations using the SOP model (Wagner, 1981) model, as specified in Eq. 19. The *top graph* shows V in acquisition and extinction for 50 simulated trials with each; the *bottom graphs* (**D**) to (**F**) show Vs across simulations moments for the first (**D**) and 50th (**E**) acquisition trial and the first extinction trial (**F**). The corresponding assumed A1 and A2 processes for USs and the A1 processes for CSs are indicated in (**A**) to (**C**). CS and US durations are indicated by the *black bars*.

The conditioned effect of a CS is to activate A2 elements of the US if $V_{cs} > 0$, or to antagonize the activation of such elements if $V_{cs} < 0$. There is no mechanism in the model for the extinction of a conditioned inhibitor, only its counterconditioning by excitatory learning.

Figure 7.15 is a simulation of the SOP model using the rule expressed, assuming a stimulus trace of the form shown earlier for this model (Figure 7.9b). The main graph depicts the change in associative strength, on a trial-by-trial basis, for 50 trials of acquisition (CS+US) followed by 50 trials of extinction (CS alone). The bottom graphs depict the changes in associative strength in the first trial of acquisition (Figure 7.15a), at the 50th trial, at which time V was asymptotic (Figure 7.15b), and on trial 51, the first extinction trial (Figure 7.15c), across simulation moments. Above each of these graphs are the CS and US functions according to which the change in associative strength is computed on a moment-by-moment basis. As is evident, on the first trial there is no change in V until the moment of application of the US, at which time the concurrent processing of CS_{A1} with US_{A1} results in a net increment in V for that trial. At asymptote, however, the onset of the CS results in a decrement in V across the extent of the CS, because, before the US, there is concurrent processing of CS_{A1} with US_{A2} (the latter being elicited by the onset of the CS). Thus, although there is an increment in V with application of the US, the net effect is no change in V for that trial. When the US is omitted in extinction, the pre-US extinction effectively results in a net decrement in V.

The dual-state stimulus representation of SOP is fundamental to how the model characterizes the stimulus trace, as was shown in Figure 7.9b. The dual-state representation also allows the model to deal with two phenomena of classical conditioning that are not accounted for by other real-time models, in addition to the time-dependent priming effects that motivated the model. For one, SOP asserts that the output of a secondary US node, US_{A2}, distinguishable from the primary US node, drives the CR. Thus, the model allows that the CR may not mimic, and may even appear antagonistic to, the prominent initial UR. Common examples of this are when a naive rat is exposed to foot shock: the prominent UR is activity but the CR, elicited by a tone that signaled the shock, is freezing (Blanchard & Blanchard, 1969; Bolles & Collier, 1976; Bolles & Riley, 1973), or when a pharmacological agent, such as morphine, is used as a US: a primary component of the UR is sedation whereas the notable CR is hyperactivity (Paletta & Wagner, 1986). Another advantage of the dual-state assumption of SOP is that the model makes the unique predictions that the backward pairing of a CS and a US will result in excitatory conditioning with sufficiently short intervals (Terry & Wagner, 1985), and in inhibitory conditioning at longer intervals, a prediction for which there is considerable empirical support (Champion & Jones, 1961; Maier et al., 1976; Moscovitch & LoLordo, 1968; Plotkin & Oakley, 1975; Spooner & Kellogg, 1947; Wagner & Larew, 1985).

Time-Derivative Models

The model first proposed by Sutton and Barto (1981; Barto & Sutton, 1982) can be characterized as assuming that reinforcement is the *time derivative* of the composite activity of an adaptive unit. The Sutton and Barto, or *SB,* model came out of adaptive system theory and followed a history of attempts to construct adaptive networks of neuron-like elements based on Hebbian connectivity rules (Anderson et al., 1977; Grossberg, 1975; Marr, 1969). For example, Grossberg (1968, 1969, 1975) postulated a real-time neural network in which connectivity changes occurred by pairing of presynaptic and postsynaptic signals in combination with other network effects, such as memory decay and network interactions. Such networks had been moderately successful in describing the recognition, processing, associative storage, and retrieval of spatial patterns, but they had been less successful in describing temporal processing (Sutton & Barto, 1981). The intent of the SB model was to address this deficiency, as well as to account for the same kinds of stimulus competition effects that had been the domain of the Rescorla–Wagner model.

The SB model assumes that a change in the activity in a "neuron-like" adaptive unit (like that shown in Figure 7.3b) is necessary for modification of associative connection weights. With reference to Figure 7.3b, activity on any input pathway is assumed to cause an immediate change in the adaptive unit's output, y, and to make the connection from the input pathway to the adaptive unit eligible for modification for the duration of the *eligibility trace*, $\bar{x}_i$ (which is initiated by some nonzero values of an input CS for some period of time; refer to Figure 7.9c). Reinforcement at time t is the difference between the output, y, at time t and $\bar{y}$, a weighted average of the values of the output variable over some time preceding t, that is, $\bar{y} = \lambda + \Sigma X_i V_i$ at time $t - 1$. It is assumed that $X_i = 1$ if the CS is present, and 0 if it is not, that $\lambda = 1$ if the US is present and 0 if it is not, and that $\Sigma X_i V_i = $ the weighted sum of all CS inputs to the adaptive unit. Change in associative strength, V_i, for each pathway, then is described as

$$\Delta V_i = \beta \bar{x}_i [y - \bar{y}], \tag{21}$$

where β is a rate parameter and $\bar{x}_i$ is the eligibility trace generated by the physical stimulus x.

Figure 7.16 shows an acquisition and extinction function derived with the rules of this model. In acquisition, V reaches some asymptote and then declines to 0 with extinction. Detailed pictures of how the model computes are shown in the small graphs for the first trial (Figure 7.16a), trial 50 (Figure 7.16b), at which point V was asymptotic, and trial 51 (Figure 7.16c), the first extinction trial. Each small graph shows $(y - \bar{y})$, as well as the CS eligibility trace and the duration of the CS and relatively long duration of the US, across simulation moments. In the first trial, at the outset of which $V = 0$, $(y - \bar{y}) = 0$, until the moment of application of the US, at which time $(y - \bar{y}) > 0$, and, because the CS eligibility trace then is large, there is some increment in V. The term $(y - \bar{y})$ is < 0 when the US is terminated, but since the CS eligibility trace also is negligible at this point, there is no corresponding change in V, so that the net outcome for this trial is a positive increment in V. (It is necessary, for this exposition, that the US input be long enough that the eligibility trace has decayed appreciably at its termination.) Later in training (trial 50), after the CS has reached its asymptotic connection weight, the term $(y - \bar{y})$ is positive with the onset of the CS. Because this simulation assumed that the CS and US overlapped by a single moment, the difference term also is positive with the onset of the US, but negative with the offset of the well-conditioned CS. Again, the negative difference with the offset of the US is ineffective, because the eligibility trace has decayed, so that there is no net change in weights on this trial. With extinction (trial 51), the positive difference value generated at CS onset is counterbalanced by an equally negative difference value at CS offset, but because the eligibility trace is stronger at the point of CS offset than at CS onset, there is a net decrement in connection weights.

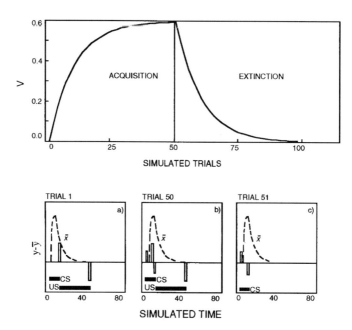

FIGURE 7.16. Simulations using the Sutton and Barto (1981) model. The *top graph* shows cumulative V in acquisition and extinction for 50 simulated trials with each; the *bottom graphs* (**A**) to (**C**) show (y − ȳ) across simulation moments for the first (**A**) and 50th (**B**) acquisition trial and the first (**C**) extinction trial. The CS eligibility trace is shown with *black bars* that indicate the stimulating CSs and USs.

Under some conditions of CS and US presentation, the SB model computes similarly to the Rescorla–Wagner model. The central idea of the SB model is that reinforcement at any time is given by the time derivative of the output of the adaptive unit. If all CSs have simultaneous onsets and offsets, and their offsets coincide with the onset of the US, effective reinforcement is provided only at the time of the offset of the CSs and the onset of the US (the prior onsets of the CSs produce reinforcement but there is no change in connection weights because the eligibility traces are essentially still 0). The CS offsets produce a decrement in V, equal to $-\Sigma V_i$, and the simultaneous US onset produces an increment in V that is equal in strength to the input of the US, that is, $+V_{US} = \lambda$. The net change then is identical to that specified by the Rescorla–Wagner rule ($\lambda - \Sigma V_i$).

The Sutton and Barto model (1981, 1990) served as an important bridge between the domain of animal learning and the domain of neural network models and adaptive systems research. The SB model accounts for many of the same stimulus selection effects as does the Rescorla–Wagner model, with some interesting differences (Sutton & Barto, 1981, 1990). The model has been attractive also because it makes no assumptions about "teaching units," as already noted; the difference term (y − ȳ), by which weight changes are computed, is indifferent as to whether the difference is generated by a CS or by a US. The model thereby strongly predicts *higher order conditioning* (Pavlov, 1927), where a CS_A, previously paired with a US, can result in acquisition of a CR when it subsequently is paired with a novel CS_B, in the arrangement $CS_B - CS_A$.[11]

As noted, in the instance of simultaneous CS offset and US onset, the SB model makes the same predictions about changes in associative strengths as does a real-time extension of the Rescorla–Wagner model. In other temporal arrangements, however, the SB model makes quite different predictions. Notably, the model predicts inhibitory conditioning whenever the CS overlaps and terminates with the US (a common procedure that is well known to produce excitatory conditioning), since the CS has a more favorable eligibility trace at CS offset than at CS onset. To avoid this implication, Sutton and Barto (1990) proposed their temporal difference (TD) model, which incorporated a change in the reinforcement term from the SB model.

Unlike the SB model, the TD model distinguishes between primary and secondary reinforcers, and specifies that US signal input is a direct reinforcement term (similar to the Rescorla–Wagner model). The TD model specifies that

$$\Delta V_i = \alpha_i \beta \bar{x}_i [\lambda + \gamma(\Sigma X_i V_i) - \bar{V}_i], \qquad (22)$$

where $\Sigma X_i V_i = \Sigma X_{i(t)} V_{i(t-1)}$, $\bar{V}_i = \Sigma X_{i(t-1)} V_{i(t-1)}$, λ is the primary reinforcement input signal, equal to 1 when the stimulus is present and to 0 when it is not, X_i is the CS input signal, equal to 1 when the CS is present and to 0 when it is not, $\bar{x}_i$ is the CS eligibility trace, and α and β are positive constants. The term γ is a discount parameter, applied to the predicted output of the adaptive unit at time t, made on the basis of $X_{(t)} V_{i(t-1)}$. Note that the connection weights in the preceding time step are used to estimate the value of V for the current time step, so that γ can be conceptualized as a "penalty" for using $V_{j(t-1)}$ as the estimate of $V_{j(t)}$ (Moore et al., 1998).

Many of the real-time models described in this chapter assume very little about the stimulus trace that is generated by the US. For most real-time models, the US is not represented by a stimulus, but rather as a "teacher unit," that simply allows for the computation of an error term, where the unit assumes values of either 1 or 0 (Schmajuk, 1997; Schmajuk & Moore, 1985, 1988; Sutton & Barto, 1981, 1990). One exception is SOP (Wagner, 1981). Another is the time-derivative, *SBD* model of Moore et al. (1986), which specifies that input from the US unit does not fall instantly to 0 with the offset of the US, but that there is some decaying trace that influences post-US computations of CS–US connections. Further elaboration of the SBD model is the *VET* model (Desmond, 1990; Desmond and Moore, 1988; Moore et al., 1989), that was described earlier for its

---

[11] The Rescorla–Wagner model supposes that any stimuli can produce conditioning so that CS–CS associations are as possible as CS–US associations. Sensory preconditioning and second-order conditioning could be attributed to such associations. This is different from the approach of Sutton and Barto.

characterization of the CS as a tapped delay-line. The primary impetus for VET was to increase the power of the tapped delay-line CS construct to predict the various timing characteristics of the rabbit conditioned NM responses. The strategy was to change the representation of the US.

In the VET model, the US activates two computational units, V (association) and E (temporal *expectation* of reinforcement) to which each CS element is assumed to form connections. In addition, the V unit receives input from the E unit. The CS–E connections can only increment at the single moment at which the CS element turns on. The CS–V associations increment according to a delta rule, weighted by two eligibility traces, one for each element of the CS, h_{ijk}, and an overall trace, $\bar{x}_{ij}$, that is globally available to all V and E connections, as well as by the output of the E unit, r, according to the rule:

$$\Delta V_{ijk} = \theta (\lambda - \Sigma V_{ijk} x_{ijk}) h_{ijk} \bar{x}_{ijk} r,$$
$$\Delta E_{ijk} = \theta [\lambda - r] \Delta x_{ijk} \bar{x}_{ij}, \quad (23)$$

where θ is a learning rate parameter, $0 < \theta \leq 1$; λ is the reinforcement signal, $0 < \lambda \leq 1$; h_{ijk} is an eligibility trace for each V_{ijk} connection, $0 < h_{ijk} \leq 1$; $\bar{x}_{ij}$ is an overall CS eligibility parameter for onset and offset processes, $0 \leq \bar{x}_{ij} \leq 1$; $\Delta x_{ijk} = 1$ when the CS turns on and is otherwise 0; and r is the output of E and represents the temporal expectation of reinforcement, equal to 1 at the moment the US is expected, and otherwise 0.

A Hidden-Unit Model

Schmajuk and DiCarlo (1992; see also Lamoureaux et al., 1988; Schmajuk & Buhusi, 1997; Schmajuk et al., 1998), proposed a hidden-unit representation of the stimulus (see Figure 7.7b). In this model, a learning rule must be specified for modifying the connection strengths between input units and hidden units, VH_{ij}, between input units and the output unit, VS_i, and between hidden units and the output unit, VN_j. The connections with the output are adjusted on a moment-by-moment basis according to a modified delta rule, in which λ (which equals 1 in moments of reinforcement and in 0 moments of nonreinforcement) is contrasted with the summed activity that is propagated to the output unit, B_j. That is,

$$\Delta VS_i = [\tau_{S_i} \beta(\lambda - B_{us})][1 - |VS_i|],$$
$$\Delta VN_j = [\tau_{n_j} \beta(\lambda - B_{us})][1 - |VN_j|], \quad (24)$$

where $B_{us} = (\Sigma \tau_{S_i} VS_i + \Sigma \tau_{n_j} VN_j)$ τ_{S_i} is the trace activity of the input unit, as defined in Eq. 14, and τ_{n_j} is the activity of the hidden unit, which depends on the activation propagated to it, σ_{H_j}, and the assumed operating characteristics of the unit. It is assumed that

$$\sigma_{H_j} = \Sigma \tau_{S_i} VH_{ij}, \quad (25)$$
$$\tau_{n_j} = \sigma_h^n / \beta^n + \sigma_h^n, \quad h = 0 \text{ if } \sigma_h = 0, \quad (26)$$

so that the hidden-unit activity is a sigmoidal function of its input.

The connections between input units and hidden units are modified according to a backpropagation rule where the overall error, $\lambda - B_{us}$, is weighted by the fractional contribution of the hidden units involved. That is,

$$\Delta VH_{ij} = [\Sigma \tau s_i \, \beta [1/(1 + e^{-kanjVN_j(\lambda - B_{us})}) - .50)][1 - |V_{ij}|]. \tag{27}$$

Using this model, it is customary to assume that the input–hidden-unit weights have random positive and negative starting values. The fact that the output characteristic of the hidden units is sigmoidal serves to make the hidden unit more effectively activated by multiple input units. The assumption also is made that the stimulus trace of the hidden units is weighted more heavily than the trace of the input units, which gives the hidden units an advantage over the input units for establishing connections with the US unit.

Some Applications to Real-Time Issues

Priming

SOP is, in part, a further elaboration of some of the assumptions of Wagner's earlier priming model (Wagner, 1976, 1978), which was referred to earlier. However, the SOP distinction between two states of stimulus representational activity (A1 and A2), and their different temporal parameters, allows the theory to account for how stimulus priming effects are time dependent in a way that a single molar trace (e.g., Schmajuk, 1997) and a trial-level (Wagner, 1978) model cannot. The decremental effects of priming CSs (Best & Gemberling, 1977; Kalat & Rozin, 1973; Wagner, 1978) and USs (Terry, 1976) are appropriately predicted by SOP to have some intermediate interval of greatest effectiveness, corresponding to the time when the representative A1 elements are sufficiently inhibited by their corresponding A2 elements that they cannot be reactivated by another presentation of the eliciting stimulus. Several simulations of self-generated priming are shown in Figure 7.17. The top graph depicts the A1 and A2 processing of a target stimulus when it was presented alone. The second graph shows the A1 processing of the target stimulus when it was preceded by an identical, priming stimulus at varying intervals of time (the priming stimulus is not shown). What can be seen in Figure 7.17b is that when the priming stimulus occurred shortly (10 simulation moments) before the target stimulus, there is an additive effect, so that target stimulus representation was greater than if it had not been primed. When the priming stimulus was presented at some intermediate times (20 or 40 simulation moments) before the target stimulus, the representation was diminished, and, finally, if the interval was long enough (60 simulation moments), the priming was ineffective.

CR Timing

The molar trace models that have been described predict response output in terms of the molar input traces weighted by acquired connection strengths. Thus, their prediction is that, as response amplitude or vigor increases with training, the CR comes to occur to the CS before the time of presentation of the US, the latency of

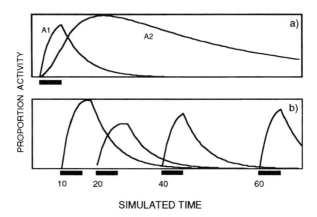

FIGURE 7.17. Simulation of the priming effects predicted by SOP. (**A**) A1 and A2 node activity elicited when a target stimulus is presented alone. (**B**) Processing in that same target stimulus node when the target is preceded, or primed, by presentation of the same stimulus at 10, 20, 40, or 60 simulation moments.

CR onset decreases with training, and the CR ceases following CS offset. Application of a strength mechanism to a simple nonmonotonic stimulus trace similar to that used by Schmajuk and DiCarlo (1992; see Eq. 14) is illustrated with the curves in Figure 7.18a: with a small amount of training ($V = 0.20$), it may be assumed that response output is small and, with increased training ($V = 0.60$), that response output is larger, with a corresponding increase in response amplitude and decrease in response latency proportional to the amount of training. Many models of conditioning have followed this characterization (Gormezano, 1972; Hull, 1943; see Desmond, 1990, for a discussion). SOP as a molar trace model specifies the CR as the output of the US_{A2} node, which increases in amplitude and decreases in latency as a function of training. These characteristics are shown in Figure 7.18b: the CS trace shown in the inset was used to simulate CRs to CSs with different acquired association strengths, $Vs = 0.05, 0.20$, and 0.40. Although these molar trace models do predict some temporal features of the CR, they do not predict the way that CRs can come to be timed to the locus of US delivery. For example, application of a US at two different ISIs should result in some singular strength of conditioning and some singular CR topography. The observations are otherwise.

It is generally agreed that, in order to account for the manner in which the timing of conditioned responding is determined by the duration of the CS–US interval, some temporally distributed componential representation of the CS is necessary. In fact, CR shaping routinely follows from those real-time models that employ some variation on the Rescorla–Wagner rule along with such a representation. To illustrate this, we applied a real-time version of the Rescorla–Wagner learning rule, generalized to an elementistic representation of the CS, to the pattern of elemental activity that was shown previously in Figure 7.11d and is given again in Figure 7.19a. The pattern of elemental activity is assumed to be consistent from

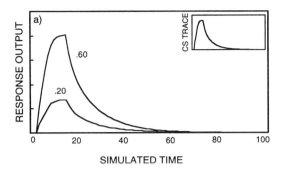

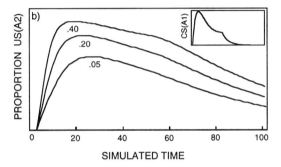

FIGURE 7.18. Timing in two molar models. (**A**) Response output assumed by the model of Schmajuk (1997), using the stimulus trace (Eq. 14) as shown in the *inset* for two associative strengths, 0.20 and 0.60. (**B**) Response output assumed by the model of Wagner (1981) using the SOP trace shown in the *inset,* for three associative strengths, 0.05, 0.20. and 0.40.

trial to trial. Changes in associative strength for each element of the CS were calculated moment by moment according to the rule

$$\Delta v_i = s_i \beta (\lambda_j - \Sigma s_i v_i), \tag{28}$$

where v_i is the associative strength of an individual element, $s_i = 1$ if the element is active and 0 if it is not, $\lambda_j = 1$ if the US is present and 0 if it is not, and $\beta^+ > \beta^-$, as otherwise specified for the real-time Rescorla–Wagner model (Eq. 16). We have borrowed the terms v_i and s_i from Blough (1975), who proposed a similar application of the Rescorla–Wagner learning rule to a set-theoretic representation of the CS.

Using this rule, the stimulus trace shown in Figure 7.19a, and a 1-moment binary signal US, we investigated how well we could predict CR timing under conditions of training with three different single and one double CS–US interval. The three single CS–US intervals were 2, 16, or 64 moments, and the double CS–US intervals were 16 or 64 moments (half the trials at each), In training, it was assumed that the CS signal terminated with the US. In testing, which was conducted after 10 simulated trials and after 320 simulated trials, the CS signal was 100 moments in duration. Two different CS traces were generated (using the same activation and decay parameters), and the set of simulations was conducted for each

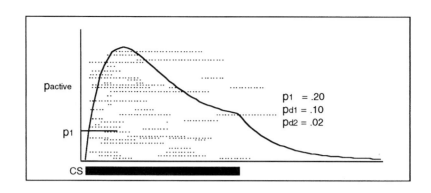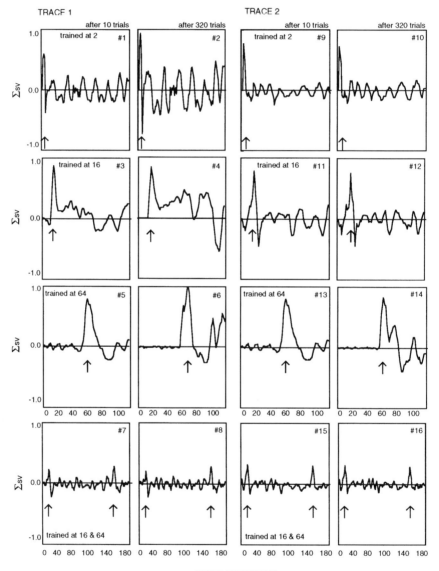

trace. Variations between simulations using the two traces were expected to the degree that the activity course of individual stimulus elements, of which the CS trace was composed, should exhibit variance within the bounds set by the parameters indicated.

The small graphs in Figure 7.19 show the results of these simulations, using the momentary $\Sigma s_i v_i$ as the performance measure. The stimulus trace is reproduced at the top of the figure. The first two columns of graphs show the result of simulations with the Trace #1 after 10 trials and then after 320 trials. The second two columns show the result of simulations with Trace #2 also after 10 trials and then after 320 trials. Each row of graphs depicts results after a specified CS–US interval: a 2-moment CS–US interval was used in the first row, 16 in the second and 64 in the third. The fourth row depicts results after the double 16/64 training interval. US application is indicated by the arrows.

This simple application of the Rescorla–Wagner model to a real-time problem fares relatively well. The outcome appears to be in good accord with the empirical data in terms of CR timing; that is, it computes that the acquired CRs peak at about the usual time of the US. The model predicts a systematic change in the shape of the CR, which can be seen best in this example in late versus early tests with the 64-moment interval. We found it interesting that training with shorter intervals, which produced not only appropriately timed CRs but also a relatively high level of associative strength, resulted in "multiple CRs" in test trials when the CS duration was extended beyond the training interval, as shown. We have observed this effect under several circumstances in rabbit eyelid conditioning studies performed in our laboratory: for example, after training rabbits with a 1050-ms CS–US interval with an auditory, light, or vibrotactual CS and a 50-ms paraorbital shock US, tests with CSs of 5 s in duration produce not only the initial CR but a second, and often a third and fourth eyelid closure, occurring at irregular intervals (Brandon et al., 2000a).

Of the various other models using componential representations of the CS that were described here (Desmond, 1990; Desmond & Moore, 1988; Gluck et al., 1990; Grossberg & Schmajuk, 1987, 1989), each has been shown to provide a

FIGURE 7.19. The *top graph* shows activity of the individual elements of a CS_{A1} node in SOP. These elements were assumed to be generated by the CS whose duration is indicated by the *black bar,* given the p1, pd1, and pd2 parameters listed. The *bottom graphs,* 1 to 16, show the results of simulations of acquisition training with various CS–US intervals, using this elementistic SOP trace with a real-time version of the Rescorla–Wagner learning rule (Eq. 28). Two stimulus traces were generated, and then each was trained at three single CS–US intervals (2, 16, or 64) or with two intervals (16 and 64). Each pair of graphs shows "CRs," which are momentary $\Sigma s_i v_i$ across a common test CS interval, in the absence of a US. The first of each pair of graphs shows a CR after 2 training trials, and the second, after 320 training trials. Thus, *#1* shows, for trace 1, a CR after 10 trials with a 2-moment CS–US interval; *#2* shows the same after 320 trials; graph *#7* shows a CR after training at two intervals, 16 and 64, after 10 trials, and *#8,* the same after 320 trials. The *arrows* indicate where the US was applied in training.

good account of the important characteristics of CR timing that have been identified: that is, a CR whose onset precedes the US and succeeds the onset of the CS, and whose peak approximates the moment of US application. These models account for the data on ISI shifts and for double-peaked CRs, as they were intended to do. Our investigations have indicated that these various ways of representing the CS are comparably powerful, and that what appears critical is a representation which is sufficiently complex and varied across time.

The ISI Function

Several current quantitative models of classical conditioning have followed the same rationale as that offered by Hull for derivation of the ISI function; that is, they have specified a particular CS trace, the value of which at any designated interval is assumed to determine the level of conditioning that accrues. The logic is that when the trace is small (at very short intervals or very long intervals), there will be little conditioning because ΔV is weighted by the value of the trace that is effective at the moment of US application (Desmond, 1990; Hull, 1943; Mauk & Donegan, 1997; Schmajuk, 1997; Wagner, 1981). The particular CS trace proposed sometimes is designed to account for numerous effects other than the ISI function. For example, we have indicated how the CS trace assumed in SOP was calculated to account for priming phenomena. Another theoretical strategy has been to build into a model an eligibility trace that produces the ISI function of the Pavlovian CR under consideration (Desmond & Moore, 1988; Klopf, 1988; Moore et al., 1986; Sutton & Barto, 1990; Moore et al., 1998). Obviously, the latter approach is less satisfying if there is no other rationalization for the eligibility function.

A molar stimulus trace approach to derivation of the ISI function was used by Mazur and Wagner (1982), and we describe it here to show how such an approach works and how it fails. For the simulations represented in Figure 7.20, we assumed a molar SOP stimulus trace with the same parameters as that used to generate the elementistic SOP trace in Figure 7.9b; that is, a trace where stimulus onset produces a period of recruitment that is followed by adaptation for the duration of the external stimulus, and then some decay to baseline beginning with the offset of the stimulus. This trace is shown in Figure 7.20a. The real-time version of the Rescorla–Wagner learning rule (Eq. 16) was used, with training CS–US delay intervals of 0 (simultaneous CS and US), 1, 2, 4, 8, 16, and 64 moments, and the CS terminated with US delivery. At varying points in training, the resulting Vs were assessed, or a "CR" was computed, assuming it to be the maximum value of $\Sigma \tau_i V_i$, where τ_i was the weight of the trace active at any moment. (In test trials, there was no US.) Importantly, where CRs were computed, they were computed separately for test trials where each CS was of the same duration as in training (a confounding of training and testing interval), as well as tests with a common, 64-moment CS (an unconfounded test of the different training intervals). The various outcomes of these simulations are shown in the subsequent functions.

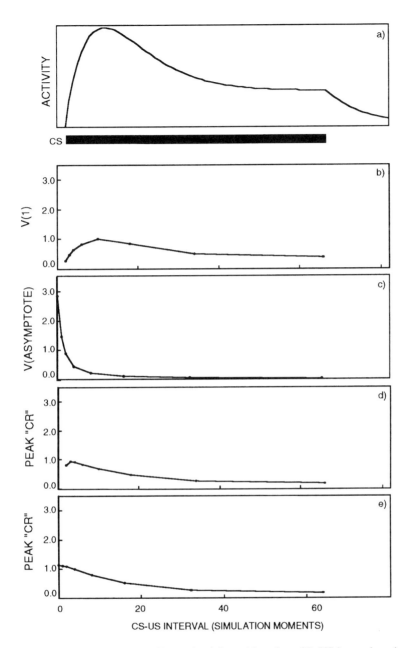

FIGURE 7.20. Simulations of the effects of training with various CS–US intervals, using a molar SOP CS trace (**A**) and the real-time Rescorla–Wagner learning rule as specified in Eq. 16. For each function, the CS–US intervals in training were 0, 1, 2, 4, 8, 16, 32, or 64 moments. The function in (**B**) is the V reached for each interval after 1 training trial, and the function in (**C**) is that reached after 200 training trials. In (**D**) 200 trials again were evaluated, but this time in terms of peak $\Sigma \tau_i V_i$ in CS-alone test trials, where the CS was the same duration as that used in training. (**E**) Same as (**D**) except that test trials were made with a common (64-moment) test interval.

The first simulation function (Figure 7.20b) shows Vs computed for this CS trace after training for only one trial. As is apparent, the nonmonotonic function obtains: acquired V reflects simply the salience of the trace present at each CS–US conditioning interval. The second function (Figure 7.20c) shows testing after training at each interval sufficient for the Vs to reach stable asymptotes (here, 320 trials). Now the function shows a simple decreasing V, the longer the CS–US interval. Although with few training trials the weak CS trace at short intervals accrued little conditioning relative to that shown by training with a CS trace that had time to grow larger, with more training smaller traces acquired higher asymptotic levels than larger traces (recall the two traces in Figure 7.14a, where the smaller trace was slower to reach asymptote but eventually reached a higher asymptote than the larger trace). Small traces of the longer CS–US intervals, as compared to short intervals, are disadvantaged by the greater opportunity for pre-US nonreinforcement to decrease responding.

Figure 7.20d shows simulations that may best approximate the empirical data: the measure of conditioning was the peak amplitude of the "CR." Here, however, the CS–US interval in training and testing were confounded; that is, the points were generated via test trials with CS durations equal to those used in training. Thus, from this function alone, it is not possible to know whether test performance varied by CS–US interval because of dependencies of learning, performance, or both. The model actually indicates that the nonmonotonic interval function in Figure 7.20d is primarily a "performance" function; when peak CRs were assessed over a common (64-moment) interval, the inverted U-shaped function disappeared, as is seen in Figure 7.20e.

Figure 7.21 shows the same series of functions for a different kind of CS trace, one used in the real-time models of Schmajuk (1997), among others, and is reproduced in the top graph. This trace shows the same kind of recruitment as the SOP trace, but no adaptation phase. With learning computed with the real-time version of the Rescorla–Wagner rule, Eq. 16, after one training trial (Figure 7.21b), there is better conditioning at intermediate intervals than at short intervals, but, because conditioning on the first trial simply reflects the value of the CS trace at the time of reinforcement, there is equally good conditioning at all the longer intervals. The subsequent functions tell the same story as the SOP trace simulations: with training to asymptote, Vs show a decreasing function with ISI duration (Figure 7.21c); with test intervals confounded with training interval (Figure 7.21d), peak CR shows a nonmonotonic function; this function is lost (Figure 7.20e) with test trials run at a common interval.

Our conclusion from these simulations, and similar simulations using molar stimulus trace functions and the learning rules of various other real-time models (Schmajuk, 1997; Sutton & Barto, 1991; Wagner, 1981), is that the ISI pattern shown in Figures 7.20 and 7.21 is quite general. It is a characteristic of further weighted models that use some variation on the delta rule (we include, here, SOP), that a small CS trace (such as that which occurs at the outset of the CS) will reach higher asymptotic V values than will a larger trace (such as that which occurs later). It is a characteristic of weighted (e.g., Schamjuk, 1997; Wagner, 1981) and unweighted (e.g., TD) models that long-duration CSs have more opportunity

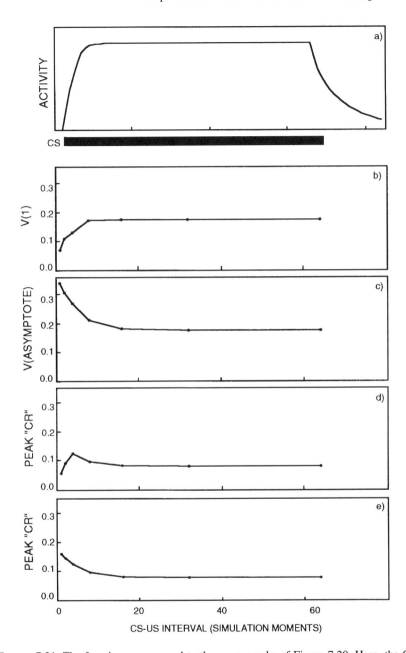

FIGURE 7.21. The functions correspond to the same graphs of Figure 7.20. Here, the CS trace and learning rule of Schmajuk (1997) were applied (see Eqs. 14 and 17) at CS–US intervals of 0, 1, 2, 4, 8, 16, 32, or 64 moments. The function in (**B**) is the V reached for each interval after 1 training trial, and the function in (**C**) is that reached after 200 training trials. In (**D**) 200 trials again were evaluated, but this time in terms of peak $\Sigma \tau_i V_i$ in CS-alone test trials, where the CS was the same duration as that used in training. Same as (**D**) except that test trials were made with a common (64-moment) test interval.

to lose associative strength. Thus, a short CS accrues a high level of excitatory strength, whereas successively later elements lose net strength in proportion to their distance from the US. Together, these processes generate a monotonic decreasing ISI function.

The planar trace model offered by Desmond (1990) purports to derive the nonmonotonic ISI function, and it is informative to see how this is accomplished. As noted already, the model is a variation of the network model of Desmond et al. (1986; also Desmond & Moore, 1988; Moore et al., 1989). Specifications about the nature of the planar array CS result in the number of elements eligible for modification increasing and then decreasing over time (Desmond, 1990, p. 445, figure 16). What is germane here is that the weight of each element in the array is arbitrarily limited, not by the dynamics of a general learning rule but by a separate parameter. A summation effect then operates: over the initial CS period, the longer the stimulating trace is on, the greater the number of elements recruited and the greater the summed output tendency. Eventually, as a result of the propagation decay properties of the planar trace, fewer elements are eligible for learning, and summed output tendency decreases. These two assumptions, that the number of elements activated is a U-shaped function of CS duration and that there is an arbitrary limit on the associative weight each may obtain, allows for computation of poorer CR elicitation at shorter and longer intervals than at intermediate intervals.

We have derived the nonmonotonic ISI function with a similarly constrained learning rule, applied to another elementistic view of the CS trace. Like Desmond (1990), the central idea was that the CS trace involved varying numbers of elements and that individual element associative strength was bounded by some parameter Λ, where $\Lambda < \lambda$, so that the ISI function reflected the number of elements conditioned. To illustrate how this works, we show here the results of training the same elementistic, correlated SOP trace as was used in earlier simulations (Figure 7.19) with three different learning rules and derivation of the ISI function for each.

The first model examined was the real-time version of the Rescorla–Wagner rule, generalized to an elementistic representation of the CS as in Eq. 28. An elementistic SOP trace was trained with CS–US delay intervals of 0 (simultaneous CS and US), 1, 2, 4, 8, 16, 64, and 128 moments. At varying points in training, the peak "CR" was measured over a test interval by computing the maximal value of $\Sigma s_i v_i$ across a CS duration of 130 moments. (In these test trials, there was no US.) Importantly, these were unconfounded tests; that is, a common test duration was applied after training at each interval. As can be seen in the top graph of Figure 7.22, the real-time Rescorla–Wagner model predicted a nonmonotonic function after one training trial, but a flat function at asymptote, as the maxium $\Sigma s_i v_i$ approached λ for all intervals.

The second model used a real-time version of the simple linear operator rule (e.g., Hull, 1943); that is,

$$\Delta v_i = s_i \beta (\lambda_j - s_i v_i), \tag{29}$$

where the parameters are the same as for Eq. 28. In this case, each v_i is constrained to approach λ, but there are different numbers of elements at different intervals.

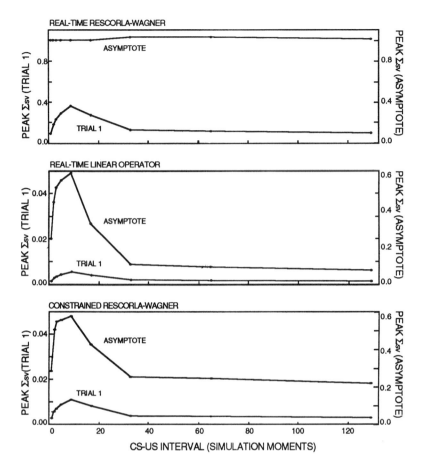

FIGURE 7.22. Simulations of the effects of training with various CS–US intervals, using an elementistic SOP CS trace (as shown in Figure 7.19A) and three different real-time models. For each function, the CS–US intervals in training were 0, 1, 2, 4, 8, 16, 32, or 64 or 128 moments. Each graph shows peak $\Sigma s_i v_i$ generated in common-interval (130-moment) CS-alone test trials, either after 1 training trial or when Vs were asymptotic. The *first function* shows the derivations of the real-time version of the Rescorla–Wagner model (Eq. 28); the *second* shows the derivations of a real-time linear operator rule (Eq. 29); the *third graph* shows the derivations of a constrained real-time Rescorla–Wagner model (Eq. 30). Note that the ordinate scales differ among the models and that different ordinate values are used for trial 1 versus asymptotic Vs for the second and third functions.

This captures, we think, the essential conditions employed by Desmond (1990) in application of the planar array model to the ISI function derivation, as described previously. Simulation results for this model are shown in Figure 7.22b. As can be seen, both after one trial and at asymptote, an inverted U-shaped function is generated. The model operates so that $\Sigma_s V$ is greater, the greater the number of elements present at the time of reinforcement to acquire excitatory connection strength.

The third model used a constrained Rescorla–Wagner rule, applied to an elementistic representation of the CS; that is,

$$\Delta v_i = s_i \beta (\lambda_j - \Sigma s_i v_i), \tag{30}$$

where v_i of the individual elements was constrained to some value Λ, where $\Lambda < \lambda_j$. As seen in Figure 7.22c, the ISI function also was an inverted U shape at asymptote as well as after one trial.

All three models showed the appropriate nonmonotonic function before asymptote; only the linear operator model and the constrained Rescorla–Wagner model showed this function at asymptote. All three models also showed some conditioning with simultaneous presentations of CS and US. Both these issues await further empirical analyses: that is, the extent to which the nonmonotonic function obtains pre- and postasymptotically, and whether simultaneity is a sufficient condition for acquisition, if the proper test assessments are made.

One of the challenges to real-time models of classical conditioning has been to develop a model that derives both the appropriate ISI function as well as appropriate CR-timing characteristics. We examined the CR-timing derivations of the same three models for which ISI functions were evaluated in Figure 7.22; the simulation outcomes are shown in Figure 7.23. In each case, the elementistic SOP CS trace was reinforced at moment 2, 16, or 64. (This is a partial replication of some of the data shown earlier, in Figure 7.19; we include it here for purposes of comparison.) The left-hand column of plots shows that application of the Rescorla–Wagner learning rule resulted in well-timed "CRs" (i.e., peak $\Sigma s_i v_i$), which developed over training. This model predicts, then, as already noted, good timing; it does not predict a nonmonotomic ISI function. The middle columns of Figure 7.23 show timing for the linear operator model. A comparison of performance after 10 trials with performance at asymptote reveals that this model also predicts accurately timed CRs but without any inhibition of early pre-US responding with training. The only effect of additional training is to change the amplitude of the peak output CR; it does not shape the CR. The right-hand columns of Figure 7.23 depict the constrained Rescorla–Wagner model. The model shows appropriate timing of peak CRs and, as well, the development of inhibition of delay. Note that, for the longer intervals of 16 and 64 moments, there is less pre-US responding after 320 trials than after 10 trials.

The Overlap Problem Again

It is useful to view the momentary reinforcement of different parts of a theoretical elementistic stimulus trace such as that used in the timing simulations shown above, as an instance of the AX+, BX− problem: there are some elements (X) that are reinforced and nonreinforced, there are some elements (A) that are uniquely reinforced, and some elements (B) that are uniquely nonreinforced. The Rescorla–Wagner rule computes that, with a sufficiently rich trace (that is, with sufficient number of elements), the B elements, which can become as inhibitory as the delta rule allows, protect the X elements from the extinction they would otherwise

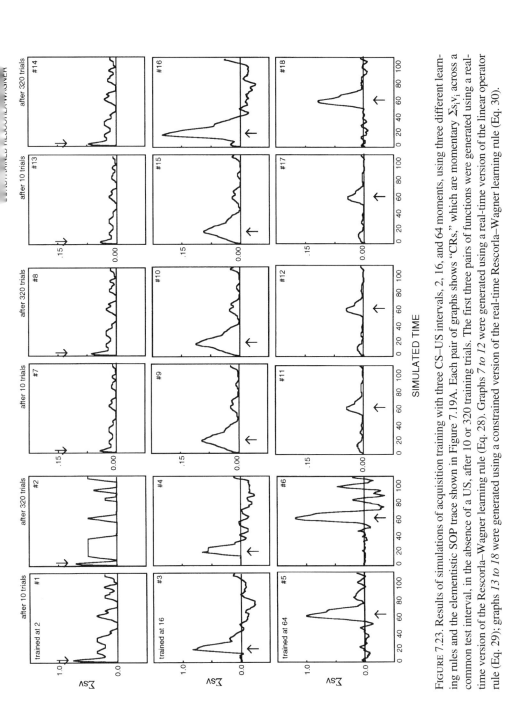

FIGURE 7.23. Results of simulations of acquisition training with three CS–US intervals, 2, 16, and 64 moments, using three different learning rules and the elementistic SOP trace shown in Figure 7.19A. Each pair of graphs shows "CRs," which are momentary $\Sigma_i s_i V_i$, across a common test interval, in the absence of a US, after 10 or 320 training trials. The first three pairs of functions were generated using a real-time version of the Rescorla–Wagner learning rule (Eq. 28). Graphs *7 to 12* were generated using a real-time version of the linear operator rule (Eq. 29); graphs *13 to 18* were generated using a constrained version of the real-time Rescorla–Wagner learning rule (Eq. 30).

acquire with long CS durations, so that eventually, $\Sigma V_{AX} = \lambda$, and $\Sigma V_{BX} = 0$, irrespective of the CS–US interval. Thus, at asymptote, the CS–US interval function is flat. In some respects, we are confronted again with anomalous theoretical predictions for the AX+, BX− problem because of a too-powerful delta rule.

We showed here one way of addressing this issue, that is, by constraining individual Vs to some value $< \lambda$, but perhaps a more theoretically attractive approach is to assume that the individual elements of the trace vary in effective probability of detection, perhaps because of varying interference or noise (cf. Mauk & Donegan, 1997). That is, one solution to this problem is a trial by trial, or moment by moment, variability in A, B, and X, so that each can fail to be detected. Then, the problem becomes AX+, BX−, A+, B−, X±, —±, with specifiable probabilities. As shown earlier with the detection model version of the Rescorla–Wagner model (Wagner, 1972), it is possible to show then that, at asymptote, $\Sigma V_{AX} < \lambda$ and $\Sigma V_{BX} > 0$. Another possibility is to assume a trial by trial, or moment by moment, variability in A, B, and X, so that A and B are less than perfectly correlated with reinforcement. Then, the problem becomes AX±, BX±, when the probability of AX+ is greater than the probability of BX+, or the probability of AX− is less than the probability of BX−. Depending on the specified probabilities, it is then possible to show that, at asymptote, the expected $\Sigma V_{AX} < \lambda$ and the expected $\Sigma V_{BX} > 0$.

Occasion Setting

Recall that, in simultaneous feature positive (AX+, X−) and feature negative (AX−, X+) discriminations, A acquires substantial excitatory and inhibitory loadings, respectively, but in the sequential instances of this same problem, where the onset of A precedes that of X, A does not appear to have comparable excitatory inhibitory loadings, although the discriminative behavior obtains. Several computational real-time models of Pavlovian conditioning have been applied to the phenomenon of occasion setting (Brandon & Wagner, 1998; Schmajuk et al., 1998; Zackheim et al., 1998). Despite markedly different assumptions about the structure of the assumed CS representations, most models use a similar strategy to account for why, when a subject is asked to solve a feature positive, AX+, X− or a feature negative, AX−, X+, problem, the associative weights acquired by the various elements A and X appear to depend on their temporal arrangement.

Brandon and Wagner (1998; see also Wagner & Brandon, 2001) proposed that the temporal arrangement of A and X is critical because it determines which context-dependent and context-independent elements are present at the time of reinforcement. The rules for generation of the replaced elements model described earlier (Eq. 15) were used in the simulations of the compound AX in simultaneous and successive arrangements, shown in Figure 7.24. The compound AX was paired with a US, whereas X alone was not. It was assumed that the context-independent elements (a_i and x_i) were weighted more heavily than the context-dependent elements, so that they generated a higher composite activity. The fig-

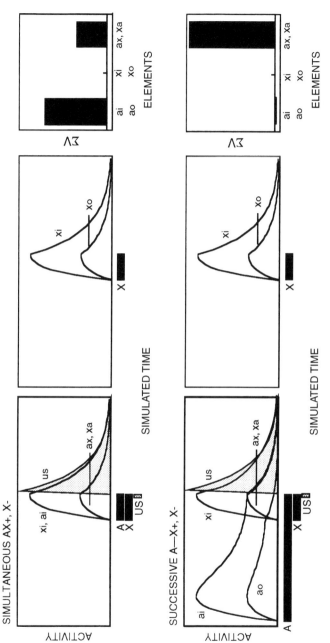

FIGURE 7.24. Stimulus traces and associative weights assumed to account for the differential performance in simultaneous (*top graphs*) versus successive (*bottom graphs*) feature positive training, AX+, X−, by the real-time replaceable elements model of Wagner and Brandon (2001). The traces shown were generated by CSs A and X and by the US. The associative weights (ΣV) were those accrued as a function of the simultaneous and successive training shown.

ure shows how, in each instance of a reinforced trial, the model assumes there to be the configural elements, a_x and x_a, elicited by the conjunction of A and X, that were in competition with both a_i and x_i. When the AX compound was reinforced in the simultaneous case, these traces also overlapped in large part with the US trace. In the successive case, however, the a_i trace was relatively degraded by the time of X and US presentation, so that the configural elements a_x and x_a had the opportunity to gain more associative strength. The results of simulations of AX+, X− after simultaneously and successively arranged AX cues are shown in the bar graphs of Figure 7.24: as can be seen, a_i acquired considerable excitatory response tendency when the arrangement was simultaneous, and more than a_x and x_a, whereas after the successive arrangement, a_x and x_a acquired greater excitatory tendencies than a_i, which had virtually no associative loading. The asymptotic loading to the x_i elements was negligible for both conditions. What was necessary for this outcome was not only to assume the existence of the context-dependent cues, a_x and x_a, but the temporal justification for those elements having either a disadvantageous (the simultaneous) or an advantageous (the successive) relationship to the elements of the feature that were equally predictive of reinforcement and nonreinforcement.

The real-time, hidden unit models of Schmajuk and Buhusi (1997) and Schmajuk et al. (1998) (see also Lamoureux et al., 1998) also explain occasion setting in terms of what units are presumed to be active at the time of reinforcement. The models use the rules already described for the model of Schmajuk and DiCarlo (Eqs. 24 to 28). The prediction is that in a simultaneous feature positive problem, AX+, X−, because the feature A and the target X are in the same temporal position, on AX+ trials, A and X each acquires an equal amount of associative strength. The nonreinforcement of X eventually results in $VS_a > VS_x$. In the serial arrangement, A—X+, X−, the feature, A, is sufficiently distant from the application of the US that VS_a is essentially 0. VS_x grows in strength on A—X+ trials, because X is proximal to the US. Because X eventually overpredicts the US on X− trials, the hidden units acquire associative weights appropriate to a reduction in the output error by generating strong inhibition of the output on X− trials and weak inhibition on A—X+ trials; this results in the generation of inhibitory hidden unit–US connections to correct for the error. At asymptote, responding on X− trials is small because excitatory X–US connections are offset by inhibitory hidden unit–US connections (where X activates the hidden unit), and responding on A—X trials is strong because the inhibitory influence of the hidden units on the US is ameliorated by the inhibitory action of A on the hidden units.[12]

---

[12] Our simulations of this model for the serial feature postive problem have shown that this output pattern is dependent on the starting values chosen for the input–hidden-unit weights, and on inclusion of an input node representing the context whose trace is active at a positive constant value. Specifically, the starting weights between CS inputs and hidden units, and between context input and hidden units, have to include at least one substantially positive value (e.g., +.25).

Multiple Process Real-Time Models

Probably the first real-time quantitative models that proposed multiple interactive learning modules, both to determine learning as well as performance in classical conditioning, were those of Grossberg and colleagues (Grossberg, 1975, 1982, 1991; Grossberg & Levine, 1987; Grossberg & Schmajuk, 1987). Networks were characterized as modular, the modules in some instances being similarly modified by the stimulating environment, and in other instances being unequally sensitive to certain aspects of that environment. Importantly, the output of these modules was interdependent.

The interacting modules in the current models proposed by Grossberg and colleagues (Grossberg, 1982; Grossberg & Levine, 1987; Grossberg & Schmajuk, 1987) are a *cognitive-reinforcement circuit* and a *sensory-cognitive circuit* (Figure 7.3d). Conditioning consists of two-way connections between *sensory (S) representations* and *drive (D) representations,* where the former encodes properties of external stimuli in a short-term memory (STM) storage, and the latter encodes the collective outcome of sensory, reinforcement, and homeostatic processing. Connections also are established from sensory representations to *motor representations,* at which action commands are generated. Pairing a CS sensory representation with activation of a drive representation, via a US, results in strengthened connections between the sensory representation of the CS and the drive representation of the US, which makes the CS a *conditioned reinforcer.* In turn, the drive-to-sensory representation feedback pathway is strengthened, which results in a shift in attention toward motivationally significant and US-compatible sensory representations, which in this case are those of the CS with which the US was paired. This shift of attention occurs because the sensory representations compete among themselves for a limited capacity short-term memory.

AESOP, which is an acronym for an *affective extension of SOP,* also is a multiple process model. It was prompted in part by the data on occasion setting in classical conditioning (Brandon & Wagner, 1991; Wagner & Brandon, 1989). The essential difference between SOP and AESOP is in how the latter conceptualized the US representation, as is shown in Figure 7.3e. In AESOP, the US representation is dichotomized, in a manner borrowed from Konorski (1967; see also Mackintosh, 1983). One aspect of the US representation is assumed to code the emotive attributes of the US, and activation here produces emotive responses (such as the Conditioned Emotional Response, or CER). The other aspect of the US representation is assumed to code for sensory/perceptual aspects of the US, resulting in primarily skeletal-motor responses (such as an eyeblink or a leg flexion). CSs are assumed to acquire independent associations to each aspect of the US, on the basis of the same learning rules as specified for SOP (Eq. 19). However, it is assumed that the decay parameters, pd1 and pd2, for the US-emotive elements are smaller than those for the US-sensory elements, thus distinguishing the activity dynamics of the two aspects of the US. One outcome of this assumption is that US-emotive conditioning can occur at longer CS–US intervals than can US-sensory conditioning, in line with many data in the classical conditioning literature showing, for example, that differential heart rate conditioning is best with a CS–US interval of 2.25 s, whereas differential NM conditioning is best with a CS–US interval of 750 ms (Van Dercar & Schneiderman, 1967; see also Schneiderman, 1972; Yehle, 1968).

There is an asymmetry assumed in AESOP, which is that whereas the primary outcome of the elicitation of the US-sensory representations is a skeletal-motor UR, there are two outcomes of eliciting US-emotive representations: one is to elicit an emotive UR or CER, and the other is to modulate the ongoing processing of stimulus nodes that are independently activated. The modulation by elicitation of the US-emotive representation was incorporated into the model by allowing that the emotive CR acted to increase the effective p1 parameter for all nodes concurrently activated (Brandon & Wagner, 1998), according to the rule:

$$p_{1,i} = p_{0,i}(1 + kp_{A2,e}), \quad 0 \leq p_1 \leq 1, \tag{31}$$

where $k \leq 1$ and $p_{A2,e}$ is the theoretical p_{A2} of the CER; this has the effect of increasing the rate of conditioning, as well as the vigor, of the CR (cf. Gewirtz et al., 1998).

These modulation properties were assumed to operate in situations of occasion setting: for example, in a sequential A—X+, X— problem, the long CS–US interval for the feature A may be effective enough for differential CER conditioning but too long for motor CR conditioning, whereas the shorter CS–US interval for X is effective for the conditioning of the motor CR (see Brandon & Wagner, 1998, and Wagner & Brandon, 1989, for simulations of these effects). It is then possible that the CER controlled by A will modulate the CR elicited by X.

Concluding Comments

This overview of computational models in classical conditioning obviously could not describe all the models nor any of the models in much detail, nor was it feasible to report many of the empirical observations that stand in evaluation of various models. What we have tried to do is broadly categorize the theoretical issues and critical empirical observations that have prompted different model types in analyses of classical conditioning. We have looked for continuity across models and unique contributions within models. Where possible, we asked how previously held assumptions, applied to later models, either were useful or led to erroneous predictions, using our own models as demonstrations.

At present we can recommend no "best" model that is able to address all the available data in the domain of classical conditioning. There remain several issues that pose continuing problems for many of the models described here, and, as such, represent incentives for the development of more comprehensive models. One such issue is associated with the widespread use of the delta rule in computing changes in CS–US associations, which has certain consequences in the real-time domain. The use of a real-time version of the Rescorla–Wagner learning rule may, as demonstrated, allow a model to retain the predictive powers of the original, trial-level model, while extending its use to the domain of timing effects. However, as also demonstrated, models using the unrestrained delta rule predict an inverted nonmonotonic ISI function for only a relatively small number of trials. With a sufficient number of training trials, the predicted ISI function becomes a monotonically decreasing function (assuming a molar CS trace) or becomes a flat function (assuming a componential trace). Unfortunately, we do not know the extent to which the nonmonotonic ISI function obtains for unconfounded tests following delay conditioning, and whether this depends on asymptotic or pre-asymptotic assessment; nor do we know whether this function obtains for most Pavlovian CRs or just some of them. There are, however, many empirical questions that need to be resolved. For example, delta rule models compute inhibition of delay, but it is unclear how ubiquitous this phenomenon is in classical delay conditioning.

Another issue of importance in the evaluation of existing models and in the development of new models concerns the representation of stimuli involved in the learning situation. There has been an increasing acceptance of the notion that models of classical conditioning should include a complex, componential representation of the stimulus, as it appears that several of the most problematic phenomena of classical conditioning, such as timing, configural learning, and occasion setting effects, require more than unitary representations of the CS and US. The idea of representing CSs and USs as aggregations of components is not new (Estes, 1950; Konorski, 1967), but in recent years there has been renewed interest in incorporating complex stimulus representations into computational models originally developed with simpler assumptions about stimulus representation (McLaren et al., 1989; McLaren & Mackintosh, 2000).

As is apparent from some of the issues described here, the combination of stimulus representation and learning rule can be relatively arbitrary; that is, a single learning rule can be applied to various representations with equal facility. An exception is SOP, for which stimulus representation is derived from the same suppositions about state dynamics as are the rules for excitatory and inhibitory learning. However, as shown, the majority of phenomena of Pavlovian conditioning, including those that appear especially sensitive to temporal variables, can be accounted for with the assumption of an SOP trace with a real-time version of the Rescorla–Wagner learning rule.

We offered a description of the evolution of computational theories of associative learning in terms of the evolution of the investigation of a common problem that we have described as $AX+$, $BX-$. The statistical associationist theory description of this problem in terms of common and unique stimulus elements (Estes, 1959b; Konorski, 1948), provided a powerful way of understanding how the discrimination could be difficult, why some generalization between the reinforced and nonreinforced stimulus might occur, and why the discrimination eventually could be at least partially resolved. However, this view of the stimulus predicted imperfect resolution of the problem. The Rescorla–Wagner learning rule offered one manner of deducing errorless discrimination behavior in the AX+, BX− problem, while maintaining this otherwise useful elementistic conception of the stimulus; alternatives were new views of the stimulus representation, offered by Estes (1959a) and Pearce (1987), which had in common the notion that the identities of the elements are virtually lost in the two configurations.

The fate of the common elements, X, in the AX+, BX− problem was an issue again in the challenges of apparent stimulus selection effects, as can be seen in the relative validity experiments by Wagner et al. (1968): if the only requisite for the acquisition of connection strength was that a stimulus occur concurrently with reinforcement, then X should be learned about as well in the condition AX+, BX− as in the condition AX±, BX±. Or, if as proposed by Hull (1943), if equal likelihoods of reinforcement and nonreinforcement "neutralize" cues, then again X should be equally neutralized in these two conditions. What the experimental data showed was that how much was learned about X depended upon the relative reliability of A to predict reinforcement, and that of B to predict nonreinforcement, as predicted by a delta rule application either to the efficacy of the US (Rescorla & Wagner, 1972) or to the efficacy of the CS (Pearce & Hall, 1980).

We have entertained the possibility here that attempts to account for some of the temporal variables in classical conditioning pose the same AX+, BX− problem. If we consider that, across trials, some temporal aspects of a CS (A) that occur when the US is presented are uniquely correlated with reinforcement, and some (B) that occur when the US is not presented are uniquely correlated with nonreinforcement, then a delta rule computation of the contribution of the common, X elements results in good CR timing but a flat ISI function. If the delta rule is constrained, then a U-shaped ISI function can obtain. The extent to which this characterization can be offered for all elemental conceptualizations of the CS trace is yet to be resolved.

The AX+, BX− problem arose again when reports of occasion setting necessitated some change in learning rules or notions of stimulus representation: why was it that the excitatory and inhibitory properties that were apparent for A and B, respectively, for the simultaneous problem, were apparently negligible for the sequential problem, that is, A—X+, B—X−? Responses to the challenges posed by occasion setting have been to change our theories of stimulus representation, for example, the hidden-unit model of Schmajuk et al. (1992), the AESOP model of Wagner and Brandon (1989), and the further extension of AESOP in Brandon and Wagner's replaceable elements model (1998). It has been notable to us how our explanations of occasion setting have precedence in earlier work: that A and B might contribute to performance to X via their emotive properties is an instantiation of Konorski's (1967) proposition that emotive modulation is not confined to interactions of classical with instrumental conditioned behaviors, but must also occur within the classical conditioning paradigm itself. The "context-dependent" aspect of stimulus representation in our replaceable elements model is a computational version of the stimulus traces posited by Hull (1943, 1945, 1952) to occur as a result of the afferent neural interaction of input stimulus traces.

References

Adrian, E.D. (1928). *The basis of sensation.* New York: Norton.
Allan, L.G. (1993). Human contingency judgements: rule based or associative? *Psychological Bulletin, 114,* 435–448.
Amsel, A. (1958). The role of frustrative nonreward in noncontinuous reward situations. *Psychological Bulletin, 55,* 102–119.
Anderson, J.A., Silverman, J.W., Ritz, S.W., and Jones, R.S. (1977). Distinctive features, categorical perception, and probability learning: some applications of a neural model. *Psychological Review, 85,* 413–451.
Anderson, J.R. (1985). *Cognitive psychology and its implications* (2nd ed.). New York: Freeman.
Anderson, J., and Bower, G. (1973). *Human associative memory.* New York: Winston.
Atkinson, R.C., and Estes, W.K. (1963). Stimulus sampling theory. In R.D. Luce, R.R. Bush, and E. Galanter (Eds.), *Handbook of mathematical psychology* (pp. 121–268). New York: Wiley.
Bahçekapili, H.G. (1997). *An evaluation of Rescorla & Wagner's elementistic model versus Pearce's configural model in discrimination learning.* Unpublished doctoral dissertation. New Haven, CT: Yale University.
Barto, A.G., and Sutton, R.S. (1982). Simulation of anticipatory responses in classical conditioning by a neuron-like adaptive element. *Behavioural Brain Research, 4,* 221–235.
Barto, A.G., Anderson, C.W., and Sutton, R.S. (1982). Synthesis of nonlinear control surfaces by a layered associative search network. *Biological Cybernetics, 43,* 175–185.
Bellingham, M., Gillette Bellingham, K., and Kehoe, E.J. (1985). Summation and configuration in patterning schedules in the rat and rabbit. *Animal Learning & Behavior, 13,* 152–164.
Best, M.R., and Gemberling, G.A. (1977). The role of short-term processes in the CS preexposure effect and the delay of reinofrcement gradient in long-delay taste-aversion learning. *Journal of Experimental Psychology: Animal Behavior Processes, 3,* 253–263.
Bitterman, M.E. (1964). The CS–US interval in classical and avoidance conditioning. In W.F. Prokasy (Ed.), *Classical conditioning* (pp. 1–19). New York: Appleton-Century-Crofts.
Blanchard, R.J., and Blanchard, D.C. (1969). Crouching as an index of fear. *Journal of Comparative and Physiological Psychology, 67,* 370–375.
Blough, D.S. (1975). Steady state data and a quantitative model of operant generalization and discrimination. *Journal of Experimental Psychology: Animal Behavior Processes, 1,* 3–21.
Bolles, R.C., and Collier, A.C. (1976). The effect of predictive cues on freezing in rats. *Animal Learning & Behavior, 4,* 6–8.
Bolles, R.C., and Riley, A.L. (1973). Freezing as an avoidance response: another look at the operant-respondent distinction. *Learning & Motivation, 4,* 268–275.
Boneau, C.A. (1958). The interstimulus interval and the latency of the conditioned eyelid response. *Journal of Experimental Psychology, 56,* 464–471.
Boyd, T.L., and Lewis, D.J. (1976). The effects of single-component extinction of a three component serial CS on resistance to extinction of the conditioned avoidance response. *Learning and Motivation, 7,* 517–531.

Brandon, S.E., and Wagner, A.R. (1991). Modulation of a discrete Pavlovian conditioning reflex by a putative emotive Pavlovian conditioned stimulus. *Journal of Experimental Psychology: Animal Behavior Processes, 17,* 299–311.

Brandon, S.E., and Wagner, A.R. (1998). Occasion setting: influences of conditioned emotional responses and configural cues. In N. Schmajuk (Ed.), *Occasion setting: associative learning and cognition in animals* (pp. 343–382). Washington, DC: American Psychological Association.

Brandon, S.E., Myers, K., Vogel, E.H., and Wagner, A.R. (2000a). "Multiple peaked rabbit eyelid conditioned responses following training with a single CS–US interval." Poster presented at the meetings of the Eastern Psychological Assoc., Baltimore, MD.

Brandon, S.E., Vogel, E.H., and Wagner, A.R. (2000b). A componential view of configured cues in generalization and discrimination in Pavlovian conditioning. *Behavioral Brain Research, 110,* 67–72.

Bush, R.R., and Mosteller, F. (1951). A mathematical model for simple learning. *Psychological Review, 58,* 313–323.

Bush, R.R., and Mosteller, F. (1955). *Stochastic models for learning.* New York: Wiley.

Champion, R.A., and Jones, J.E. (1961). Forward, backward, and pseudo-conditioning of the GSR. *Journal of Experimental Psychology, 62,* 58–61.

Chiang, C.-Y. (1993). A generalization of the Rescorla–Wagner model of conditioning and learning. *British Journal of Mathematical & Statistical Psychology, 46,* 207–212.

Coleman, S.R., and Gormezano, I. (1971). Classical conditioning of the rabbit's (*Oryctolagus cuniculus*) nictitating membrane response under symmetrical CS–US interval shifts. *Journal of Comparative & Physiological Psychology, 77,* 447–455.

Daly, H.B., and Daly, J.T. (1982). A mathematical model of reward and aversive nonreward: its application in over 30 appetitive learning situations. *Journal of Experimental Psychology: General, 111,* 441–480.

Davis, M. (1970). Interstimulus interval and startle response habituation with a "control" for total time during training. *Psychonomic Science, 20,* 39–41.

Denny, M.R. (1971). Relaxation theory and experiments. In F.R. Brush (Ed.), *Aversive conditioning and learning* (pp. 235–295). New York: Academic Press.

Desmond, J.E. (1990). Temporally adaptive responses in neural models: The stimulus trace. In M. Gabriel and J. Moore (Eds.), *Learning and computational neuroscience: foundations of adaptive networks.* (pp. 421–456). Cambridge, MA: MIT Press.

Desmond, J.E., and Moore, J.W. (1988). Adaptive timing in neural networks: the conditioned response. *Biological Cybernetics, 58,* 405–415.

Desmond, J.E., Blazis, D.E., Moore, J.W., and Berthier, N.E. (1986). Computer simulations of a classically conditioned response using neuron-like adaptive elements: response topography. *Society for Neuroscience Abstracts, 12,* 516.

Donegan, N.H. (1981). Priming-produced facilitation or diminution of responding to a Pavlovian unconditioned stimulus. *Journal of Experimental Psychology: Animal Behavior Processes, 7,* 295–312.

Donegan, N.H., and Wagner, A.R. (1987). Conditioned diminution and facilitation of the UR: a sometimes opponent-process interpretation. In I. Gormezano, W.F. Proaksy, and R.F. Thompson (Eds.), *Classical conditioning* (pp. 339–369). Hillsdale, NJ: Erlbaum.

Dubin, W.J., and Lewis, D.J. (1973). Influence of similarity of components of a serial conditioned stimulus on conditioned fear in rats. *Journal of Comparative and Physiological Psychology, 85,* 304–312.

Ebel, H.C., and Prokasy, W.F. (1963). Classical eyelid conditioning as a function of sustained and shifted interstimulus intervals. *Journal of Experimental Psychology, 65,* 52–58.

Egger, M.D., and Miller, N.E. (1962). Secondary reinforcement in rats as a function of information value and reliability of the stimulus. *Journal of Experimental Psychology, 64,* 97–104.

Estes, W.K. (1950). Toward a statistical theory of learning. *Psychological Review, 57,* 94–104.

Estes, W.K. (1959a). Component and pattern models with Markovian interpretations. In W.K. Estes and R.R. Bush (Eds.), *Studies of mathematical learning theory* (pp. 9–52). Palo Alto, CA: Stanford University Press.

Estes, W.K. (1959b). The statistical approach to learning theory. In S. Koch (Ed.), *Psychology: a study of a science, Vol. 2* (pp. 380–491). New York: McGraw-Hill.

Estes, W.K. (1969). Outline of a theory of punishment. In B.A. Campbell and R.W. Church (Eds.), *Punishment and aversive behavior* (pp. 57–82). New York: Appleton-Century-Crofts.

Estes, W.K., and Hopkins, B.L. (1961). Acquisition and transfer in pattern versus component discrimination learning. *Journal of Experimental Psychology, 61,* 322–328.

Fowler, H., Kleiman, M.C., and Lysle, D.T. (1985). Factors affecting the acquisition and extinction of conditioned inhibition suggest a "slave" process. In R.R. Miller and N.E. Spear (Eds.), *Information processing in animals: conditioned inhibition* (pp. 113–150). Hillsdale, NJ: Erlbaum.

Frey, P.W., and Ross, L.E. (1968). Classical conditioning of the rabbit eyelid response as a function of interstimulus interval. *Journal of Comparative and Physiological Psychology, 65,* 246–250.

Frey, P.W., and Sears, R.J. (1978). Model of conditioning incorporating the Rescorla–Wagner associative axiom, a dynamic attention process, and a catastrophe rule. *Psychological Bulletin, 85,* 321–340.

Gaioni, S.J. (1982). Blocking and nonsimultaneous compounds: comparisons of responding during compound conditioning and testing. *Pavlovian Journal of Biological Sciences, 17,* 16–29.

Gardner, H. (1985). *The mind's new science.* New York: Basic Books.

Gewirtz, J.C., Brandon, S.E., and Wagner, A.R. (1998). Modulation of the acquisition of the rabbit eyeblink conditioned response by conditioned contextual stimuli. *Journal of Experimental Psychology: Animal Behavior Processes, 24,* 106–117.

Gluck, M.A. (1992). Stimulus sampling and distributed representations in adaptive network theories of learning. In A.F. Healy, S.M. Kosslyn, and R.M. Shiffrin (Eds.), *From learning theory to connectionist theory, Vol. 1* (pp. 169–199). Hillsdale, NJ: Erlbaum.

Gluck, M.A., and Bower, G.H. (1988) From conditioning to category learning: an adaptive network model. *Journal of Experimental Psychology, 117,* 227–247.

Gluck, M.A., Reifsnider, E.S., and Thompson, R.F. (1990). Adaptive signal processing and the cerebellum: models of classical conditioning and VOR adaptation. In M.A. Gluck and D.E. Rumelhart (Eds.), *Neuroscience and connectionist theory* (pp. 131–185). Hillsdale, NJ: Erlbaum.

Gormezano, I. (1972). Investigations of defense and reward conditioning in the rabbit. In A.H. Black and W.F. Prokasy (Eds.), *Classical conditioning. II* (pp. 151–181). New York: Appleton-Century-Crofts.

Gormezano, I., and Kehoe, E.J. (1981). Classical conditioning and the law of contiguity. In P. Harzem and M.D. Zeiler (Eds.), *Advances in analysis of behaviour, Vol. 2. Predictability, correlation, and contiguity* (pp. 1–45). New York: Wiley.

Gormezano, I., Kehoe, E.J., and Marshall, B.S. (1983). Twenty years of classical conditioning research with the rabbit. *Progress in Psychobiology & Physiological Psychology, 10,* 197–277.

Grossberg, S. (1968). Some physiological and biochemical consequences of psychological postulates. *Proceedings of the National Academy of Sciences of the United States of America, 60,* 758–765.

Grossberg, S. (1969). Some networks that can learn, remember, and reproduce any number of complicated space-time patterns, I. *Journal of Mathematics and Mechanics, 19,* 53–91.

Grossberg, S. (1970). Neural pattern discrimination. *Journal of Theoretical Biology, 27,* 291–337.

Grossberg, S. (1975). A neural model of attention, reinforcement and discrimination learning. *International Review of Neurobiology, 18,* 263–327.

Grossberg, S. (1976). Adaptive pattern classification and university recording: 1. Parallel development and coding of neural feature detectors. *Biological Cybernetics, 23,* 121–134.

Grossberg, S. (1982). *Studies of mind and brain.* Dordrecht: Reidel.

Grossberg, S. (1991). A neural network architecture for Pavlovian conditioning: reinforcement, attention, forgetting, timing. In M.L. Commons, S. Grossberg, and J.E.R. Staddon (Eds.), *Neural network models of conditioning and action* (pp. 149–180). Hillsdale, NJ: Erlbaum.

Grossberg, S., and Levine, D.S. (1987). Neural dynamics of attentionally modulated Pavlovian conditioning: blocking, interstimulus interval, and secondary reinforcement. *Applied Optics, 26,* 5015–5030.

Grossberg, S., and Schmajuk, N.A. (1987). Neural dynamics of attentionally modulated Pavlovian conditioning: conditioned reinforcement, inhibition, and opponent processing. *Psychobiology, 15,* 195–240.

Grossberg, S., and Schmajuk, N.A. (1989). Neural dynamics of adaptive timing and temporal discrimination during associative learning. *Neural Networks, 2,* 79–102.

Gulliksen, H., and Wolfle, D.L. (1938). A theory of learning and transfer. I. *Psychometrika, 3,* 127–149.

Hearst, E., and Franklin, S.R. (1977). Positive and negative relations between a signal and food: approach-withdrawal behavior. *Journal of Experimental Psychology: Animal Behavior Processes, 3,* 37–52.

Hilgard, E.R., and Marquis, D.G. (1940). Conditioning and learning. New York: Appleton-Century-Crofts.

Hoehler, F.K., and Leonard, D.W. (1976). Double responding in classical nictitating membrane conditioning with single-CS, dual-ISI pairing. *Pavlovian Journal of Biological Science, 11,* 180–190.

Holland, P.C. (1983). Occasion-setting in Pavlovian feature positive discriminations. In M.L. Commons, R.J. Herrnstein, and A.R. Wagner (Eds.), *Quantitative analyses of behavior: discrimination processes, Vol. 4* (pp. 182–206). New York: Ballinger.

Holland, P.C. (1984). Differential effects of reinforcement of an inhibitory feature after serial and simultaneous feature negative discrimination training. *Journal of Experimental Psychology: Animal Behavior Processes, 10,* 461–475.

Holland, P.C. (1985). The nature of conditioned inhibition in serial and simultaneous feature negative discriminations. In R.R. Miller and N.E. Spear (Eds.), *Information processing in animals: conditioned inhibition* (pp. 267–297). Hillsdale, NJ: Erlbaum.

Holland, P.C. (1986). Transfer after serial feature positive discrimiantion training. *Learning and Motivation, 17,* 243–268.

Holland, P.C., and Ross, R.T. (1981). Within-compound associations in serial compound conditioning. *Journal of Experimental Psychology: Animal Behavior Processes, 7,* 228–241.

Hull, C.L. (1930). Simple trial-and-error learning: a study in psychological theory. *Psychological Review, 37,* 241–256.

Hull, C.L. (1935). The conflicting psychologies of learning—a way out. *Psychological Review, 42,* 491–516.

Hull, C.L. (1937). Mind, mechanism, and adaptive behavior. *Psychological Review, 44,* 1–32.

Hull, C.L. (1943). *Principles of behavior: an introduction to behavior theory.* New York: Appleton-Century-Crofts.

Hull, C.L. (1945). The discrimination of stimulus configuration and the hypothesis of afferent neural interaction. *Psychological Review, 52,* 133–142.

Hull, C.L. (1950). Behavior postulates and corollaries—1949. *Psychological Review, 57,* 173–180.

Hull, C.L. (1951). *Essentials of behavior.* New Haven: Yale University Press.

Hull, C.L. (1952). *A behavior system.* New Haven: Yale University Press.

James, J.J., and Wagner, A.R. (1980). One-trial overshadowing: evidence of distributive processing. *Journal of Experimental Psychology: Animal Behavior Processes, 6,* 188–205.

Jenkins, H.M. (1985). Conditioned inhibition of key pecking in the pigeon. In R.R. Miller and N.E. Spear (Eds.), *Information processing in animals: conditioned inhibition* (pp. 327–353). Hillsdale, NJ: Erlbaum.

Jenkins, H.M., and Harrison, R.H. (1960). Effect of discrimination training on auditory generalization. *Journal of Experimental Psychology, 59,* 246–253.

Jones, J.E. (1962). Contiguity and reinforcement in relation to CS–US intervals in classical aversive conditioning. *Psychological Review, 69,* 176–186.

Kalat, J.W., and Rozin, P. (1973). "Learned safety" as a mechanism in long delay taste-aversion learning in rats. *Journal of Comparative and Physiological Psychology, 83,* 198–207.

Kamin, L.J. (1965). Temporal and intensity characteristics of the conditioned stimulus. In W.F. Prokasy (Ed.), *Classical conditioning* (pp. 118–147). New York: Academic Press.

Kamin, L.J. (1968). 'Attention-like' processes in classical conditioning. In M.R. Jones (Ed.), *Miami Symposium on the Prediction of Behavior: Aversive Stimulation* (pp. 9–33). Miami: University of Miami Press.

Kamin, L.J. (1969). Predictability, surprise, attention and conditioning. In B.A. Campbell and R.M. Church (Eds.), *Punishment and aversive behavior* (pp. 279–296). New York: Appleton-Century-Crofts.

Kehoe, E.J. (1988). A layered network of associative learning: learning to learn and configuration. *Psychological Review, 95,* 411–433.

Kehoe, E.J. (1990). Classical conditioning: fundamental issues for adaptive network models. In M. Gabriel and J. Moore (Eds.), *Learning and computational neuroscience: foundations of adaptive networks* (pp. 389–420). Cambridge, MA: MIT Press.

Kehoe, E.J., and Gormezano, I. (1980). Configuration and combination laws in conditioning with compound stimuli. *Psychological Bulletin, 87,* 351–378.

Kehoe, E.J., and Graham, P. (1988). Summation and configuration in negative paterning of the rabbit's conditioned nictitating membrane response. *Journal of Experimental Psychology: Animal Behavior Processes, 14,* 320–333.

Kehoe, E.J., and Morrow, L.D. (1984). Temporal dynamics of the rabbit's nictitating membrane response in serial compound conditioned stimuli. *Journal of Experimental Psychology: Animal Behavior Processes, 10,* 205–220.

Kehoe, E.J., and Napier, R.M. (1991). In the blink of an eye: real-time stimulus factors in delay and trace conditioning of the rabbit's nictitating membrane response. *Quarterly Journal of Experimental Psychology, 43B,* 257–277.

Kehoe, E.J., Marshall-Goodell, B., and Gormezano, I. (1987). Differential conditioning of the rabbit's nictitating membrane response to serial compound stimuli. *Journal of Experimental Psychology: Animal Behavior Processes, 13,* 17–30.

Kelso, S.R., Ganong, A.H., and Brown, T.H. (1986). Hebbian synapses in hippocampus. *Proceedings of the National Academy of Sciences of the United States of America, 83,* 5326–5330.

Kimble, G.A., and Ost, J.W.P. (1961). A conditioned inhibitory process in eyelid conditioning. *Journal of Experimental Psychology, 61,* 150–156.

Kimmel, H.D. (1965). Instrumental inibitory factors in classical conditioning. In W.F. Prokasy (Ed.), *Classical conditioning* (pp. 148–171). New York: Appleton-Century-Crofts.

Klopf, A.H. (1972). *The hedonistic neuron: a theory of memory, learning, and intelligence.* New York: Hemisphere.

Klopf, A.H. (1988). A neuronal model of classical conditioning. *Psychobiology, 16,* 85–125.

Klopf, A.H. (1989). Classical conditioning phenomena predicted by a drive-reinforcement model of neuronal function. In J.H. Byrne and W.D. Berry. (Eds.), *Neural models of plasticity: experimental and theoretical approaches* (pp. 105–132). New York: Academic Press.

Kohler, W. (1930). La perception humain (trans. M. Henle). *Journal de Psychologie Normale et Pathologique, 27,* 5–30.

Kohler, W. (1938). Simple structural functions in the chimpanzee and in the chicken. In W.D. Ellis (Ed.), *A source book of gestalt psychology.* New York: Harcourt, Brace.

Konorski, J. (1948). *Conditioned reflexes and neuron organization.* Cambridge, England: Cambridge University Press.

Konorski, J. (1967). *Integrative activity of the brain.* Chicago: University of Chicago Press.

Krechevsky, I. (1938). A study of the continuity of the problem-solving process. *Psychological Review, 45,* 107–33.

Lamoureux, J.A., Buhusi, C.V., and Schmajuk, N.A. (1998). A real-time theory of Pavlovian conditioning: simple stimuli and occasion setters. In N.A. Schmajuk and P.C. Holland. (Eds.), *Occasion setting: associative learning and cognition in animals* (pp. 383–424). Washington, DC: American Psychological Association.

Lashley, K.S. (1942). An examination of the 'continuity theory' as applied to discrimination learning. *Journal of General Psychology, 26,* 241–265.

Levinthal, C.F. (1973). The CS–US interval function in rabbit nictitating membrane response conditioning: single vs multiple trials per conditioning session. *Learning and Motivation, 4,* 259–267.

Levinthal, C.F., and Papsdorf, J.D. (1970). The classically conditioned nictitating membrane response: the CS–US interval function with one trial per day. *Psychonomic Science, 21,* 296–297.

Levinthal, C.F., Tartell, R.H., Margolin, C.M., and Fishman, H. (1985). The CS–US interval (ISI) function in rabbit nictitating membrane response conditioning with very long intertrial intervals. *Animal Learning & Behavior, 13,* 228–232.

Logan, F.A. (1956). A micromolar approach to behavior theory. *Psychological Review, 63,* 63–73.

Lovejoy, E.P. (1965). An attention theory of discrimination learning. *Journal of Mathematical Psychology, 2,* 342–362.

Lubow, R.E. (1973). Latent inhibition. *Psychological Bulletin, 79,* 398–407.

Lubow, R.E., and Moore, A.U. (1959). Latent inhibition: the effect of nonreinforced preexposure to the conditioned stimulus. *Journal of Comparative & Physiological Psychology, 52,* 415–419.

Mackintosh, N.J. (1965). Selective attention in animal discrimination learning. *Psychological Bulletin, 64,* 124–150.

Mackintosh, N.J. (1975). A theory of attention: variations in the associability of stimuli with reinforcement. *Psychological Review, 82,* 276–298.

Mackintosh, N.J. (1983). *Conditioning and associative learning.* New York: Oxford University Press.

Macrae, M., and Kehoe, E.J. (1995). Transfer between conditional and discrete discriminations in conditioning of the rabbit nictitating membrane response. *Learning and Motivation, 26,* 380–402.

Maier, S.F., Rapaport, P., and Wheatley, K.L. (1976). Conditioned inhibition and the UCS-CS interval. *Animal Learning & Behavior, 4,* 217–220.

Marks, L.E. (1974). *Sensory processes.* New York: Academic Press.

Marr, D.A. (1969). A theory of cerebellar cortex. *Journal of Physiology, 202,* 437–470.

Martin, I., and Levy, A.B. (1965). Efficiency of the conditioned eyelid response. *Science, 150,* 781–783.

Mauk, M.D., and Donegan, N.H. (1997). A model of Pavlovian eyelid conditioning based on the synaptic organization of the cerebellum. *Learning & Memory, 3,* 130–158.

Mauk, M.D., and Ruiz, B.P. (1991). The timing of conditioned eyelid responses: differential conditioning using multiple inter-stimulus intervals. *Behavioral Neuroscience, 106,* 666–681.

Mazur, J.E., and Wagner, A.R. (1982). An episodic model of associative learning. In M. Commons, R. Herrnstein, and A.R. Wagner (Eds.), *Quantitative analyses of behavior, Vol. 3. Acquisition* (pp. 3–39). Cambridge, MA: Ballinger.

Medin, D.L. (1975). A theory of context in discrimination learning. In G.H. Bower (Ed.), *The psychology of learning and motivation, Vol. 9* (pp. 109–165). San Diego: Academic Press.

Millenson, J.R., Kehoe, E.J., and Gormezano, I. (1977). Classical conditioning of the rabbit's nictitating membrane response under fixed and mixed CS–US intervals. *Learning and Motivation, 8,* 351–366.

Moore, J.W. (1991). Implementing connectionist algorithms for classical conditioning in the brain. In M.L. Commons, S. Grossberg, and J.E.R. Staddon (Eds.), *Neural network models of conditioning and action* (pp. 181–199). Hillsdale, NJ: Erlbaum.

Moore, J.W., and Stickney, K.J. (1980). Formation of attentional-associative networks in real time: role of the hippocampus and implications for conditioning. *Physiological Psychology, 8,* 207–217.

Moore, J.W., and Stickney, K.J. (1982). Goal tracking in attentional-associative networks: spatial learning and the hippocampus. *Physiological Psychology, 10,* 202–208.

Moore, J.W., Desmond, J.E., Berthier, N.E., Blazis, D.E.J., Sutton, R.S., and Barto, A.G. (1986). Simulation of the classically conditioned nictitating membrane response by a neuron-like adaptive element: response topography, neuronal firing, and interstimulus intervals. *Behavioural Brain Research, 21,* 143–154.

Moore, J.W., Desmond, J.E., and Berthier, N.E. (1989). Adaptively timed conditioned responses and the cerebellum: a neural network approach. *Biological Cybernetics, 62,* 17–28.

Moore, J.W., Choi, J., and Brunzell, D.H. (1998). Predictive timing under temporal uncertainty: the TD model of the conditioned response. In D.A. Rosenbaum and C.E. Collyer (Eds.),

Timing of behavior: neural, computational, and psychological perspectives (pp. 3–34). Cambridge, MA: MIT Press.

Moscovitch, A., and LoLordo, V.M. (1968). Role of safety in the Pavlovian backward fear conditioning procedures. *Journal of Comparative and Physiological Psychology, 66*, 673–678.

Myers, K.M., Vogel, E.H., Shin, J., and Wagner, A.R. (2001). A comparison of the Rescorla–Wagner and Pearce models in a negative patterning and a summation problem. *Animal Learning & Behavior, 29*, 36–45.

Newlin, R.J., and LoLordo, V.M. (1976). A comparison of pecking generated by serial, delay, and trace autoshaping procedures. *Journal of the Experimental Analysis of Behavior, 25*, 227–241.

Paletta, M.S., and Wagner, A.R. (1986). Development of a context-specific tolerance to morphine: support for a dual-process interpretation. *Behavioral Neuroscience, 100*, 611–623.

Parker, D.B. (1985). *Learning-logic (TR-47)*. Cambridge, MA: Massachusetts Institute of Technology, Center for Computational Research in Economics and Management Science.

Patterson, M.M. (1970). Classical conditioning of the rabbit's (*Oryctolagus cuniculus*) nictitating membrane response with fluctuating ISI and intracranial CS. *Journal of Comparative and Physiological Psychology, 2*, 193–202.

Pavlov, I.P. (1927). *Conditioned reflexes* (Trans. by G.V. Anrep). London: Oxford University Press.

Pearce, J.M. (1987). A model for stimulus generalization in Pavlovian conditioning. *Psychological Review, 94*, 61–73.

Pearce, J.M. (1994). Similarity and discrimination: a selective review and a connectionist model. *Psychological Review, 101*, 587–607.

Pearce, J.M., and Hall, G. (1980). A model for Pavlovian learning: variations in the effectiveness of conditioned but not unconditioned stimuli. *Psychological Review, 87*, 532–552.

Pearce, J.M., and Redhead, E.S. (1993). The influences of an irrelevant stimulus on two discriminations. *Journal of Experimental Psychology: Animal Behavior Processes, 19*, 180–190.

Pfautz, P.L., and Wagner, A.R. (1976). Transient variations in responding to Pavlovian conditioned stimuli have implications for mechanisms of "priming." *Animal Learning & Behavior, 4*, 107–112.

Plotkin, H.C., and Oakley, D.A. (1975). Backward conditioning in the rabbit (*Oryctolagus cuniculus*). *Journal of Comparative and Physiological Psychology, 88*, 586–590.

Prokasy, W.F., and Papsdorf, J.D. (1965). Effect of increasing the interstimulus interval during classical conditioning of the albino rabbit. *Journal of Comparative and Physiological Psychology, 60*, 249–252.

Razran, G. (1957). The dominance-contiguity theory of the acquisition of classical conditioning. *Psychological Bulletin, 54*, 1–46.

Rescorla, R.A. (1968). Probability of shock in the presence and absence of CS in fear conditioning. *Journal of Comparative and Physiological Psychology, 66*, 1–5.

Rescorla, R.A. (1972). "Configural" conditioning in discrete-trial bar pressing. *Journal of Comparative and Physiological Psychology, 79*, 307–317.

Rescorla, R.A. (1973). Evidence for the "unique stimulus" account of configural conditioning. *Journal of Comparative and Physiological Psychology, 85*, 331–338.

Rescorla, R.A. (1979). Conditioned inhibition and extinction. In A. Dickinson and R.A. Boakes (Eds.), *Mechanisms of learning and motivation* (pp. 83–110). Hillsdale, NJ: Erlbaum.

Rescorla, R.A. (1980). Simultaneous and successive associations in sensory preconditioning. *Journal of Experimental Psychology: Animal Behavior Processes, 6,* 207–216.
Rescorla, R.A. (1985). Inhibition and facilitation. In R.R. Miller and N.E. Spear (Eds.), *Information processing in animals: conditioned inhibition* (pp. 299–326). Hillsdale, NJ: Erlbaum.
Rescorla, R.A., and Wagner, A.R. (1972). A theory of Pavlovian conditioning: variations in the effectiveness of reinforcement and nonreinforcement. In A.H. Black and W.F. Proasky (Eds.), *Classical conditioning. II* (pp. 64–99). New York: Appleton-Century-Crofts.
Revusky, S. (1971). The role of interference in association over a delay. In W.K. Honig and P.H.R. James (Eds.), *Animal memory* (pp. 155–213). New York: Academic Press.
Robbins, D. (1970). Stimulus selection in human discrimination learning and transfer. *Journal of Experimental Psychology, 84,* 282–290.
Ross, R.T., and Holland, P.C. (1981). Conditioning of simultaneous and serial feature-positive discriminations. *Animal Learning & Behavior, 9,* 292–303.
Rudy, J.W. (1974). Stimulus selection in animal conditioning and paired-associate learning: variations in the associative process. *Journal of Verbal Learning & Verbal Behavior, 13,* 282–296.
Rumelhart, D.E., and McClelland, J.L. (1986). *Parallel distributed processing: explorations in the microstructures of cognition. Vol. 1.* Cambridge: MIT Press.
Rumelhart, D.E., Hinton, G.E., and Williams, G.E. (1986). Learning internal representations by error propagation. In D.E. Rumelhart and J.L. McClelland (Eds.), *Parallel distributed processing: explorations in the microstructure of cognition: Vol. 1. Foundations* (pp. 318–362). Cambridge, MA: Bradford MIT Press.
Saavedra, M.A. (1975). Pavlovian compound conditioning in the rabbit. *Learning and Motivation, 6,* 314–326.
Schmajuk, N.A. (1997). *Animal learning and cognition: a neural network approach.* Cambridge: Cambridge University Press.
Schmajuk, N.A., and Buhusi, C.V. (1997). Stimulus configuration, occasion setting, and the hippocampus. *Behavioral Neuroscience, 111,* 1–24.
Schmajuk, N.A., and DiCarlo, J.J. (1992). Stimulus configuration, classical conditioning, and hippocampal function. *Psychological Review, 99,* 268–305.
Schmajuk, N.A., and Moore, J.W. (1985). Real-time attentional models for classical conditioning and the hippocampus. *Physiological Psychology, 13,* 278–290.
Schmajuk, N.A., and Moore, J.W. (1988). The hippocampus and the classically conditioned nictitating membrane response: a real-time attentional-associative model. *Psychobiology, 16,* 20–35.
Schmajuk, N.A., Lam, Y.W., and Gray, J.A. (1966). Latent inhibition: a neural network approach. *Journal of Experimental Psychology: Animal Behavior Processes, 22,* 321–349.
Schmajuk, N.A., Lamoureux, J.A., and Holland, P.C. (1998). Occasion setting: a neural network approach. *Psychological Review, 105,* 3–32.
Schneiderman, N. (1966). Interstimulus interval function of the nictitating membrane response under delay versus trace conditioning. *Journal of Comparative and Physiological Psychology, 62,* 397–402.
Schneiderman, N. (1972). Response system divergencies in aversive classical conditioning. In A.H. Black and W.F. Proasky (Eds.), *Classical conditioning. II: Current theory and research* (pp. 313–376). New York: Appleton-Century-Crofts.
Schneiderman, N., and Gormezano, I. (1964). Conditioning of the nictitating membrane of the rabbit as a function of CS–US interval. *Journal of Comparative and Physiological Psychology, 57,* 188–195.

Shanks, D.R. (1993). Human instrumental learning: a critical review of data and theory. *British Journal of Psychology, 84,* 319–354.
Sherrington, C.S. (1929). Ferrier lecture. *Proceedings of the Royal Society, B, 55,* 332.
Skinner, B.F. (1938). *The behavior of organisms.* New York: Appleton-Century-Crofts.
Smith, M.C. (1968). CS–US interval and US intensity in classical conditioning of the rabbit's nictitating membrane response. *Journal of Comparative and Physiological Psychology, 66,* 679–687.
Smith, M.C., Coleman, S.R., and Gormezano, I. (1969). Classical conditioning of the rabbit's nictitating membrane response at backward, simultaneous, and forward CS–US intervals. *Journal of Comparative and Physiological Psychology, 69,* 226–231.
Spence, K.W. (1936). The nature of discrimination learning in animals. *Psychological Review, 43,* 427–449.
Spence, K.W. (1937). The differential response in animals to stimuli varying within a single dimension. *Psychological Review, 44,* 430–44.
Spence, K.W. (1956). *Behavior theory and conditioning.* New Haven: Yale University Press.
Spence, K.W. (1960). *Behavior theory and learning.* Englewood Cliffs, NJ: Prentice-Hall.
Spence, K.W., and Taylor, J. (1951). Anxiety and strength of the UCS as determiners of the amount of eyelid conditioning. *Journal of Experimental Psychology, 42,* 183–188.
Spooner, A., and Kellogg, W.N. (1947). The backward-conditioning curve. *American Journal of Psychology, 60,* 321–334.
Sutherland, N.S. (1964). Visual discrimination in animals. *British Medical Bulletin, 20,* 54–59.
Sutherland, N.S., and Mackintosh, N.J. (1971). *Mechanisms of animal discrimination learning.* New York: Academic Press.
Sutton, R.S., and Barto, A.G. (1981). Toward a modern theory of adaptive networks: expectation and prediction. *Psychological Review, 88,* 135–170.
Sutton, R.S., and Barto, A.G. (1990). Time-derivative models of Pavlovian reinforcement. In M. Gabriel and J. Moore (Eds.), *Learning and computational neuroscience: Foundations of adaptive networks* (pp. 497–534). Cambridge, MA: MIT Press.
Terry, W.S. (1976). The effects of priming US representation in short-term memory on Pavlovian conditioning. *Journal of Experimental Psychology: Animal Behavior Processes, 2,* 354–370.
Terry, W.S., and Wagner, A.R. (1975). Short-term memory for "surprising" vs. "expected" unconditioned stimuli in Pavlovian conditioning. *Journal of Experimental Psychology: Animal Behavior Processes, 1,* 122–133.
Terry, W.S., and Wagner, A.R. (1985). The effects of US priming on CR performance and acquisition. *Bulletin of the Psychonomic Society, 23,* 249–252.
Thorndike, E.L. (1898). Animal intelligence: an experimental study of the associative processes in animals. *Psychological Review Monograph Supplements, 2,* no. 4 (whole no. 8).
Thorndike, E.L. (1932). *The fundamentals of learning.* New York: Teachers College.
Tieu, K.H., Keidel, A.L., McGann, J.P., Faulkner, B., and Brown, T.H. (1999). Perirhinal-amygdala circuit-level computational model of temporal encoding in fear conditioning. *Psychobiology, 27,* 1–25.
Titchener, E.B. (1896). *An outline of psychology.* New York: Macmillan.
Trabasso, T., and Bower, G.H. (1968). *Attention in learning: theory and research.* New York: Wiley.
Uhl, C.N. (1964). Effect of overlapping cues upon discrimination learning. *Journal of Experimental Psychology, 67,* 91–97.

Van Dercar, D.H., and Schneiderman, N. (1967). Interstimulus interval functions in different response systems during classical discrimination conditioning of rabbits. *Psychonomic Science, 9*, 9–10.
Van Hamme, L.J., and Wasserman, E.A. (1994). Cue competition in causality judgements: the role of nonpresentation of compound stimulus elements. *Learning and Motivation, 25*, 127–151.
Vogel, E.H. (2001). *A theoretical and empirical analysis of inhibition of delay in Pavlovian conditioning.* Unpublished doctoral dissertation, Yale University, New Haven, CT.
Wagner, A.R. (1969a). Stimulus validity and stimulus selection. In W.K. Honig and N.K. Mackintosh (Eds.), *Fundamental issues in associative learning* (pp. 90–122). Halifax: Dalhousie University Press.
Wagner, A.R. (1969b). Stimulus-selection and a "modified continuity theory." In G.H. Bower and J.T. Spence (Eds.), *The psychology of learning and motivation, Vol. 3* (pp. 1–40). New York: Academic Press.
Wagner, A.R. (1969c). Incidental stimuli and discrimination learning. In R.M. Gilbert and N.S. Sutherland (Eds.), *Animal discrimination learning* (pp. 83–111). London: Academic Press.
Wagner, A.R. (1971). Elementary associations. In H.H. Kendler and J.T. Spence (Eds.), *Essays in neobehaviorism: a memorial volume to Kenneth W. Spence* (pp. 187–213). New York: Appleton-Century-Crofts.
Wagner, A.R. (1972). *The detection model.* Unpublished manuscript.
Wagner, A.R. (1976). Priming in STM: an information-processing mechanism for self-generated or retrieval-generated depression in performance. In T.J. Tighe and R.N. Leaton (Eds.), *Habituation: perspectives from child development, animal behavior, and neurophysiology* (pp. 95–128). Hillsdale, NJ: Erlbaum.
Wagner, A.R. (1978). Expectancies and the priming of STM. In S.H. Hulse, H. Fowler, and W.K. Honig (Eds.), *Cognitive aspects of animal behavior* (pp. 177–209). Hillsdale, NJ: Erlbaum.
Wagner, A.R. (1979). Habituation and memory. In A. Dickinson and R.A. Boakes (Eds.), *Mechanisms of learning and motivation: a memorial to Jerzy Konorski* (pp. 53–82). Hillsdale, NJ: Erlbaum.
Wagner, A.R. (1981). SOP: a model of automatic memory processing in animal behavior. In N.E. Spear and R.R. Miller (Eds.), *Information processing in animals: memory mechanisms* (pp. 5–47). Hillsdale, NJ: Erlbaum.
Wagner, A.R. (1992). Some complexities anticipated by *AESOP* and other dual-representation theories. In H. Kimmel (Chair), *Pavlovian conditioning with complex stimuli.* Symposium conducted at the XXV International Congress of Psychology, Brussels, Belgium (July).
Wagner, A.R., and Brandon, S.E. (1989). Evolution of a structured connectionist model of Pavlovian conditioning (AESOP). In S.B. Klein and R.R. Mowrer (Eds.). *Contemporary learning theories: Pavlovian conditioning and the status of traditional learning theory* (pp. 149–189). Hillsdale, NJ: Erlbaum.
Wagner, A.R., and Brandon, S.E. (2001). A componential theory of Pavlovian conditioning. In R.R. Mowrer and S.B. Klein (Eds.), *Handbook of contemporary learning theories* (pp. 23–64). Hillsdale, NJ: Erlbaum.
Wagner, A.R., and Larew, M.B. (1985). Opponent processes and Pavlovian inhibition. From R.R. Miller and N.E. Spear (Eds.), *Information processing in animals: conditioned inhibition* (pp. 233–265). Hillsdale, NJ: Erlbaum.

Wagner, A.R., and Rescorla, R.A. (1972). Inhibition in Pavlovian conditioning: application of a theory. In M.S. Halliday and R.A. Boakes (Eds.), *Inhibition and learning* (pp. 301–336). San Diego: Academic Press.

Wagner, A.R., Logan, F.A., Haberlandt, K., and Price, T. (1968). Stimulus selection in animal discrimination learning. *Journal of Experimental Psychology, 76,* 171–180.

Wagner, A.R., Rudy, J.W., and Whitlow, J.W. (1973). Rehearsal in animal conditioning. *Journal of Experimental Psychology, 97,* 407–426.

Wasserman, E.A., Franklin, S.R., and Hearst, E. (1974). Pavlovian appetitive contingencies and approach versus withdrawal to conditioned stimuli in pigeons. *Journal of Comparative and Physiological Psychology, 86,* 616–627.

Weidemann, G., and Kehoe, E.J. (1997). Transfer and counterconditioning of conditional control in the rabbit nictitating membrane response. *The Quarterly Journal of Experimental Psychology: Comparative & Physiological Psychology, 50B,* 295–316.

Whitlow, J.W. (1975). Short-term memory in habituation and dishabituation. *Journal of Experimental Psychology: Animal Behavior Processes, 1,* 189–206.

Whitlow, J.W., and Wagner, A.R. (1972). Negative patterning in classical conditioning: summation of response tendencies to isolable and configural components. *Psychonomic Science, 27,* 299–301.

Widrow, G., and Hoff, M.E. (1960). Adaptive switching circuits. *Institute of Radio Engineers, Western Electronic Show and Convention, convention record, Part 4* (pp. 96–104).

Wilson, R.S. (1969). Cardiac response: determinants of conditioning. *Journal of Comparative & Physiological Psychology Monographs, 68(1, Pt. 2),* 1–23.

Woodbury, C.B. (1943). The learning of stimulus patterns by dogs. *Journal of Comparative Psychology, 35,* 29–40.

Yehle, A.L. (1968). Divergent response systems in rabbit conditioning. *Journal of Experimental Psychology, 77,* 468–473.

Yeo, A.G. (1974). The acquisition of conditioned suppression as a function of interstimulus interval duration. *Quarterly Journal of Experimental Psychology, 26,* 405–416.

Young, D.B., and Pearce, J.M. (1984). The influence of generalization decrement on the outcome of a feature-positive discrimination. *Quarterly Journal of Experimental Psychology, 36B,* 331–352.

Zeaman, D., and House, B.J. (1963). The role of attention in retardate discrimination learning. In N.R. Ellis (Ed.), *Handbook of mental deficiency: psychological theory and research* (pp. 159–223). New York: McGraw-Hill.

Zimmer-Hart, C.L., and Rescorla, R.A. (1974). Extinction of Pavlovian conditioned inhibition. *Journal of Comparative and Physiological Psychology, 86,* 837–845.

Name Index

Anrep, G.V., 4

Bacon, Francis, 1
Brown, Thomas, 189
Bykov, K.M., 48

Cason, Hulsey, 7
Coleman, Stephen R., xii

Descartes, R., 174
Desmond, J.E., 96
Dodge, Raymond, 8
Doty, R.W., 94

Ebbinghaus, H., 15

Gantt, W.H., 7
Gormezano, Isidore, xiii–xv, 8–9
Grant, David, xiv, 7, 8
Guthrie, E.R., 4

Hall, Marshall, 3
Hilgard, E.R., 8
Hull, Clark, 233, 238

Loucks, R.B., 94

Moore, J.W., 96, 254, 255

Pavlov, Ivan P., x, 1–5, 9, 94, 171–173, 238

Reid, Thomas, 2

Schmajuk, N.S., 255
Sechenov, I.M., 174
Spence, Kenneth W., xiv, 7, 233, 238
Stickney, K.J., 254–255

Thompson, Richard, 96
Tsukahara, N., 131–132

Voogd, J., 101

Wagner, A.R., 255
Watson, John B., 4
Wells, H.G., 3
Woodruff-Pak, D.S., 160
Wundt, Wilhelm, 1

Subject Index

Accessory abducens nucleus, 31
ACe (central nucleus of amygdala), 56
Acetylcholine (ACh), 155
Acidic substance, 48
Acoustic startle reflex, 69
Acquisition, term, xi
Acquisition curves, 172
Acquisition trials, xi
Active avoidance, 53
AD (Alzheimer's disease), 36, 154–155
Adaptive unit, 243
Adaptive unit model, 263
Adaptive utility of conditioned fear, 52–55
AESOP, 295–296
Afferent limb, 86
Afferent neural interaction, principle of, 240
Afferent systems to cerebellum, 99–101
Afterhyperpolarization (AHP), 158
Age-related memory disorders (ARMD), 148–149
Aging, animal models of disorders of, 155–156
AHP (afterhyperpolarization), 158
AIP (anterior interpositus nucleus), 105
Air puff, 17
Alpha conditioning, 17
Alpha responses, 17, 22, 177
Alpha value, 66
Aluminum chloride, 161
Alzheimer's disease (AD), 36, 154–155
Amnesiac mutant of *Aplysia*, 25
AMPA knockout, 39
Amygdala
  central nucleus of (ACe), 56
  in fear conditioning, 55–56
  lateral nucleus of, 56
  role of, in nucleus accumbens switching mechanism, 70–71

Amygdaloid neurons, 57
Animal models of disorders of aging, 155–156
Animals across life span, conditioning in, 149–154
Anterior interpositus nucleus (AIP), 105
Anticipation, 3
Antidementia compounds, 156–159
Antisense, 37
Anxiety, *see* Conditioned fear
Aplysia, model system, 20–23
Appetitive motivation, 69
Apple/FIRST system, xv
Arachidonic acid, 27
ARMD (age-related memory disorders), 148–149
Associability, 248
Association(s), 4, 6
Association psychology, 1
Associative contributions, nonassociative contributions versus, 216
Associative learning, 14–16
  defined, 15
Associative tradition, 175
Asymptote, 172
Attentional-associative model, 66
  application of, to psychopathology, 72
Attention deficits, 72
Attention mechanism, 242
Auditory cortex, 63
Autonomic conditioning viewed from biobehavioral control systems perspective, 51–52
Autoshaping, 9, 19
Aversive conditioning, 53
Avoidance, active, 53

Background cues, 213

313

Backward trials, 191–192
BAS (Behavioral Approach System), 69
Behavior, phasic inhibition of, 53
Behavioral Approach System (BAS), 69
Behavioral Facilitation System (BFS), 69
Behavioral methods, fundamental, in classical conditioning, 171–218
Behavioral psychology, 5
Behaviorism, 3, 4
Beta-amyloid, 36
BFS (Behavioral Facilitation System), 69
BIC (brachium of the inferior colliculus), 58–61
Biconditional discrimination, 249
Biobehavioral control systems perspective, autonomic conditioning viewed from, 51–52
Biological evolution, 3
Bipolar patients, 72
Blocking, 134–135, 207–208, 242
Blood pressure, increased, 52
B photoreceptors, 27
Brachium of the inferior colliculus (BIC), 58–61
Bradycardia, 52
Bradycardia CR, 54
British Empiricists, 15

Cabbage mutant of *Aplysia*, 25
Calcium-calexcitin-ryanodine receptor cascade, 36
Calcium/calmodulin II kinase, 38
Calcium current, reduction in, 20
Calexcitin, 29
CA1 pyramidal cells, 31
Cardiac output, 52
Cardiac-somatic linkage, 54
Cardiovascular system, 49
Causality, 3
CCC (cerebellar cortical conditioning) model, 110–113
Cellular correlates versus causes of learning and memory, 36–37
Cellular mechanisms of classical conditioning, 14–40
Cellular models of learning and memory, 37–39

Central cholinergic modulation, conditioned arousal and, 62
Central excitatory state, 24
Central nervous system (CNS) pathways, 46
Central nucleus of amygdala (ACe), 56
CER (Conditioned Emotional Response), 237
Cerebellar afferent, 97
Cerebellar cortex, 96
  zones and microzones, 101
Cerebellar cortical conditioning (CCC) model, 110–113
Cerebellar cortical convergence to cerebellar nuclei, 107
Cerebellar efferent pathways, lesions of, 109
Cerebellar function, theories of, 104–105
Cerebellar learning, cerebellar lesions and, 113–116
Cerebellar learning models, testing, 117–123
Cerebellar lesions
  cerebellar learning and, 113–116
  conditioning and, 109
  excitability changes and, 108
  eyeblink/NMR conditioning and, 105–109
  NMR conditioning and, 105–109
Cerebellar nuclei, cerebellar cortical convergence to, 107
Cerebellar slice preparation, 33
Cerebellum, 31
  afferent systems to, 99–101
  basic anatomy and connections of, 96–97
  conditioning and, 113–117
  eyeblink control regions of, 103–104
  neuronal architecture of, 97–99
  overview of anatomy and physiology of, 96–104
Cerebral cortex, 48
  role of, in fear conditioning, 62–63
CER (conditioned emotional response) preparation, 8
c-Fos protein, 60
Cholinergic modulation, central, conditioned arousal and, 62
Cholinergic system, 149
Classical conditioning, ix, xii, 1, 16–20
  applications and extensions to clinical neuroscience, 147–164

Subject Index   315

basic procedures and findings, 171–182
cellular mechanisms of, 14–40
characterized, 46
circumstances and themes in history of, 1–9
computational theories of, 232–298
control procedures, 16–18
defined, ix–x
defining characteristics, 16
emergence of new response in, 18–20
as exemplar of learning, 172–173
fundamental behavioral methods in, 171–218
informational effects in, 7
key variables in, 189–203
neuroscientists and, ix–xii
taking place during instrumental learning, 69–72
Climbing fibers, 100
  US information and, 128–130
Climbing fiber stimulation, 38
Clinical neuroscience, classical conditioning applications and extensions to, 147–164
CNQX, 117
CNS (central nervous system) circuitry for learned emotional responses, 63
  underlying classical fear conditioning, 55–63
CNS (central nervous system) pathways, 46
Cognitive enhancers, 156–159
Cognitive-reinforcement circuit, 295
Cognitive-style theory, 7
Compensatory responses, 19
Componential stimulus representation, 234
Componential trace models, 263–265
Compound conditioned stimuli, 203–215
Compound conditioning, 204–208
  explicit differentiation between compound and its components, 208–211
  stimulus compounding and, 206
Compound stimuli
  serial, 217–218
  simultaneous, 217
Computational theories of classical conditioning, 232–298
Conditional cues, 213
Conditional discriminations, 134

serial, 213–215
Conditioned arousal, central cholinergic modulation and, 62
Conditioned emotional response (CER), 237
Conditioned emotional response (CER) preparation, 8
Conditioned fear, 47, 53; see also Fear conditioning
  adaptive utility of, 52–55
  development of classically conditioned somatic CRs and, 54–55
  motivational properties of, 53–54
Conditioned incentive stimuli, 70, 71
Conditioned inhibition (s/R), 181, 209, 235–236
Conditioned motivation, 6
Conditioned reflex, 4
Conditioned reinforcer, 70, 295
Conditioned stimuli
  compound, 203–215
  serial, 211–215
Conditioned stimulus-response (S-R) connections, 4
Conditioned stimulus variables, 195–197
Conditioning
  in animals across life span, 149–154
  cerebellar lesions and, 109
  cerebellum and, 113–117
  compound, see Compound conditioning in humans across life span, 153–154
  real-time models of, see Real-time models
  term, xi
  transfer of, 216–217
Conditioning-induced changes, 60
Conditioning-induced plasticity, 130–132
Configural model, 250–251
Configurations, 239
Contextual cues, 213
Continuity view, 241
Continuous reinforcement, 201
Control systems perspective, biobehavioral, autonomic conditioning viewed from, 51–52
Coping response, 49
Corneal afferents, 89
Cortex, 31
Cortical folia, 96–97
Cortical plasticity, 112–113

Counterconditioning, 186–188
Covert CRs, 6
CR (conditioned response), x, 6, 172
  characterized, 47
  covert, 6
  discrete, functional utility of, 48–52
  functional utility and neurobiology of, 46–74
  latency of, CS–US interval and, 51
  lost and recovery of, 185–188
  original, acceleration of extinction of, 187
  retention, extinction, and recovery of, 216
  retention of, in rabbits across life span, 152–153
CR acquisition, 172, 199–200, 201
  basic determinants of, 216
  demonstrations of, 176–182
  rates of, 53
CREB (cyclic AMP response element binding) protein, 21
CR emergence, UR after, 199
CR extinction, 202
CR maintenance, 201–202
CR onset latency, 193
CR peak latency, 193–194
CR strength, 8
CR timing, 192–194, 217, 257–258, 280–284
CR timing effects, 254
CS (conditioned stimulus), 16, 171
  interspersed, 201–202
  lobule HVI and, 110
  reflex modification of UR by, 198–199
CS-alone, 178
CS-alone trial, 172
CS-CR preparations, 8
CS durations, 195
  test, 195–196
CS intensity, 196–197
CS-IR preparations, 8
CS offset, role of, 196
CS pathways, 123–126
  information processing in, 126–130
CS preexposure, 200–201
CS priming effects, 255–257
CS procedures, serial, 191

CS processing, theories of variations in, 246, 247–249
CS–US contiguity, 216
CS–US interval, 9, 189–194
  latency of CR and, 51
  maximum effective, 191
  minimum effective, 190–191
CS–US pairings, 18, 172
CS–US trial, 172
Cyclic AMP-mediated phosphorylation, 20–21
Cyclic AMP response element binding (CREB) protein, 21

DA (dopamine), 67–68
Defense reaction, 69
Defensive behaviors, 53
Definitions, generality of, 173–174
Delay, inhibition of, 51, 263
Delay conditioning, 150–151, 189, 258
  trace conditioning versus, 195
Delta rule, 246
Detection model, 253
Developmental disorders, 162
Diacylglycerol, 27
Differential conditioning as nonassociative control, 180
Discrete autonomic CR, 47
Discrimination reversal, 133
Distribution of training, 194–195
Dog, 2
Dopamine (DA), 67–68
Drive (D) representations, 295
Drosophila, model system, 23–25
Dunce mutant of *Aplysia,* 25

Effective UR, 50
Effective US, 50
Efferent limb, 86
Eligibility trace, 263, 275
Emotions, 4
Empirical real-time issues, 255–260
Enhanced learning, 37
Epinephrine, 50
EPSPs (excitatory postsynaptic potentials), 31–32

Excitability changes, cerebellar lesions and, 108
Excitation, xi, 181
  in fear preparations, 192
Excitatory conditioning, xi
Excitatory postsynaptic potentials (EPSPs), 31–32
Excitatory tendency (sE_R), 235
Exclusive-OR (XOR) problem, 210
Expected change, 253
Experimental psychology, 2
Extinction, xi, 185–186
Extratrial stimulus effects, temporal dependence of, 254
Eyeblink, 31
Eyeblink arrangement, human, 7
Eyeblink control Purkinje cells, learning in, 110–111
Eyeblink control regions of cerebellum, 103–104
Eyeblink/NMR conditioning, cerebellar lesions and, 105–109
Eyeblink response, mammalian, 88
Eyelid blink in rabbits, 89–94
Eyelid closure, 18

Fear conditioning, *see also* Conditioned fear
  amygdala in, 55–56
  classical, CNS circuitry underlying, 55–63
  neurobiology of, 55
  role of cerebral cortex in, 62–63
  sensory pathways in, 56–62
Fear preparations, excitation in, 192
Feature cues, 213
Feature extraction, 54
Feature negative discrimination, 209
Feature positive discrimination, 210–211
Fight/flight response, 53
Filtering, 64
Finger withdrawal, 18
FIRST operating system, xv
Food, skin shock and, 18
Foot contraction, 25–26
Forgetting, 188

Functional anatomy of skeletal muscle conditioning, 86–136
Functionalism, 8

GABA (gamma-aminobutyric acid), 27
Galantamine, 157
Galvanic Skin Response (GSR), 258
Gamma-aminobutyric acid (GABA), 27
Gated signal spectrum, 265
Generality of definitions, 173–174
Generalization, 182–183
Generalization decrement, 206
Generalization gradient, 183
Generalization test, 240
General model types, 254–255
General transfer, 183
Gestalt patterns, 239
Gill withdrawal reflex, 20
Glucose, 50
Gnostic unit, 234
Goal-directed behaviors, 19
Golgi cells, 99
Golovan mutant of *Aplysia,* 25
Gormezano box, xiv–xv
GSR (Galvanic Skin Response), 258

Habit strength (sH_R), 235–236
Habituation, 14
Heart rate, 17
Heart rate conditioning, 49, 191
Hermissenda, model system, 25–30
Hidden-unit model, 253, 278–279
Hippocampal lesions, 66–67
Hippocampus, 31, 66
  in NMR conditioning, 132–135
Homeostatic response, 50
Human eyeblink arrangement, 7
Humans across life span, conditioning in, 153–154
HVI lobule, 33
  CS and US information converges in, 110
Hypotheses, 2

Inactivation response, 99
Inferential physiology, 2

Inferior olive, 31, 99
Informational effects in classical conditioning, 7
Information processing in CS pathway, 126–130
Infraorbital nerve, 89
Inhibition, xi, 181
  conditioned (s/R), 181, 209, 235–236
  of delay, 51, 263
  latent (LI), 65, 134, 200–201
    as model for selective attention, 65–69
Instrumental approach behaviors, 19
Instrumental conditioning, xii, 6, 70, 174
Instrumental learning, classical conditioning taking place during, 69–72
Instrumental-response (IR) performance, 8
Insulin secretion, 49
Intact sensory pathways, 37
Intermittent reinforcement, 201
Internal events, 46
Interneurons, 20
Interpositus nucleus, mossy fiber input to, 128
Interstimulus interval (ISI) function, 254, 258–259, 284–291
Intertrial interval (ITI), xi, 172
Interval
  between onset of CS and US, xi
  between trials, xi, 172
IR (instrumental-response) performance, 8
Irradiation, xi
ISI (interstimulus interval) function, 254, 258–259, 284–291
ITI (intertrial interval), xi, 172

Jaw movement, 18
Jaw-opening response, 88

Knee jerk, 18
Knockdown experiment, 39
Knockouts, 37
Korsakoff's disease, 160

Laboratory setting, x

Latent inhibition (LI), 65, 134, 200–201
  as model for selective attention, 65–69
Latheo mutant of *Aplysia,* 25
Law of Contiguity, 15, 189
Laws of Association, 15
  Secondary, 189
Layered network, 252–253
Learn, learning to, 183–184
Learned arousal, 62
Learned emotional responses, CNS circuitry for, 63
Learning, 236
  causes of, cellular correlates versus, 36–37
  cellular models of, 37–39
  classical conditioning as exemplar of, 172–173
  in eyeblink control Purkinje cells, 110–111
  to learn, 183–184
  neurobiological analysis of, 87–88
Learning curves, 172
Learning equation, 242
Learning rules, 235–238, 242–249, 265–279
  time-derivative, 255
Learning theory, 7
  quantitative, 233
  statistical, 236
Least-mean-squares (LMS) algorithm, 246
Leg flexion, 18
Leg flexion withdrawal reflex, 87
Levator palpebrae (l.p.) muscle, 91
LI, *see* Latent inhibition
Life span
  animals across, conditioning in, 149–154
  humans across, conditioning in, 153–154
  rabbits across, retention of conditioned response in, 152–153
Limb flexion responses, 88
Linear control system model, 52
Linotte mutant of *Aplysia,* 25
LMS (least-mean-squares) algorithm, 246
Long-term depression (LTD), 33, 37, 38–39
Long-term memory, 173
Long-term potentiation (LTP), 37, 38–39, 151

LTD (long-term depression), 33, 37, 38–39
LTP (long-term potentiation), 37, 38–39, 151

Mackintosh model, 247
Mammalian eyeblink response, 88
Mammalian order, 2
Manic episodes, 72
Marr–Albus theory, 110–112
Massed trials, xi
Materialism, 3
MCP (middle cerebellar peduncle), 124–126
Mechanism, 3
Medial geniculate nucleus, *see* mMG
Mediating response model, 6
Mediation-based theory, 6
Membrane excitability, 33
Memory
  causes of, cellular correlates versus, 36–37
  cellular models of, 37–39
Memory disorders, age-related (ARMD), 148–149
Mesopontine cholinergic cell groups, 62
Metrifonate, 157
Middle cerebellar peduncle (MCP), 124–126
Military Medical Academy (St. Petersburg), 1
Mind, 3
mMG (medial geniculate nucleus), 57
  as selective attention gating mechanism, 64–65
Model preparations, 176
Model systems, 14–15, 20–34
  Aplysia, 20–23
  Drosophila, 23–25
  Hermissenda, 25–30
  molecular cascade common to, 35–36
  rabbit eyeblink/nictitating membrane response, 30–34
Model systems approach, 147–148
Molar stimulus trace, 234
Molar trace models, 260–263
Molecular cascade common to model systems, 35–36

Momentary models, 255, 265–274
Mossy fiber-granule cell synapses, 98
Mossy fiber input to interpositus nucleus, 128
Motivational variables, 238
Motor output pathway, 19
Motor representations, 295
Motor responses, skeletal, 53
Multiple process real-time models, 295–296

NAC, *see* Nucleus accumbens
Nalyot mutant of *Aplysia*, 25
Naturalism, 3
Nefiracetam, 158
Negative feedback mechanism, 52
Negative patterning, 210
Negative transfer, 184
Neobehaviorism, 6
Nervous system, 86
Neural interaction, afferent, principle of, 240
Neural substrates, 19
  of skeletal muscle conditioning, 94–96
Neurobiological analysis of learning, 87–88
Neurobiology of fear conditioning, 55
Neurological damage, 159–161
Neuronal architecture of cerebellum, 97–99
Neuronal recordings, 118
Neurons, 86
Neurophysiology, 5
Neuroscience, clinical, classical conditioning applications and extensions to, 147–164
Neuroscientists, ix, xii
  classical conditioning and, ix–xii
Neurotoxicity, 161–162
New Psychology, 2
Nicotine, 157
Nictitating membrane (NM), 88
Nictitating-membrane response (NMR)
  conditioning, *see* NMR conditioning
  model system, 30–34
  preparations, 9
  in rabbits, 89–94
Nimodipine, 158
NM (nictitating membrane), 88

NMDA overexpression, 37
NMDA receptors, 56
NMR, *see* Nictitating-membrane response
NMR conditioning
  cerebellar lesions and, 105–109
  hippocampus in, 132–135
Noise factors, 238
Nonassociative contributions, associative contributions versus, 216
Nonassociative control(s), 175
  differential conditioning as, 180
  paired conditions versus, 178
Nonassociative factors, 16
Nonassociative learning, 14
Noncontinuity view, 241
Nonreinforced trial, 176
Nonspecific autonomic CR, 47
Nonspecific transfer, 183–184
Nonvoltage-dependent potassium current, 21
Nuclear plasticity, 112–113
Nucleo-olivary inhibition, 130
Nucleus accumbens (NAC), 67
  function of, 68–69
  switching mechanism, role of amygdala in, 70–71

Objectivism, 3, 8
Occasion-setters, 213
Occasion setting, 213, 254, 259–260
Odor avoidance procedure, 24–25
Old Psychology, 3
Olivary/cerebellar stimulation studies, 129
Olivary lesion/inactivation studies, 129
Olivo-cortico-nuclear module, 101–103
100% reinforcement, 201
Operant conditioning, xii
Orbicularis oculi (o.o.) muscle, 91
Overlap problem, 239–240, 292–295
  new version of, 253–254
  solved, 244–246
Overprediction, 207
Overshadowing, 204–205, 242

Paired conditions, nonassociative controls versus, 178
Paired training, 178

Partial reinforcement, 201–202
Partial reinforcement extinction effect, 202
Partial warning, 203
Pearce–Hall model, 249
Pearce model, 250
Percentage response, 179
Perception, 233
Performance hypothesis, 114–115
Performance rules, 238–239
Periocular afferent, 90
Phasic inhibition of behavior, 53
Phenylephrine, 52
Phorbol ester, 31
Phosphorylation, cyclic AMP-mediated, 20–21
Phototropic response, 25
Physiology, 4–5
Picrotoxin, 38
PIN (posterior intralaminar nucleus), 57
PKC (protein kinase C), 27–29
Polymodal stimulation, 60
Pontine nuclei, 120
Pontine stimulation, 124
Positive patterning, 210–211
Positive transfer, 184
Positivism, 3
Postencephalitis syndrome, 160
Posterior intralaminar nucleus (PIN), 57
Potassium channel function, 33
Potassium current, nonvoltage-dependent, 21
Potentiated startle procedure, 179
Predictive homeostasis, 49
Premotor "blink" area, 93–94
Primary lemniscal pathway, 57
Primary negative reinforcer, 70
Primary positive reinforcer, 70
Primary reinforcer, 70
Priming, 279
Priming effects, 255–257
Priming model, 248
Principle of afferent neural interaction, 240
Proboscis extension, 23–24
Protein kinase C (PKC), 27–29
Pseudoconditioning, 24, 177–178
Psychology, 2
Psychology programs, ix

Psychopathology, application of attentional-associative model and switching model to, 72
Pupillary reflex, 86
Purkinje cell activity, 33
Purkinje cells, 98–99
　eyeblink control, learning in, 110–111
　recordings from, 121–123
Pyramidal cells, CA1, 31

Quantitative learning theory, 233

Rabbits
　across life span, retention of conditioned response in, 152–153
　eyelid blink and nictitating membrane response in, 30–34, 89–94
Radish mutant of *Aplysia*, 25
Reacquisition, 185–186
Reactive homeostasis, 49
Real-time issues
　applications to, 279–295
　empirical, 255–260
Real-time models, 254–295
　of conditioning, 190
　multiple process, 295–296
Reflex(es), 3, 174
　skeletal muscle, 86–87
Reflex center, 86
Reflex modification of UR by CS, 198–199
Reflex tradition, 174–175
Reinforced trial, 176
Reinforcement, xi–xii, 176
　schedules of, 200–203
Reinforcement learning, xii
Reinforcer, 176
Relations, 4
Relative validity effects, 242
Replaceable elements model, 251
Rescorla–Wagner model, 207, 237, 242–245, 265
　with replaceable elements, 251–252
　with unique cues, 249–250
Rescorla–Wagner rule, constrained, 290
Response likelihood, 179

Retardation test, 209–210
Retention interval, 188
Retractor bulbi (r.b.) muscle, 91
RT-PCR, 30
Rutabaga mutant of *Aplysia*, 25
Ryanodine receptor (RYR), 29
Ryanodine receptor synthesis, 35
RYR (ryanodine receptor), 29

Salivary-conditioning procedure, 7
Salivary system, 2
Salivation, 18, 48
Savings, 188
SC (superior colliculus), 58–61
Schaeffer collaterals, 31, 38
Schedules of reinforcement, 200–203
Schizophrenia, 72
Scientific developments, xv
Scientism, 3
Scopolamine, 155–156
S current, 21
SDL (state-dependent learning), 116
Secondary Laws of Association, 189
Secondary lemniscal system, 57
Secondary reinforcer, 70
Second-order conditioning, 19, 212
Selective attention, 63–64
　latent inhibition as model for, 65–69
Selective attention gating mechanism, mMG as, 64–65
Selective attention mechanisms, 63–72
Sensitization, 14
Sensory-cognitive circuit, 295
Sensory pathways in fear conditioning, 56–62
Sensory preconditioning, 134, 212
Sensory (S) representations, 295
sE_R (excitatory tendency), 235
Serial compound stimuli, 217–218
Serial conditional discriminations, 213–215
Serial conditioned stimuli, 211–215
Serial CS procedures, 191
Serotonin, 21
Set point, 49
Short-term memory, 173
Short-term stimulus trace, 260
sH_R (habit strength), 235–236

Signalization, 1–2
Signal-to-noise ratio, changing, 64
Sign-directed behaviors, 19
Silent learning, 15
Simultaneous compound stimuli, 217
Siphon withdrawal, 21–22
Siphon withdrawal reflex, 20
s/R (conditioned inhibition), 181, 209, 235–236
Skeletal motor responses, 53
Skeletal muscle conditioning
    functional anatomy of, 86–136
    neural substrates of, 94–96
Skeletal muscle reflexes, 86–87
Skin shock, food and, 18
Somatic–cardiac link, 49
Somatic CRs, development of classically conditioned, conditioned fear and, 54–55
Somatomotor conditioned responses, 46
SOP assumptions, 272
SOP model, 260–262
Spaced trials, xi
Specific transfer, 182–183
Spinal cord reflex center, 86
Spinal trigeminal nucleus, 31
Spontaneous configuration, 205–206
Spontaneous recovery, xi, 185
S-R (stimulus-response) association, 5
S-R (conditioned stimulus-response) connections, 4
S (sensory) representations, 295
S-S (stimulus-stimulus) association, 5
Startle reflex, acoustic, 69
State-dependent learning (SDL), 116
Statistical learning theory, 236
Stimulating trace, 263
Stimuli, x, 174
Stimulus associability value, 66
Stimulus compounding, 204
    compound conditioning and, 206
Stimulus effects, extratrial, temporal dependence of, 254
Stimulus generalization, xi
Stimulus representation, 233–234, 249–253, 260–265
    componential, 234
Stimulus-response (S-R) association, 5

Stimulus-response reinforcer contingencies, 175–176
Stimulus selection, 240
Stimulus selection mechanisms, 64
Stimulus-stimulus (S-S) association, 5
Stimulus trace, 189
    short-term, 260
Stimulus variables
    conditioned, 195–197
    unconditioned, 197–200
Summation, 204
Summation test, 209
Superior colliculus (SC), 58–61
Supraorbital nerve, 89
Sutton and Barto model, 276–277
Switching model, 67
    application of, to psychopathology, 72
Synaptic efficacy, 60
Synaptic excitability, 33
Synaptic plasticity, 38
Synaptic signals, 36

Tacrine, 157
Tail shock, 20
Take-away messages, 215–218
Tapped delay-line, 263
Target response, 17
Target stimulus, 213
TD (temporal difference) model, 277
Telencephalic structures, 95
Temporal dependence of extratrial stimulus effects, 254
Temporal difference (TD) model, 277
Temporal specificity, 215
Test trials, 172
Thalamus, 62
Thorndikean learning, 6
Time derivative, 274
Time-derivative learning rules, 255
Time-derivative models, 274–278
Tone, 17
Trace conditioning, 133, 150–151, 189, 258
    delay conditioning versus, 195
Trace interval, 189
Trace models, componential, 263–265
Trace theory, 192
Training